MOON
SHOT

MOON SHOT

The Inside Story of America's Race to the Moon

By Alan Shepard and Deke Slayton
with Jay Barbree and Howard Benedict

INTRODUCTION BY NEIL ARMSTRONG

Turner Publishing, Inc.
ATLANTA

Published by Turner Publishing, Inc.
A Subsidiary of Turner Broadcasting System, Inc.
1050 Techwood Drive, N.W.
Atlanta, Georgia 30318

Distributed by Andrews and McMeel
A Universal Press Syndicate Company
4900 Main Street
Kansas City, Missouri 64112

First Edition 10 9 8 7 6 5 4

Printed in the U.S.A.

Library of Congress Cataloging-in-Publication Data
Shepard, Alan B. (Alan Bartlett), 1923-
    Moon Shot : the inside story of America's race to the moon / by
Alan Shepard and Deke Slayton with Jay Barbree and Howard Benedict.
    —1st ed.
        p.   cm.
    Includes index.
    ISBN 1-878685-54-6 : $21.95 ($28.50 Can.)
    1. Project Apollo (U.S.)—History. 2. Space flight to the moon—History.
I. Slayton, Donald K., 1924-1993. II. Title.
TL789.8.U6A58177 1994
629.45'4'0973—dc20   94-3027
                                                                  CIP

For

BOBBIE

JO

JOY

LOUISE

The wives

CONTENTS

UNA INCOGNITA. The unknown moon. A silent sentinel. For all of man's history it had hung overhead, remote, unreachable, unknowable.

Marching across the heavens each day and circling our earth monthly, the moon has fascinated scientists and inspired poets. Its changing shape provides a perpetual clock-calendar in the sky, a marker for planting, for holidays, for religious celebrations. So near and yet so far, men and moon intertwining for millennia, but never touching.

In the twentieth century, two distinctly different technologies emerged: the digital computer and the liquid-fueled rocket. Two great world powers, ideological adversaries, each recognized that the rocket, which could operate in a vacuum, and the computer, which could enable precision navigation, might break the barrier to space travel.

Both the Soviet Union and the United States believed that technological leadership was the key to demonstrating ideological superiority. Each invested enormous resources in evermore spectacular space achievements. Each would enjoy memorable successes. Each would suffer tragic failures. It was a competition unmatched outside the state of war. Finally, and unpredictably, the competitors would join in a cooperative effort that would contribute to the demise of the Cold War that enveloped them.

The moon's isolation of nearly 5 billion years would soon end. Early in the space age, man-made probes flew near the moon. Others soon crashed into the lunar surface. Robot craft landed and transmitted pictures and scientific measurements back to earth laboratories. The stage was set for a visit by man.

The Soviets established an impressive number of "firsts": first to place a satellite in orbit, first to send a probe to the moon, first to place a human in space, first to orbit two manned craft simultaneously, first to have a human exit his craft in space. But it would be the Americans who would accomplish the seemingly impossible, sending men to the moon and returning them safely to earth.

History will remember the twentieth century for two technological developments: atomic energy and space flight. One threatened the extinction of society, one offered a survival possibility. If earth were ever threatened by man-made or natural catastrophe, space flight, could, just possibly, provide protection or escape.

Alan Shepard and Deke Slayton knew the practical aspects and the visceral feelings of flight. Both were experienced airplane test pilots. Test pilots have the responsibility for finding errors in airplane design. They may discover them during flight, but they would much prefer to identify the problems before going aloft. As two of the seven initial American astronauts, this search for perfection served them well.

Deke and Alan were at the heart of the manned space program. Deke was responsible for the selection of flight crews and their preparedness to fly in space. He took an intense interest in the well-being of his flock, protecting, supporting, and encouraging them. They were test pilots, and he understood them. He was a superb boss.

Alan, as chief of the Astronaut Office, was responsible for day-to-day operations. Astronauts were needed for spacecraft tests, for design reviews, for newspaper interviews. With equanimity, he distributed these seemingly limitless tasks to a very limited number of "his boys." He was an impenetrable barrier to inappropriate or untimely requests. He was "the man in the middle" and handled it well.

Moon Shot is their story. Much more than the story of their flights in space, it details their central role in the most exciting adventure in history. Howard Benedict and Jay Barbree, two of the world's most experienced space journalists, reported the triumphs and the tragedies from the dawn of the space age. They are exceptionally well qualified to recall and record the remarkable events and emotions of the time.

Deke Slayton did not survive to see this book in print. He was taken from us much sooner than we would have liked. But he would not have us mourn. It was not his way. He played the hand he was dealt, good or bad, with equal enthusiasm. We are in his debt.

Luna is once again isolated. Two decades have passed without footfalls on its dusty surface. No wheeled Rovers patrol the lunar highlands. Silent ramparts guard vast territories never yet visited by man. Unseen vistas await the return of explorers from earth.

And they will return.

Neil Armstrong
February, 1994

PROLOGUE

THEY DESCENDED THROUGH rolling clouds building on a hot summer afternoon over southeast Texas. The first tremors of heat rising from the earth jostled their aircraft, and they grew impatient to plant their feet on solid ground. Arriving from all points of the compass in airliners, corporate jets, military aircraft, even private propeller planes, they were members of one of the world's most elite fraternities.

They were astronauts. The first in the race to orbit the earth, to walk in space, and the first and only to walk on the moon. They were men and women with the names of Shepard, Leonov, Glenn, Borman, Stafford, and dozens more—all coming to Houston in June 1993 to pay their last respects to Deke Slayton, NASA's former chief of flight operations, a fellow astronaut, friend, and mentor who had recently died of brain cancer.

Looking down across the land and seeing the neat, campus-like Johnson Space Center, where the seeds for the early space journeys had been nurtured, some were surprised at the familiar landscape. They had expected change after all these years, but at first glance the surroundings seemed just as they remembered them. As the wings of their planes banked, they looked upon the buildings that had been constructed so many years ago on land donated by Rice University. First created as the Manned Spacecraft Center, no one had known when the site was selected in 1961 how long the nation would support men in space. Only two Americans and two Russians had by then ridden rockets away from the earth. But those early missions had been enough to trigger a moon race between two superpowers seeking national prestige in a world where many nations were choosing sides, and technical and scientific superiority was crucial to how the United States was perceived on the world stage. The goal was to put a man on the moon, and in the 1960s both Washington and Moscow grabbed for the brass ring, marshaling enormous engineering and industrial might for the massive challenge this goal represented.

America's National Aeronautics and Space Administration would enlist more than twenty thousand companies and more than four hundred thousand people to do the job. At the very apex of this pyramid of talent and ambition were seven outstanding military pilots with the new title of astronaut, men of strong will and purpose, skilled in flight, who would lead the way to another world. These men, idolized and hailed as heroes, were called the Mercury Seven. Their names became household words: Scott Carpenter, Gordon Cooper, John Glenn, Gus Grissom, Wally Schirra, Alan Shepard, and Deke Slayton.

There were critics of the great adventure who condemned President John F. Kennedy's quest for national honor as a waste of money. They complained that the cost of reaching the moon was at the expense of Americans with down-to-earth needs. Those who had a clear vision of the future dismissed the carping and grumbling. The technical effort that would go into landing on the moon would have countless other applications: it would accelerate development of technology to build better, smaller, and less expensive computers and create miracles in electronics, raise global weather forecasting from a medieval art to a science, revolutionize communications, and use the vantage of weightlessness to create new medicines, machines, and cures for the diseases and plagues of mankind.

But many scientists, looking at the greater stage of history, pushed away all such reasons and justifications. Man goes higher and higher, they said quietly, because of basic instinct. From that time when the first thinking being struggled to stand upright, when he began to rationalize and think in abstract terms, when he remembered the past so he could better see the future, he was driven by this inborn instinct to grasp for heights, to reach outward.

The astronauts were America's first messengers as the world reached for unprecedented heights beginning in the 1960s. And when a young president with great clarity of vision said they should go to the moon, they went.

Now, decades later, those same astronauts were returning to Houston to salute Deke Slayton, the man who helped them realize dreams they had all shared, the man they had chosen to lead them. Their aircraft rolled out and leveled as they began their final approach, the runways rising to meet them. Each astronaut saw the blurred outline of his or her plane moving across the space center and realized that parts of his or her own past were gone forever. As they passed Mission Control, time stopped for many of them. They looked through walls, through metal and glass, seeing with the startling clarity of memory their own eager younger selves, seeing the years roll back to the time they first reached the moon.

CHAPTER ONE

THE LANDING

THE TWO ASTRONAUTS SWOOPED toward the lunar landscape in their landing craft—the first of their kind to descend on the moon. As they looked through their helmet visors in wonder, the lifeless face of the moon rushed toward them.

The landing ship was named *Eagle*, and within its boxy and cramped cabin Neil Armstrong and Buzz Aldrin stood with legs spread slightly, their booted feet flat against the deck flooring. Each was sealed within the protective layers of a pressurized spacesuit, and each instinctively tugged at the cinches on his body harness, secure in his tight confines.

Eagle rushed on in the grip of the gravitational pull of the moon below. Armstrong and Aldrin knew they were moving by what their instruments told them and by the gliding motion of craters past their triangular windows. Inside their capsule they were still virtually weightless, freed of the physical sensation of earth's gravity as their craft slid along its parabolic arc without resistance.

Flying backward with their faces parallel to the silent and airless surface below, they glanced at the glowing numbers of their timers. They were minutes from the moment they would ignite the engine beneath their feet and descend to the moon's surface. Time seemed to stretch endlessly.

A quarter million miles away, in Houston, a fellow astronaut named Charlie Duke studied glowing instrument panels, listened to the chatter of his moon-bound colleagues and that of those manning consoles around him. Everything he observed and heard fed him critical data about *Eagle*'s plunge toward the lunar surface.

Duke was the only one authorized to communicate with Armstrong and Aldrin at this moment. It was time to send the message.

"*Eagle*, Houston," he spoke into his microphone. His words, crackling and charged with static, raced across space at 186,300 miles per second to the two astronauts falling toward craters and dusty rocks. "If you read, you're GO for powered descent."

At that moment *Eagle* was curving downward from the backside of the moon, emerging from a twenty-two-minute period during which the moon was between the spacecraft and the earth, blocking communication with Mission Control. Duke's sputtering message was the first the astronauts heard as they came around the lunar limb.

Armstrong and Aldrin were not alone in space. A third member of the Apollo 11 crew, Michael Collins, was fifty miles above them, in lunar orbit in their command ship, *Columbia*. He had heard clearly the vital message from the control center.

"*Eagle*, this is *Columbia*." His words flashed instantly into the headsets of Armstrong and Aldrin. "They just gave you a GO for powered descent."

The two men glanced at each other. "Roger," Armstrong acknowledged. They were now headed for a waterless sea known as Tranquility.

Inside Houston's Mission Control Center, a small army of tense flight controllers sat with eyes riveted to their data consoles. The control center was configured in terraced rows of consoles and monitors, and at the back of the huge room the last row of electronic equipment was the highest. Each successive row toward the front of the room was lower than the one behind it, so that each skilled technician sitting at these consoles had a clear view not only of the lead people deep in electronic wizardry but also of the wide projection screens covering the front wall. The screens of the monitors before each man glowed and flashed with every twitch and slide of the *Eagle*.

In their midst was a man who spurned the long hair and outlandish garb that defined the youth of his decade. With an outdated crew cut adding starkness to his features, Gene Kranz seemed strangely out of place among his team. He was the final authority, the flight director whose sweeping powers would decide the what, when, and where of the first-ever descent and landing on the moon. Now, with *Eagle* dropping silently toward lunar dust and craters, no one called him by his name. He answered to the name of Flight.

Kranz leaned into his communications panel. With an easy motion he switched from the standard talk network to an auxiliary loop. Now he could be heard only within Mission Control. What Gene Kranz had to say was meant for his team only.

"Hey, gang." Heads turned. Kranz smiled. "We're really going to land on the moon today. No bullshit. We're really gonna do it." It was July 20, 1969.

Smiles, grins, and thumbs-up met his words. That's all Kranz said for the moment. He took his seat and switched his microphone so that the tense, nail-biting visitors in the viewing rooms could again hear him speak. These visiting dignitaries, including congressional officials and members of the NASA family, marveled at the sight of the man with the crisp clothes and dazzling vest of white brocade and silver thread—the good-luck vest Kranz wore with professional calm.

The calm was a front. The flight director's stomach was knotted, his heart thumped hard in his chest, and when he lifted his right hand from the cover of the moon-landing flight plan, he left behind a wet, perfect image of his perspiration-soaked palm.

Close by the flight director Kranz, Deke Slayton studied the "situation monitors." A fighter pilot, a test pilot, and one of the original seven astronauts of Project Mercury, he had been grounded.

Slayton could have been one of those men in *Eagle*. But he had never flown a Mercury spacecraft. Nor had he ever lofted at five miles a second into orbit within a Gemini, or tested an Apollo in one of its checkout flights to the moon. The men in Mission Control who looked at Deke Slayton knew all this, felt for the man, marveled at the grim steadfastness that kept him supporting astronauts after he had been chained to the ground for medical reasons. Slayton was now one of the bosses. He was the astronauts' "mother hen," and he would gladly have traded it all for just one space flight.

Deke was there as an observer. If something went wrong, he wanted to be on hand for Armstrong and Aldrin. He would slide into Charlie Duke's place as capsule communicator to provide his friends advice and soothing words from a man they worked closely with and respected.

Elsewhere in the room was another of the original Project Mercury astronauts. Alan Shepard was also there as an observer. Like his friend Slayton, he too had been grounded by the medics. Alan at least had had the good fortune of flying in space before his problem hit. In fact, in 1961, he had been the first American to fly in space. It was only a fifteen-minute flight, but it set in motion a race to the moon that was this day reaching its culmination.

Alan and Deke, two major players in America's space program, watched intently as the moment of truth approached.

The fateful words winged across space.

Eagle was GO to ignite its descent engine, and Armstrong and Aldrin locked their eyes to the glowing numbers displayed before them. They were almost at an invisible junction of height, speed, range, and time when everything would join together for commitment. When the instruments told them that they were 192 miles from their projected landing site, and were precisely 50,174 radar-measured feet above the long shadows of the moon, they would unleash decelerating thrust and begin slowing their speed for the touchdown.

Green-bright digits changed constantly, numbers flashing by in a breathless blur.

This was it. PDI.

Powered Descent Initiate.

On earth, radio listeners and television viewers held their breath. People prayed. Fingernails dug into palms.

Armstrong and Aldrin braced themselves for the shock of ignition as volatile propellants rushed through lines, sprayed into the ignition chamber of the powerful descent engine, and lit off. They waited for the punishing deceleration, a full body blow that would buckle their knees and vibrate through their ship.

They were met with silence and calm.

Had the descent engine fired? Where was the gripping tug of deceleration, the onset of force?

They glanced at each other, only an instant, and glued their eyes to their instruments. Confirmation of ignition glowed in the numbers. The phantom fire veil spewed forth gently, as planned, perfectly following the program. Yet the astronauts had expected some sensation.

At 10 percent power there was no sudden hard smash of energy as the *Eagle's* engines were brought to life with a caress. Gently the ship descended through the black sky.

The *Eagle's* electronic brain monitored the deceleration, measured the loss of velocity, judged height, and confirmed the angle of descent. The invisible hand of the computer then began to add power.

Throttle up. Full power!

Flame gushed beneath them. Glowing, gleaming plasma in a shock wave buoying them in vacuum. The *Eagle* rocked from side to side and pitched violently. The computer—sensitive, alert, instantly responsive— fired control thrusters to hold the craft steady. Hollow thuds, distant sub- dued bangs, could be heard within the capsule as the small thrusters bal- anced *Eagle* in the ultimate balancing act.

Gravity pulled at *Eagle* with a vengeance as it decelerated. Inside their capsule Armstrong and Aldrin, who had been weightless, free from the weight of the heavy pressure suits and helmets and boots and backpacks, were once again in a gravity field. Their arms sagged. Legs settled within their suits. Their feet pressed downward in their boots as they yielded to their down-rushing speed.

Neil Armstrong smiled. His eyes were tired but warm with anticipa- tion, immersed in the reality of their incredible adventure. He saw Buzz Aldrin grinning like a kid.

Good Lord, they were going to land on the moon!

Fuel pumped through the lines under full throttle. Flame spewed far ahead and beneath them. The *Eagle* was in full fury now, blasting away her weight and mass, slowing, slowing.

Headsets crackled. Charlie Duke in the control center was incredibly calm and professional as he called out: "*Eagle*, Houston. You are GO. Take it all at four minutes. You are GO to continue powered descent."

But all was not well.

Back on earth, Mission Control was thick with tension.

The highly trained flight controllers were focused on the monitors and consoles before them, tracking the curving line of the *Eagle's* landing path. They were waiting for the vertical metal probe that extended from the landing legs of the ship to touch the moon's surface and signal a successful landing.

Those manning the front row of consoles were in what was known as the "trench." This was where final decisions were made. Where ultimate responsibility lurked unseen but inescapable. Where Deke Slayton concentrated his attention.

Deke had confidence in all those who worked in the supercharged atmosphere of the trench, but he was especially keen about a twenty-six-year-old computer hotshot named Steve Bales. Like many in the control center, he was young. In terms of his experience and skills, however, he was a seasoned veteran.

No one called him by his own name. With a mission underway, he became GUIDO, the acronym for guidance officer. To the old-timers in the space flight business, Bales was pure genius, even if they sometimes referred to him as "that kid genius."

Today Bales had been early for his shift, excited, filled with anticipation and wonder at what was coming. This was the most important, demanding, and exciting day of his life, and that sobering thought stayed with him as he took his seat at the guidance officer's console and nervously began twisting a lock of his hair along the back of his head into a small pigtail.

Bales made a mental run through the list of possible signals that could sound danger alarms at any point in the epochal descent of the *Eagle*. And he knew that twenty-four-year-old Jack Garman, in the back room, was running through the same mental check list. Both were experts on the lunar lander's computers, and they shared the same knowledge and concerns.

Deep within the bowels of the *Eagle* were computers essential to measuring all the electronic and mechanical forces and factors that would determine the success of the lunar landing. A landing soft enough to safeguard the health of the two astronauts and maintain the structural integrity of the bug-eyed lander demanded a complex monitoring system. Changes in speed, rates of deceleration, shifting centers of gravity, weight, and balance, and engine thrust were all factors that were too sophisticated for the human mind to process without the aid of computers, the electronic brains designed to perform superfast computations and to ride shotgun on everything that happened aboard *Eagle*.

The brief burst of rocket gases from a thruster, for example, would register one and a half seconds later in Houston. At the speed of light, the exchange of conversation between Houston and the *Eagle* skittering toward the Sea of Tranquility required three seconds.

Every man in the tiers of Mission Control knew the computers aboard the *Eagle* also contained sensitive electronic watchdogs. Alarm systems to detect imbalance, misalignment, deviation from the exquisitely created flight plan. Only Bales and Garman were familiar with each of those alarms and what it meant. They were the only two people in that vast control system equipped to interpret any alarm emergencies aboard *Eagle*.

Everything they monitored aboard the landing craft was green and go. The tension was there, but everyone was feeling pretty good about the descent.

Suddenly *Eagle's* computers shrilled madly.

Alarm!

Emergency signals flashed within *Eagle* and one and a half seconds later on consoles in Houston. No one expected a cry of danger. Not now.

Eagle's descent engine had blazed at partial throttle for twenty-six seconds. Everything fit within the flight plan. The flight-control computer advanced the throttle to full power. For another three minutes and thirty-four seconds everything had gone well.

But at six thousand feet above the moon a yellow light flashed at the two astronauts. Buzz's voice responded immediately as he called out the numbers flashing on his flight panel and on the console before Steve Bales.

"Program alarm," Buzz snapped crisply. "It's a twelve-oh-two."

Twelve-oh-two. A warning that the ship's main computer was overloaded. So much was happening, so many performance signals were being generated that the computer could not absorb them all. It was a cry for help.

There was no panic in Mission Control. But everyone sensed an abort. A hellish maneuver that would explosively separate the upper ascent stage of the *Eagle* from the landing stage and squeeze every ounce of thrust from the ascent rocket to make it climb back to an altitude of sixty miles for docking with the command ship *Columbia*.

The flight controllers crossed their fingers, prayed, and clenched their teeth. They swallowed hard, as if their hearts had surged into their throats. All eyes were on Steve Bales.

He stared at his console. Coded numbers told him instantly what was going wrong. He wanted, needed, confirmation of safety and a better identification of the problem. Immediately he was in contact with the back room, where Jack Garman had heard the call, had checked his computers, had made his own analysis.

Garman spoke carefully, articulating his words to Bales. "It's executive overflow. If it does not occur again, we're fine."

Bales understood instantly. The computer within *Eagle* was being overtaxed with data. Bales knew that force-feeding a computer could lead to overload. But he also knew this computer. He understood what it was doing.

It was doing just what it was supposed to do. *Eagle's* computer operated on a fixed cycle of one second. He saw it was recycling on schedule and knew the hardware was probably still in good condition. Every second, the computer needed to navigate, guide, trim engine thrust, update all earlier data, and display multifaceted data to the crew. It also was performing the calculations necessary for the *Eagle* to abort the landing.

If the computer failed to do all these things in the allotted one second, it would flash the 12–0–2 overload alarm.

To hell with that, Bales judged. He and Garman had programmed the *Eagle's* computer so all the key functions were performed in the time allotted. His computer was working, and that's what mattered.

But Neil Armstrong's voice demanded a response. "Give us the reading on that twelve-oh-two program alarm," he asked.

"GUIDO?" Gene Kranz shouted the question into the loop of Mission Control. Everyone hung on the edge of the cliff.

Bales wanted more time.

Kranz didn't have time. Armstrong and Aldrin didn't have a single second to spare as they plunged downward. Kranz stared at Bales. The flight director slammed a clenched fist against his console.

Bales jerked in his seat. No time to judge, weigh, consider.

"GO!" he shouted. He closed his mike, staring at his console. "Go, damn it," he said in a hoarse whisper only to himself.

Charlie Duke showed surprise. He didn't have time to wait, either. His words came forth immediately. "We've got, uh, we're GO on that alarm, *Eagle*."

The beat speeded up. The closer *Eagle* came to lunar soil, the greater the need for instant judgment and reaction.

The astronauts were four thousand feet above moondust. Kranz keyed his mike. "All flight controllers. Coming up on GO-NO GO for landing."

He called every flight controller in the big room. Everyone responded with "GO." Until Kranz again came to Steve Bales. His guidance officer hesitated.

Kranz had no time for hesitation. Not with a ship and two men blazing onto another world a quarter million miles away.

"GUIDO, you happy?" Kranz snapped.

Bales was not pleased. Program alarms still were leaping onto the monitoring screens. Bales had to judge whether Neil and Buzz would have a good computer working for them not just now but tomorrow as well, when they were scheduled to blast off from the moon. In a flash of memory and instinct, he reviewed all the practice and simulator runs they'd made. Just weeks earlier, similar alarms had sounded in the simulator, and he and Garman had studied them, calculated the system was working all right. But that was a practice. This was it. He had a gut feeling and an absolute conviction the hardware was working properly, that he wasn't jeopardizing the lives of two men so far away. Yet, there were those alarms. . . .

"GO," he said firmly.

Charlie Duke called it out. "*Eagle*, you're GO for landing."

Three thousand feet up another alarm sounded. A 12–0–1. They made an immediate judgment call in Mission Control. Another "executive overflow."

"You're GO," said Charlie Duke.

Two thousand feet high, craters widening rapidly below, Neil called it out again.

"Twelve alarm. Twelve-oh-one alarm."

Kranz shouted to Bales. "GUIDO? What about it?"

Deke Slayton locked eyes with Steve Bales. The young man read confidence in that look. It was Steve's call. Whatever he said would . . .

"GO!" Bales snapped. "Just GO!"

Charlie Duke looked at Slayton. Deke grinned, turned his right thumb upward with a quick, firm, stabbing motion. Instinctively Deke scanned the room for Alan Shepard. He spotted him in the rear of the room. Their eyes met. Both winked, flashed the thumbs-up sign.

Charlie keyed his mike. He swallowed hard. "We're GO, *Eagle*. Hang tight, we're GO. . . ."

Thirteen hundred feet above the moon's surface, *Eagle* began its final descent. Flames gushed downward as the craft slowed. Neil Armstrong had flown his mission right along the edge of the razor. He and Buzz functioned as one mind. Now they were doing more than falling moonward. They were so close Neil had to *fly* this ship. He punched PROCEED into his keyboard. The computer would handle the immediate descent tasks. Buzz would back up both man and electronic brain so Neil could switch his eyes and senses to flying in vacuum.

Both men looked through the triangular windows to study the surface of the moon. They'd made simulated runs so many times they knew their intended landing site as well as they did familiar airfields back home. Almost immediately they noticed that they weren't where they were supposed to be.

Damn!

Eagle had overshot the landing zone, Home Plate, by four miles. A slight navigational error and a faster than intended descent speed accounted for the *Eagle* missing its planned touchdown site in the Sea of Tranquility.

Neil scowled at the surface still rising toward them. Boulders surrounded a yawning crater wider than a football field, and *Eagle* was running out of fuel and headed straight for it. There was no time to waste.

In the lunar void there was no gliding to conserve fuel. *Eagle* was only dead weight in vacuum. There also was no opportunity to orbit again for another try at landing. Their only chance to succeed was to land now.

Neil Armstrong gripped the hand controller in his fist, firm and strong, with a touch honed by years of flight in jets and rockets. He knew the "thin edge" well, both in atmosphere and in hurtling in orbit about his home planet. Now he had to fly as he'd never flown before. Knowledge, experience, skill. X-planes and spacecraft, everything coming to this one moment, this place above desolation.

Neil Armstrong needed to hand-fly the rest of the way.

His fingers alternately tightened and eased his grip on the controls. *Eagle* was sailing down at twenty feet per second. Neil nudged the power, slowing to nine feet per second.

He needed to *feel* his ship.

He attuned his senses to the rocking motions, the nudges, and the skidding motions of the sixteen small positioning thrusters that kept the *Eagle* aligned throughout its descent. A level touchdown was their ticket to safety, survival, and the return home.

Mission Control listened, mesmerized and awed, to the voices closing in on lunar soil. Neil guided the bird. Buzz watched the landing radar, called out numbers that bespoke volumes of split-second judgment and maneuvering.

"Five hundred forty feet [height above the surface], down at thirty [feet per second] . . . down at fifteen . . . four hundred feet, down at nine . . . forward . . . three hundred fifty feet, down at four. . . . three hundred feet, down three and a half . . . forty-seven forward . . . one and a half down . . . thirteen forward . . . eleven forward, coming down nicely . . . two hundred feet, four and a half down . . . ."

It sounded better than it was. Neil balanced the bug-like machine on a solid cone of fire. *Eagle* was now in a directed drifting mode. There was no place to land. Rocks, huge boulders, and deadly craters were strewn everywhere.

Mission Control was dead silent.

Neil fired the *Eagle*'s right bank of maneuvering jets. The *Eagle* scooted across rubble billions of years old.

There! There beyond a field of boulders, slightly to the left—the rocks were thinning out, revealing a smooth, flat area. The new Home Plate.

The numbers ghosted back to earth.

"Five and a half down . . . 5 percent . . . seventy-five feet . . . six forward . . . ninety seconds," Buzz chanted. "Ninety seconds."

He had been carefully watching the fuel gauge, as had Mission Control. Ninety seconds of fuel left in their tanks for the descent. The *Eagle* needed to land in ninety seconds or—

No one wanted to think about it. If the engine gulped its last surge of fuel before they touched down, this close to the moon, they would crash, falling to the surface without power.

Buzz knew there was plenty of fuel for *Eagle*'s ascent stage. But that was needed for an abort or to get off the moon after a successful landing. If the descent fuel was exhausted before the ascent engine kicked in on an abort, it would be too late.

Neil didn't bother with the ifs and could-bes. He'd flown this same approach hundreds of times in the ground simulators and in the clumsy, jet-powered lunar landing trainers.

At such moment a pilot flies. Everything else is shoved aside. Neil could *feel* what fuel they had left. His eyes and mind and hands worked beautifully with orchestrated skill. He would bring *Eagle* down and bring her down level.

It would not be easy. *Eagle* was now top-heavy, the ascent stage still crammed with fuel, the tanks of the descent stage perilously close to empty. Neil and Buzz might not have worried about setting down at a steep angle, but they needed to remain on an even keel. Otherwise, *Eagle* could hit the surface tilted, out of position to relaunch itself for its life-saving rendezvous with the command ship *Columbia*.

Slayton, Shepard, and everyone else in the control center were gritting their teeth. Anxious and concerned, all they could do was watch. The fate of the mission was in the hands of the two astronauts.

Charlie Duke sounded the warning. "Sixty seconds."

There was one minute of fuel left. In sixty short seconds the rocket power flaming beneath *Eagle* would burn out. The tanks would be empty. An abort would need to be initiated seconds before that happened if *Eagle* was not to crash.

Balancing on slashing flames and banging thrusters, Neil Armstrong calmly aimed for his new landing site.

The *Eagle*'s commander kept one thought uppermost in his mind. Fly his machine. That was it. Simple, exquisitely to the point. Fly it down the invisible corridor. *Eagle* swayed gently from side to side as the thrusters responded to Neil's commands.

Far away, down through the atmosphere and the clouds, enclosed within Mission Control, the flight controllers were almost frantic with their inability to do anything more to aid Neil and Buzz.

Deke Slayton knew they had to leave it to the pilots and "final approach." As *Eagle* skittered over boulders and across craters, only Neil Armstrong's judgment counted. He was there. He was flying. But the clock kept ticking, the precious remaining seconds slipping away like grains of sand in an hourglass. Charlie Duke looked at Deke. Charlie held up both hands, palms out. He didn't need to voice the question. Gene Kranz did it for him on the internal communications loop. Everyone heard his words.

"Capcom, this is Flight." Kranz's words to Duke were firm and steady. "You'd better remind them there ain't no damn gas stations on the moon."

Charlie nodded and keyed his mike. A timer stared at him. "Thirty seconds," he said. An entire speech in two words.

"Light's on."

This time the announcement was from Buzz Aldrin as he watched an amber light blink balefully at him from the master caution-and-warning panel. It was the low-fuel signal.

Buzz eyed another button, half afraid he might have to punch it. It read ABORT STAGE.

Neil didn't respond. There was no time left for word exchanges. All his senses, brought to needle-point sharpness at this moment, concentrated on the multiple tasks involved in settling *Eagle* on the moon. He searched the ground below, experienced eyes seeing, judging, deciding. The lunar landing vehicle had come out of the vacuum sky like a meteor. It was now transformed. It was more helicopter than spaceship or airplane, but its lifting and hovering force was fire, transparent thrust.

Eagle's descent engine burned fiercely and silently in the void.

Buzz intoned the numbers like a priest, steady and clear, voicing the final moments flashing away. He had confidence in Neil's ability. But his hand did not stray far from the ABORT STAGE button.

"Seventy-five feet," he called out.

"Six forward . . .

"Light's on . . . down two and a half . . . forty feet, down two and a half . . ."

Eagle was now slipping downward fifty feet above the moon. Men and machine embraced a new level of potential danger. This close to the surface, they had no margin for error. If their space vessel failed them, or if they ran out of fuel, they would not have time to abort. If they ran into any problem this close, there would be no time for circuits and solenoids and explosive charges to separate the *Eagle* from its lifeless descent stage, no time for fuel to stream through lines into the ascent stage's combustion chamber beneath their boots, no time for the fuel to ignite and hurl them upward before they crashed.

Time was their enemy.

"Thirty feet . . .

"Two and a half down . . ."

Then, the magic words!

"Kicking up some dust . . .

"Faint shadow . . ."

So close now! So close!

There was no turning back! The door behind them had closed.

"Four forward . . .

"Drifting to the right a little . . . "

Everyone in Mission Control, and in the visitors' viewing gallery, and throughout the vast halls of NASA, everyone anywhere who knew what was happening just above the moon was hoping, praying, straining.

Fuel flashed away.

Neil Armstrong flew *Eagle* with the smooth touch of an experienced pilot landing a winged aircraft at his home airfield.

Still the fuel blazed away. Millions of hearts pounded madly.

Then, these words from Buzz Aldrin on the moon. "Contact light!"

"Okay, engine stop . . . descent engine command override off . . ."

In Houston, Capcom Charlie Duke was choking with relief. But he still needed voice confirmation. He wanted to hear the words.

"We copy you down, *Eagle*," he radioed. Then waited.

Three seconds for the voices to rush back and forth, earth to moon and moon back to earth.

Those three incredible seconds, and then came the call.

"Houston . . . "

Neil Armstrong had landed so smoothly that Buzz wasn't taking any chances. Were they really down? Stopped? He studied the lights on the landing panel to be certain the wonder was real.

Four lights gleamed brightly on the panel. Four marvelous lights welcoming them to another world where no human being had ever been. The lights banished all doubt. Four round landing pads at the end of the *Eagle's* legs rested, level, in lunar dust.

Neil allowed himself the luxury of a long, deep breath as he stared through his helmet visor at the alien world before him. He was surprised at how quickly the dust, hurled away by the final thrust of the engine, had settled back on the surface. Within seconds the moon looked as if it had never been disturbed by the strange machine now resting on the firm lunar soil. Buzz stared at the rocks and shadows of the moon, marveled at the horizon that curved into velvety blackness just a mile away.

Neil's voice was calm, confident, clear.

"Houston, Tranquility Base here. The *Eagle* has landed."

It was 4:17:42 P.M. EDT, Sunday, July 20, 1969.

Charlie Duke spoke above the bedlam of cheering and applause in Mission Control.

"Roger, Tranquility. We copy you on the ground. You've got a bunch of guys about to turn blue. We're breathing again. Thanks a lot."

"Thank you . . . "

Deke was exuberantly pounding on the back of and shaking hands with anyone he could reach. His eyes sought out his friend Alan Shepard, who was doing the same. Intuitively Alan looked at Deke. Again they exchanged a thumbs-up. There would be time later to celebrate.

The words flashed across space to Houston.

"That may have seemed like a very long final phase." Neil Armstrong was all business again. "The auto targeting was taking us into a football-field-sized crater, with a large number of big boulders and rocks for about one or two crater diameters around it, and it required flying manually over the rock field to find a reasonably good area."

Charlie Duke had a grin like a mule eating briars. "Roger, we copy. It was beautiful from here, Tranquility," he sang out. "Be advised there's lots of smiling faces in this room, and all over the world."

Desolation rolled in from all sides the longer they looked out at the moon. No birds, no wind, no clouds, no blue sky. Only rocks and shadows and craters and dust.

Neil and Buzz turned to face each other. Buzz grinned, reached over, and clasped Neil's hand. Then, overwhelmed by the enormity of the moment, they slapped each other about their backs as a sound never before heard on the moon burst through their headsets.

Charlie Duke had opened his microphone, and the wild and tumultuous celebration inside Mission Control poured and vibrated and rumbled across a quarter million miles and brought smiles to the faces of the only two living creatures on the moon.

Half a continent away, at Cape Canaveral, Florida, several members of the team that had launched Armstrong, Aldrin, and Collins on their epochal journey cheered as radio messages confirmed the *Eagle's* safe landing on the Sea of Tranquility.

One engineer turned to a fellow worker. "I want you to do something for me," he said. "Can you hold your breath for a full minute?"

"Sure, why?" the man replied.

"Take a deep breath now," the engineer instructed as he tripped his stopwatch and told the man to start counting in his head.

Suddenly the engineer slammed his hand against the back of his coworker. "Hey! Why'd you do that? I was only up to sixteen seconds! That was nothing."

"That's the point," the engineer grinned at his fellow workers. "That's all the fuel they had left in that lander. They had nothing."

THE BEGINNING

THERE ARE SOME SOUTHERN TOWNS that are cocooned in time, content to let the industrial and technological age pass by. In the mid-1950s one such community was Huntsville, Alabama. It was like many other towns of its vintage and size, moving with a courtly glide, its major contribution to its citizens a courthouse centered in the town square. Like many of its sister communities, the square held a monument to the Confederacy. Here was the sacred ground to recall days of past sacrifices.

The future loomed barely ten miles west of Huntsville. The future was Redstone Arsenal, an unlovely complex along Alabama Highway 72 in the thick of the north Alabama clay hills and tall pines that stretched on to the Tennessee River. Here the U.S. Army loaded explosive materials into artillery shells, bombs, and other weapons that helped America secure a hands-down triumph in World War II. With the war over, however, activity at the Redstone Arsenal ceased. The Army closed the facility, and Huntsville returned to its tranquil times.

Five years later, in 1950, the arsenal came back to life as hundreds of engineers, technicians, specialists, scientists, and their support personnel descended. Their number included a 118 men who had come with their families from the center of Europe. The most prized rocket team of the infamous Third Reich.

They came to run-down Redstone Arsenal to work, to Huntsville to live, and their purpose was to construct a rocket laboratory that would propel the Western world into the second half of the twentieth century. They represented Hitler's finest, recruited by the U.S. government from a nation where only a few short years earlier Alabama's young men had fought and died. They were German citizens by birth who had been offered American citizenship and a new home amid the quiet cotton fields of rural Alabama. Having designed, constructed, tested, and launched deadly missiles for the Reich, including the V-1 and V-2 rockets whose explosive force had terrorized London during the Blitz, these scientists and engineers were now commissioned to design, construct, test, and launch long-range missiles for the United States. Arriving in Huntsville, they were confident they could exceed their past performance.

Nobody questioned their expertise. The American military was without any missile skills and considered these Germans to be the most valuable booty from the defeated Third Reich. They had been recruited through Operation Paperclip, a secret U.S. Army program created to scour Germany for rocket, atomic, and aircraft specialists who could be brought to America and kept together as a team.

The lead German scientist was Dr. Wernher von Braun, a brilliant propulsion engineer with a dynamic, commanding presence. He was a visionary who from his youth had dreamed of developing rockets to explore outer space. Many of his fellow scientists and engineers shared his vision and had established rocket clubs in pre-war Berlin. With the advent of war, these engineers had been forced to build weapons of destruction for Adolf Hitler. When von Braun's V-2 rocket first hit London, he remarked to some of his colleagues, "The rocket worked perfectly except for landing on the wrong planet."

With Germany crumbling, with the Americans and their European allies advancing from the west and the Russians from the east, von Braun called his top men to a secret meeting in a farmhouse near the launch site at the research center of Peenemünde on Germany's Baltic coast.

"Germany has lost the war," von Braun announced. "But let us not forget that it was our team that first succeeded in reaching outer space. We have never stopped believing in satellites, voyages to the moon, and interplanetary travel. We have suffered many hardships because of our faith in the great peacetime future of the rocket. Now we have an obligation. Each of the conquering powers wants our knowledge. The question we must answer is: To what country shall we entrust our heritage?" The answer was unanimous. They all wanted to surrender to America. In America they might be able to fulfill their dream of exploring space.

"We don't want to fall into the hands of the Russians," said one of the team members, Konrad K. Dannenberg. "We have had enough of a totalitarian society."

The Armament Ministry in Berlin directed von Braun to destroy all classified material relating to his missile research, fearing that it would be captured. He disobeyed, hiding his research documents in an abandoned mine in the Harz Mountains, where he planned to retrieve them later. He and several of his top scientists and technicians were then moved by SS troops to an area south of Munich, where they suspected they would be murdered by their own countrymen to keep the Allies from obtaining their missile skills.

But in the confusion of Germany's collapse, the rocket men were able to surrender to the American Army near the Bavarian ski resort of Oberjoch in May 1945. The Americans were delighted to have found the German scientists, and von Braun and 117 of his key team members were sent to the United States under contract to the Army to build rockets. Once the Germans arrived in the U.S., however, the country hardly knew what to do with them. The world was at peace, and Congress was not of a mind to appropriate much money for rocket research, much less space exploration. So von Braun and his team, lonely and discouraged, were deposited at Fort Bliss, Texas, and left to tinker with captured V-2s and instructed to teach rocketry to those in the Army who were interested.

"The United States had no ballistic missile program worth mentioning between 1945 and 1951," von Braun complained years later. "Those six years during which the Russians obviously laid the groundwork for their large rocket program are irretrievably lost."

Although the United States recruited the cream of the German rocket scientists, the Soviets captured many of those left behind and began their own missile program.

In 1950 the fortunes of the Germans at Fort Bliss changed when the Army received confirmation of Soviet rocket activity and immediately decided to establish a rocket research and development center.

Huntsville, Alabama, became the new home for the German team, and while the Army brass promised a warm reception from the local community, there were still too many empty beds, broken hearts, and still-fresh grave sites of Alabama soldiers for the people of Huntsville to welcome the Germans with any hospitality. Many were suspicious and unable to accept that the scientists had transferred their loyalties from Nazi Germany to the United States as quickly and easily as they seemed to.

Tensions eased when the Alabamans learned these men were not Nazis. And gradually this energetic, dedicated band of Germans—who had learned to speak English at Fort Bliss—won the respect and support of their stubborn hosts. Much of the credit for this turnaround went to von Braun, the charismatic leader who worked tirelessly to create goodwill within the community.

Just weeks after the arsenal reopened on June 25, 1950, North Korea invaded a startled and unprepared South Korea. Two days later President Harry S. Truman ordered U.S. troops to Korea. The Korean War energized the arsenal. The Army, under orders from Washington, directed the von Braun team to develop the country's first ballistic missile. It was required to propel a conventional or nuclear warhead two hundred miles and be mobile enough to be ferried around the battlefront by combat troops. The missile was named Redstone, after the arsenal.

Once more the German engineers and scientists were called on to build a weapon of war, their hopes for space rocket research and development derailed again. Little did they know that this slender sixty-nine-foot rocket would one day be their ticket to space.

Over the next three years the Germans and the American engineers working with them designed the Redstone from scratch, fired its engine in test stands, shaped its dynamics in wind tunnels, and verified its structure in vibrating machines. In August 1953 the first Redstone rocket roared away from the military test range at Cape Canaveral, Florida. It was a modest beginning, the Redstone flying five miles down the tracking range before crashing into the ocean.

In succeeding years Redstone missiles flew straight and true, and to von Braun they were more reliable than many of the other missiles launched from Cape Canaveral, all of which were plagued by failures and explosions.

Von Braun divided his time between Huntsville, where the Redstone was assembled, and the Cape Canaveral testing ground. Irrepressible and innovative, he wanted to set up a monitoring console on the outside walkway of the Cape's lighthouse, where from a high vantage point he would be able to observe the missile closely as it ignited and thundered into the sky. It would be like looking down the throat of a fire-spitting dragon, and while this ambition said much for von Braun's confidence in the technology he and his team had created, the unprecedented risks involved were judged to be too high by Lieutenant Colonel Asa Gibbs, who commanded the Air Force-run launch site.

Gibbs was a stocky man with a strict military sensibility who could always be found chewing on the end of a thick cigar stub. He was conservative and cautious, his ambition was to be in charge, and he did not question orders from above.

Gibbs walked up to von Braun, stood nose to nose with the German scientist, and left no doubt about what he wanted. "Wernher, you'll take down your equipment from that damned lighthouse," he growled through teeth tightly clenched around his black cigar, "and you will damn well put it back in the blockhouse where it belongs. You read me?"

"Of course," von Braun said, smiling, not quite understanding why the colonel was so upset, or appreciating that if a Redstone went awry or exploded as it lifted off, it could have meant the end of the German scientist's life.

By 1956 the Redstone had become more than a battlefield missile. Coming on line were more powerful fifteen-hundred-mile-range missiles—the Air Force Thor and Army Jupiter—and the Air Force's Atlas and Titan, the true intercontinental range brutes designed to loft nuclear warheads five thousand miles or more. Missiles this large all shared the same problem: when their warheads were hurled into space, they had to reenter the earth's atmosphere to reach their targets. Their sixteen-thousand-mile-per-hour speed would create such friction when they hit the dense air on reentry that the heat would melt the warhead. Existing materials were adequate for the warhead of the Redstone, which flew lower and slower. New protective materials had to be developed for the larger rockets, which sped toward targets at 16,000 mph. But how to test them at that speed? Thor, Jupiter, Atlas, and Titan were far from the flight-test stage. The Pentagon asked von Braun to build a rocket that could fly at sixteen thousand miles an hour, allowing it to carry out warhead reentry tests.

The Huntsville team adapted the one-stage Redstone for the job. It lengthened the rocket, modified the engine, and added two upper stages, consisting of a total of fourteen small rockets, and called the modified booster a Jupiter-C. Atop this stack they added the unarmed warhead, stuffed with recording instruments, its nose cone coated with the new protective material. It worked perfectly on the first test launch, and the dummy warhead survived.

"Do you realize what we've done?" von Braun asked his team. "We went higher than six hundred miles, we sent the warhead more than three thousand miles, and we reached a speed of sixteen thousand miles an hour—higher, farther, and faster than any rocket has flown. If we had just one more small rocket on top, we could have placed a satellite in orbit around the earth!"

The Huntsville gang, as von Braun's scientists had become known, was exuberant. Excitement swept their ranks. Von Braun and the Army asked the Pentagon for permission to add that single stage to the back-up Jupiter-C, dubbed Missile 29. Despite the resounding success of the Jupiter-C, the response to von Braun's request was anything but certain.

Two years earlier, in 1954, von Braun and other space enthusiasts from industry and academia had met in Washington to discuss the U.S. contribution to International Geophysical Year, a cooperative scientific effort through which scientists around the world would study the earth and which would be observed between July 1957 and December 1958. Von Braun said he could orbit a five-pound satellite to study the upper atmosphere by adding upper stages to the Redstone rocket. The Office of Naval Research put up eighty-eight thousand dollars, and Project Orbiter was born.

The project had a short life. A panel of scientists appointed by the White House decided that the satellite should be launched with a rocket that did not have a military origin and recommended development of a new booster called Vanguard, arguing that a rocket with non-military applications would lend more dignity to a scientific project like IGY. President Eisenhower agreed. Snorted von Braun: "I'm all for dignity, but this is a cold war tool. How dignified would our position really be if a man-made star of unknown origin suddenly appeared in our skies?"

There were reports at the time that Eisenhower and his aides wanted the world's first satellite to be orbited by an American scientific team and not by a group headed by von Braun's German war veterans. It was irrelevant to the administration that other American research centers did not yet possess the technical skills of the Germans, nor did they have the advanced hardware that had emerged from Redstone Arsenal. There was time enough for the Americans to learn, or so Washington thought.

The Eisenhower administration ignored hints from Moscow that the Soviets were quickly developing the technology that could put them in space first. Not only were they launching intercontinental range missiles to Pacific targets, but also they had announced their intention to launch their own satellite during International Geophysical Year. Washington, particularly the Pentagon, brushed off the Soviet announcement as a bluff, not believing the Soviets had developed the technology for such a feat.

But Wernher von Braun was listening. He heard the Russian broadcasts and read the detailed papers being circulated at scientific meetings around the globe. He also paid close attention to the radar reports from American sites in Iran, which confirmed beyond any doubt that Russians missiles were flying higher, faster, and farther.

Von Braun understood better than the Pentagon or the White House that if his Jupiter-C rocket could toss a satellite into orbit, then the Soviets with their five-thousand-mile-range missiles carrying heavy warheads could very well launch their own satellite.

When von Braun asked for permission to launch a satellite into space, Eisenhower denied the request, and Missile 29 was stored in a shed at the arsenal. Washington had given the Russians an open door and a free ride to lead the world into tomorrow.

If you flew nine thousand miles east from the rolling hills and piney woods surrounding Huntsville's Redstone Arsenal, you would arrive at an unspectacular piece of naked and barren landscape in the Baikonur region of the Soviet Union near the Aral Sea. Only scrub brush grows in this infertile backwater, where the few inhabitants make their way on camels.

In 1955 a small army of scientists, rocket engineers and technicians, laborers and cooks and carpenters and masons invaded this remote wasteland in Central Asia.

Living in primitive conditions until new homes and workshops could be built, the men and women of new technological Russia erected large concrete mounds, partly submerging them with earth. Enormous steel towers rose to dominate the barren landscape. Unobserved by the rest of the world, these men and women built the world's first space port, the Baikonur Cosmodrome.

They developed and tested rockets, and on the evening of October 4, 1957, they gathered around a large white rocket bathed in brilliant flood-lights. The rocket was the temple, the tiny figures scurrying around its flanks the faithful. For most of the day the rocket had experienced exas-perating technical problems that resulted in numerous countdown delays. The launch originally set for early morning was finally at hand.

The rocket was called R-7, a simple name for a momentous giant.

Inside a steel-walled room on the nearby launch pad, Sergei Korolev sat at an old wooden desk, microphone in hand, orchestrating the stop-and-go countdown. Korolev was the chief rocket engineer of the USSR, who, unlike Wernher von Braun, had the full blessing and support of his country's leader, Premier Nikita Khrushchev. His R-7, four times more powerful than the Redstone, was to send a satellite into orbit and Russia into a new page in history.

Korolev was a brilliant, simple man. He disliked fancy surroundings and, shortly after arriving at the Cosmodrome, he had built for himself a small wood frame house no better than that of any Russian peasant fam-ily. The essential difference with Korolev's house was its location: it stood halfway between the rocket assembly building and the R-7 launch pad.

Korolev had left nothing to chance. He stood side by side with mechanics and metal workers in a machine shop at the launch area, per-sonally helping to fashion and assemble what would be the first artificial satellite of the human race. Korolev followed the rule of simplicity and created a sphere of aluminum alloys with four spring-loaded whip anten-nas and two battery-powered radio transmitters that would sing their unmelodious song to the world. He fitted the satellite within a pointed metal nose cone, watched technicians installing it atop the large booster.

Once the technological glitches had been resolved, events moved rapidly and men left the launching pad for safety behind thick concrete walls. The countdown went quickly and was heard only by the launch crew and a handful of top experts and communist officials.

The unsuspecting world was about to be completely shocked.

"*Gotovnosty dyesyat minut.*"

Ten minutes and counting.

The great launch tower with workstands and umbilicals rolled back, others folded. The last power umbilicals between the launch stand and the rocket separated, falling and writhing like thick, black snakes. R-7 stood alone with its super-cold fuels venting plumes of icy fog into the night.

R-7's internal systems were alive.

The minutes were gone.

Final seconds fell like withered leaves.

"*Tri* . . .

"*Dva* . . .

"*Odin* . . ."

Korolev's voice rang out:

"*Zashiganiye!*"

Ignition!

Green-red flame created a pillow of fire that ripped into curving steel and concrete channels, blew away the darkness of the night, and sent bright-orange day flashing across the desolate landscape. The manmade light was followed quickly by a sustained roar as thunder shook all that stood for miles.

R-7 rose on a Niagara of thrust and, as it climbed into the night—a brilliant star racing against a black sky—darkness returned to the launch pad. In minutes the Russian rocket had disappeared over the Aral Sea. Korolev was far more interested in his console readouts than in the pyrotechnic wonder of his booster. The numbers were perfect. Engines had cut off on schedule, rocket stages separated as planned and, when the last engine died, protective metal flew away from the satellite. Springs pushed it free to fly in space.

The moment its power had been cut and it had been freed from the rocket, the satellite became known as Sputnik (fellow traveler). Obeying the laws of celestial mechanics, it immediately began to fall, beckoned invisibly toward the center of the earth. As fast as it fell in a wide, swooping arc, the surface of the planet below curved away beneath it, falling away at a speed of three hundred miles per minute.

Sputnik was in orbit.

Ninety minutes later, it raced over its still-steaming launch pad, its transmitter emitting a lusty *beep-beep-beep*, which blared from Baikonur's loudspeakers. Cheers and shouts of joy exploded from observers on the launch pad. Korolev turned to his associates.

"Today," he said with deep feeling, "the dreams of the best sons of mankind have come true. The assault on space has begun."

News of Sputnik swept like breaking surf across the world. Five bells clanged from Associated Press printers in newsrooms across the country, signaling a major story. Editors and reporters, jolted to attention, moved quickly to the keys pounding paper. They could hardly believe what they were reading. It was no exception in the NBC newsroom in the city of New York.

Editor Bill Fitzgerald had just put the wrap on his next scheduled newscast. He froze at his desk when the wire service machines began to clang. The bells echoed in the large newsroom as he dashed from his desk into the wire room and stood before the main AP machine. He stared with eyes wide at the incoming copy.

BULLETIN
LONDON, OCT. 4 (AP)—MOSCOW RADIO SAID TONIGHT
THAT THE SOVIET UNION HAS LAUNCHED AN EARTH
SATELLITE.
THE SATELLITE, SILVER IN COLOR, WEIGHS 184 POUNDS AND
IS REPORTED TO BE THE SIZE OF A BASKETBALL.
MOSCOW RADIO SAID IT IS CIRCLING THE GLOBE EVERY 96
MINUTES, REACHING AS FAR OUT AS 569 MILES AS IT ZIPS
ALONG AT MORE THAN 17,000 MILES PER HOUR.

"Damn!" Fitzgerald spat in disgust. Not at the news but with the realization that his fully written newscast had just gone down the toilet.

He hurried from the wire room across the wide news center and burst into Morgan Beatty's office. "Mo, we've gotta update," he yelled. "One of the damn Russian missiles got away from them, and they lost a basketball or something in space."

The veteran newscaster stared in disbelief at the agitated editor. "Give me that," he demanded, snatching the wire copy from Fitzgerald's hand.

His eyes widened as he read. "Jesus Christ, Bill, you know what this is? The Russians have just put a satellite in orbit around the earth! They've been talking about it and, dammit, they've really done it!"

Realization was sinking into Fitzgerald. He took a deep breath. "Okay, what do we do, Mo?"

"We'd better put out a hotline," Beatty said quickly. "We've got to get on the air right away." He kicked back the chair and headed for the wire room, calling over his shoulder as he left. "Get the RCA shortwave station on it! Get them on the satellite's frequency. We need the sound of that thing passing overhead!"

Sputnik hurtled through space, arcing around the world, and began a pass over the eastern United States. Its orbit took it almost directly over Huntsville, Alabama, where it was about to wreck a carefully planned evening.

More than five hundred miles below Sputnik, the Army's rocket team was enjoying cocktails at Redstone Arsenal. Top brass from Washington had joined them for an evening of business and pleasure. One of the guests, Neil H. McElroy, had just been nominated by President Eisenhower to be the secretary of defense. Wernher von Braun was delighted with the news; he judged McElroy as a man of action and quick decisions. McElroy was to replace the current defense secretary, Charles E. Wilson, who thought space flight was nonsense, and who had blocked every attempt by von Braun's team to punch a satellite into orbit.

Von Braun was eager to meet the secretary-designate. He had come loaded with charts, blueprints, slides, and reams of data, and with a magnificent meal prepared at his personal direction. Von Braun was as much a social charmer as he was a genius in rocketry. Tall, blond, square-jawed, and with an unquenchable enthusiasm for space flight, he also was experienced in dealing with bureaucratic machinery.

Wernher knew his team was the only group in America with the experience and the ability to launch a satellite into orbit. He also knew that no matter what else went on in Washington, the man who was president had also fought the legions of the Wehrmacht and seen what von Braun's V-2 rockets could do to helpless cities. Von Braun knew Eisenhower would neither forgive nor forget.

Von Braun joined with his American friend and sympathizer, Major General John B. Medaris, commander of the Army Ballistic Missile Agency at Redstone, to plead their case to launch a satellite with a souped-up Redstone rocket. They argued that Project Vanguard would fail and that the Soviet Union would embarrass the United States by orbiting the first satellite.

That evening von Braun launched into spellbinding oratory as he briefed McElroy on the potential of the Army's rockets to bring

American space flight into reality. McElroy listened with interest and understanding. Wernher and Medaris were jubilant; they felt they were getting through.

"Dr. von Braun!"

Wernher stopped in mid-sentence. They all turned to see a man running into the room.

"They've done it!" shouted Gordon Harris, the public affairs director for the rocket team.

"They've done what?" demanded von Braun.

"The Russians . . . " Harris ran up to join von Braun, Medaris, and McElroy. "They just announced over the radio that the Russians have successfully put up a satellite!"

The room froze in stunned silence. "What radio?" von Braun snapped at Harris.

"NBC." Harris sucked in air. "NBC in New York. They reported a bulletin from Moscow Radio. They've got the sounds from the satellite. The BBC has also picked up—"

"What sounds?" von Braun interrupted, his voice a steady monotone.

"Beeps," Harris told him. "Just beeps. Over and over. That's all. Beeps."

Von Braun turned to McElroy. "We knew they were going to do it," he said acidly. "They kept telling us, and we knew it, and I'll tell you something else, Mr. Secretary." A tremor of suppressed fury entered his voice. "You know we're counting on Vanguard. The president counts on Vanguard. I'm telling you right now Vanguard will never make it."

McElroy gestured in protest. "Doctor, I'm not yet the new secretary. I don't have the authority to—"

"But you will," von Braun broke in, his expression and words raw with emotion. "You will be, and when you have the authority," he said sternly, "for God's sake, turn us loose! The hardware is ready. Just give us the green light, Mr. Secretary. Just give us the green light. We can put up a satellite in sixty days."

Medaris did some quick calculations on all the work that needed to be done and told his keyed-up friend, "No, Wernher, ninety days."

"Just turn us loose," von Braun pleaded as he walked quickly from the room, not turning around. But his friends didn't miss the tears of anger and frustration in his eyes.

The world reacted with shock and no small measure of fear. A satellite racing overhead had seemed impossible until this moment. Screaming headlines repeated over and over that Sputnik was circling the earth again and again and again, at incredible speeds, unstoppable.

No one was more disturbed than the Americans. The United States had been considered the world's unchallenged technological leader, with the Soviets trailing far behind. The satellite not only marked the emergence of the Soviet Union as a technologically accomplished society, but it demonstrated for the first time that the Russian military had the rocket power to deliver nuclear weapons across continents and oceans.

American skies had been violated. During World War II no Nazi or Japanese aircraft had penetrated U.S. air space. But here was Sputnik, made in Russia, passing overhead several times a day. The Eisenhower administration was caught by surprise. In general, its spokesmen tried to minimize the significance of the Soviet achievement and the threat it represented to national security. "After all," Eisenhower told a news conference, "the Russians have only put one small ball in the air."

Soviet Premier Nikita Khrushchev saw it differently, boasting that Sputnik demonstrated the superiority of communism over capitalism. "People of the whole world are pointing to the satellite," he stated. "They are saying the U.S. has been beaten."

People everywhere were looking up at the sky, hoping to glimpse the satellite. Newspapers, radio, and television reported that Sputnik was as bright as a fourth-magnitude star, and observers lucky enough to catch the right angle at the right time—just before sunset or after sunrise, when it was dark on earth's surface but still sunlit where the new satellite traveled—could follow the incredible moving star as it passed overhead.

In Newport, Rhode Island, a young U.S. Navy test pilot, Alan Shepard, was definitely interested in seeing it.

Shepard was attending the Naval War College in Newport, the admirals' school for officers who were marked for quick promotion through the naval ranks. The young lieutenant commander would rather have been back at the Patuxent River Naval Air Station in Maryland, testing new and experimental jets. He wanted wings, not a desk. But in the eyes of his supervisors he had promise, and the top brass pointed to the Naval War College and said, "Go." So he was burning books instead of kicking in afterburners.

The paper said Sputnik could be seen that evening, and Shepard went outside to take a look at the satellite whose success gnawed at his insides. He was angry that the thing up there didn't belong to America. It instead belonged to a nation he felt was technologically inferior—a nation that couldn't even build washing machines and refrigerators that worked!

Shepard looked up into the early evening sky. A high-pressure area had stalled over Rhode Island, and the crisp, cold autumn evening was clear.

Sputnik was to appear in the Southwest and move to the Northeast, and Alan Shepard focused on an area just south of the bright evening star. No sooner had he settled into a comfortable position than he saw it. A star that blinked as it moved. "That little rascal," he said to himself as it moved closer. He stared in astonishment as Sputnik zipped overhead and disappeared in the northeastern sky. Shepard knew he was catching a glimpse of the future, and he stood with his feet planted solidly in place, not wanting that brief slice of tomorrow to leave. But all he could do was watch as Sputnik disappeared from view on its journey over Nova Scotia and across the north Atlantic and Europe to begin its sweep over its native Russia.

It appeared that Sputnik changed its flight path as it moved over the earth's surface. But it didn't. It was the earth that moved. The satellite was in its own independent orbit, moving around the globe every ninety-six minutes on a firm, invisible track fixed by earth's gravity. Earth at the equator was rotating beneath it at the rate of a thousand miles each hour.

Some three hours and two orbits after Shepard observed Sputnik, the earth rotated eastward three thousand miles beneath Sputnik's orbital track. It was ascending over the southern Pacific, moving northeast across southern California and Edwards Air Force Base. There, another young test pilot had decided to go take a look as this new manmade object raced across the heavens.

At the time Deke Slayton was a member of the highly talented Fighter Test Group at Edwards. These pilots were flying higher and faster than any others in the world. They were the best. It was nothing for them to fly upward of twelve or fourteen miles from the earth's surface and then roar off at more than twice the speed of sound.

Tonight Deke waited to see something flying higher than he'd ever gone. Forty times higher, fifteen times faster. So he stood in the desert with his dog, Ace, and they looked at the heavens. Deke was convinced this was *the* beginning. That men would soon be up there as well as soulless machines.

Ace barked furiously at something in the desert. He wanted to cut loose and chase, but Deke gripped the leash tightly and kept his face on the evening sky. He looked for anything moving, waiting until it was clear that the only things overhead were the stars and planets and what he thought might be Pegasus, the flying horse, riding over the mountains to the south. I must have been looking in the wrong place, he thought.

Ace was getting impatient, and Deke turned and took him back inside. The dog could not know that one of his own soon would be making history.

Thirty days later the Russians did it again. Sputnik 2 weighed an astonishing 1,120 pounds, and it soared as high as 1,031 miles. The numbers were overwhelming to an America struggling to get its own three-pound Vanguard satellite into orbit. But what grabbed and held the world's attention was that inside the new Sputnik was a dog—a dog named Laika.

A living, breathing creature was orbiting the earth. Laika ate and slept and woke as Russian scientists measured the animal's physiological reactions to the pressures of the launch, to the heavy gravity pulls built up by the high speeds needed to reach orbit, and then to prolonged weightlessness. Test pilots like Alan Shepard and Deke Slayton had no doubts that humans soon would follow.

The first Soviet satellite had beeped with maddening repetition. The second was sending down the sounds of a dog's heartbeat. The world was fascinated and swept up by the promise of the adventure of a new age.

The Soviets had no intention of recovering Laika, and only days after launch her air supply was exhausted and the animal slipped into hypoxic sleep and died painlessly. But while Laika was slipping into sleep, a sleeping United States was finally wakening. The American public was demanding to see a satellite with the stars and stripes in orbit.

Under pressure from the Eisenhower administration, the Vanguard team rushed its rocket to a launch pad at Cape Canaveral. The press coverage was enormous. On December 6, 1957, two months after Sputnik had been launched into space, an anxious hush fell over America as the

Vanguard countdown finally reached its single-digit seconds.

"Four, three, two, one, zero . . . " Harry Reasoner of CBS was at the Cape and broadcast an emotional, excited liftoff. It was absolutely beautiful, he said. He told his national audience he was amazed at the speed of Vanguard streaking up from its launch pad, faster than his eyes could follow.

"It was so quick," said the astounded broadcaster, "I really didn't see it."

Of course, he didn't.

Vanguard didn't go anywhere.

A few inches really didn't count when you're reaching for orbit. Vanguard spat fire, rocked, lurched forward and fell back, breaking apart and crumpling into a ghastly fireball that seared the eyes of America.

The tiny satellite wasn't quite launched. It was blasted off the top of the exploding Vanguard and rolled into hiding in the Cape's palmetto and scrub brush. It then began to broadcast its radio signals. To those covering the launch, it was a pitiful sound.

Columnist Dorothy Kilgallen asked the most appropriate question for all Americans. "Why doesn't someone go out there, find it, and kill it?"

It was a bleak day for America. The Vanguard failure generated feelings of lost confidence, wounded pride, confusion—and awe at the Soviet achievements.

Something had to be done. President Eisenhower, stung by the criticism, finally relented, gave the Army and the von Braun team the green light to take its Jupiter-C—Missile 29—out of storage and to the launch pad.

More reliable upper stage rockets were added, and at the Cape a thirty-one-pound satellite was mounted atop the stack. It carried eighteen pounds of instruments designed to measure space radiation.

Eisenhower, his science adviser James Killian, and others in the White House didn't want to be reminded that the rocket was the same damn Jupiter-C that could have placed a satellite in orbit more than a year before Sputnik. The Army was told to keep that information quiet—in fact, to change the name of the rocket, and Jupiter-C became Juno 1.

After three days of delays caused by high winds, Juno 1 was ready to lift off on January 31, 1958. At 10:45 P.M., test director Robert Moser pushed the switch to ignite the first stage. Yellow flames shot down and then splashed outward in all directions beneath the rocket. A huge pillow of dazzling fire gushed forth, and moments later thunder crashed across the Cape.

Observers, blinking at the searing flames and bathed in that marvelous roar, unleashed cheers and screams of excitement as broadcasters shouted to be heard above the thunder. Men and women cried shamelessly as Juno climbed higher and faster.

The rocket burned a fiery path into the night sky as it reached for von Braun's stars.

In the blockhouse, engineer Ernst Stuhlinger sat before his monitors, tense with excitement. Exactly on schedule he sent forth radio signals that ignited the second, then the third, and finally the fourth stage of the rocket.

Dr. Wernher von Braun, architect of the American space age, a boy dreamer who had wanted to go to the moon rather than build weapons for destruction, was refused his request to be present at Cape Canaveral when his creation roared from the pad.

"Wernher, you will be needed in Washington," General Medaris had told him. "All the key players will be at the Pentagon for a news conference once the satellite is in orbit."

"I can hold a news conference at the Cape," von Braun protested. "I need to be there. I want to be there."

"Brucker says it's important that you be there, no argument," Medaris said with finality, referring to Army Secretary Wilber Brucker.

Von Braun monitored the liftoff of his beloved rocket from a chilly communications room in the Pentagon, along with Brucker; Dr. William Pickering, head of NASA's Jet Propulsion Laboratory, where the satellite was built; and Dr. James A. Van Allen of the University of Iowa, who provided the radiation-monitoring Geiger counter for the payload. Von Braun was extremely uncomfortable in Washington. As the rocket rose, he could hear his own heart beating rapidly as he shifted in his seat and then paced the floor.

He was greatly relieved when all four stages of Juno 1 burned as planned. But no one could tell for certain if the satellite was in orbit.

The country didn't have a network of tracking stations in place, just a few, and definite confirmation of orbit would come from a tracking station in California only after the satellite had made nearly one complete revolution of the globe. Everyone would have to wait to find out if the fourth stage had burned properly, at the right angle and attitude, and with the required speed.

The minutes passed and, as the moment of confirmation neared, there was a nervous shuffling of feet in the Pentagon room. Von Braun had calculated that it would take the Juno 1 one hundred six minutes to reach the California tracking station. When the satellite signal was not received at the expected time, von Braun again began to pace about the room.

Pickering was on an open phone to the California station. "Why the hell don't you hear anything?" he shouted.

Brucker, pale and trembling, looked at von Braun. "Wernher, what happened?" he demanded.

Before von Braun could answer, Pickering's excited voice boomed through the room. "They hear her, Wernher! They hear her!"

The satellite was in a slightly higher orbit than expected, accounting for an eight-minute delay in picking up the signal.

Men and women hugged one another with unchecked joy. Wernher von Braun walked onto the stage of an adjoining auditorium filled with applauding reporters.

"It was one of the great moments of my life," he said. "I only regret we didn't do it earlier."

A jubilant America was at his feet.

In Huntsville, the town square rocked with a wild and furious celebration of fireworks, sirens, blaring horns, and the cheers of thousands of people dancing and embracing in the streets. Former Defense Secretary Charles E. Wilson, who had long throttled von Braun's program, was hanged in effigy, and the hallowed Confederate monument was topped with a mockup of the Jupiter-C missile. Except for the accents, you couldn't tell Alabaman from German, northerner from southerner, easterner from westerner, in Huntsville's thronged streets. America was celebrating itself.

Instantly Dr. Wernher von Braun became a national hero. He was on the front page of every newspaper and on the cover of many magazines. "Von Braun, forty-five, personifies man's drive to rise above the planet," *Time* magazine declared. "Von Braun, in fact, has only one interest, the conquest of space, which he calls man's greatest venture." Soon thereafter Eisenhower summoned von Braun to a white-tie dinner at the

White House and presented him with the Distinguished Federal Civilian Service Award.

The satellite von Braun's rocket had launched was named Explorer 1. It weighed only thirty-one pounds, but Dr. Van Allen's Geiger counter made the first discovery of the new era—that the earth is surrounded by huge bands of high-energy radiation composed of particles trapped in our planet's magnetic field. Scientists honored Van Allen by naming the belts after him.

With Explorer 1, the United States was catapulted into the space age and into what would become a fierce competition with the Soviet Union for dominance of this new frontier.

CHAPTER THREE

THE PILOTS

CALIFORNIA'S MOJAVE DESERT is a barren and inhospitable place, a blistering, snake-infested expanse where sand and sun whiten the bones of long-forgotten misadventurers. Powerful winds knife through the gnarled and broken Joshua trees standing like wounded sentries at Edwards Air Force Base, a great military installation on the edge of nowhere.

At the heart of this high-tech flight center, huge aircraft hangars rise like shimmering images from the boundless sea of desert. The flat, dry desert lake bed is favored by local test pilots because it stretches for miles in all directions, offering the world's longest runways.

It was at Edwards Air Force Base that the National Aeronautics and Space Administration in 1958 focused its search for America's first astronauts. Created in the frantic aftermath of Sputnik, NASA had assumed the responsibilities of the old National Advisory Committee on Aeronautics, and it also accepted the burden for investing in the great new technologies of space exploration. NASA was to develop the means to send humans into space at speeds in excess of seventeen thousand miles an hour and to return them safely to earth.

The immediate obstacle facing NASA was that there simply weren't any highly skilled test pilots with the engineering experience and discipline to risk flight in the most dangerous craft in the world. There were test pilots, of course, but before they could become astronauts and slip through the atmosphere's barriers into the vacuum beyond the planet, they had to begin not only training for space flight but to work in concert with engineers and scientists to define the parameters of space exploration.

There was no shortage of volunteers when NASA began its search for the first astronauts, but the agency rejected outright the mountain climbers, race car and speedboat drivers, parachutists and daredevils. NASA wanted stable, college-educated professionals who had been screened for security. Many military test pilots fit the bill, and Edwards, just 150 miles northeast of Los Angeles, was one place where they could be found.

It was a beautiful airfield, where pilots could come burning out of the sky and always have a place to land, where landing-speed restrictions were nonexistent and where pilots could experience the thrill of touching down faster than many airplanes could even fly.

It was almost impossible to define the particular quality that marked the special and closed fraternity of test pilots who worked at the base. Esprit de corps, pride, honor, dedication, skill, and courage were all qualities required of the men who would become astronauts, but these qualities did not go far enough in describing the particular character of the pilots, who were all highly intelligent, skilled, and knowledgeable, and who all exhibited a strong survival instinct and fierce will to succeed. Fools didn't last long at Edwards.

Deke Slayton was the lead pilot, assigned to test the new Republic F-105 Thunderchief, a fighter-bomber designed to fly and fight day and night, in clear skies and in rotten weather, as fast as twice the speed of sound. When Slayton flew, other pilots at Edwards would dump their gear in the Flight Ops Ready Room and watch and listen.

On one particular day the corps of test pilots crowded to the Ready Room when they recognized a familiar sound: the deep-throated rumble of the 105's powerful jet engine spinning faster and faster in its start-up mode. Fuel tankers, ambulances, fire trucks, and communications vehicles all were ready, and the pilots watched as the Thunderchief turned onto the runway, its nose bobbing gently as Deke Slayton braked the ship to a stop, checking his machine before the tower gave him the call.

He fed the 105 power. A deep, winding cry thundered across the flight line, echoing back from hangars, booming hoarsely as thrust increased. Deke rechecked the gauges. Satisfied with the readouts, he released the brake. The swept-wing fighter-bomber lurched forward, accelerating swiftly. From the cockpit the runway flashed by smoothly. All was strangely quiet under Deke's helmet. He brought up the nose wheel, and the 105 seemed to hang in the air as it swept ahead, the earth falling away beneath the wings. Blue sky replaced desert brown. Deke felt the gear come up and lock, the gear doors close to form a sleek shape, and in that wonderful rush of jet flight he arrowed skyward.

As the F-105 was climbing to test flight altitude, Deke's thoughts flashed like high-speed film to the time long ago when he had started flying. He recalled the announcement of war. He'd never even been up in an airplane, but somehow he knew the sky was where he belonged, and soon he was signing up for cadet training.

He went through the drill like thousands of other cadets being hammered into skilled pilots. A year after signing up, in 1942, he had earned his silver wings, and he was on his way to combat. He flew B-25 medium bombers in Europe through flak storms and attacks by swift German fighters whose view of Deke's bomber was always through a gunsight. It wasn't a time for heroics. Just fly, bomb, fight to get back to base. When the Third Reich was vanquished, Deke was sent to fight in the war still raging in the Pacific.

He went from the B-25 to the powerful A-26 Invader, a flat-bellied attack bomber. "That's when I had the real fun," he recalled. "Those low-level attacks of six Invaders flying abreast were synchronized thunder, right on the bloody deck, thank you, our props churning up salt spray from the ocean as we barreled in for the Japanese coast."

The Invaders flew in a formation so tight it seemed as if only one pilot was controlling all six planes. When Deke and his fellow pilots cut loose at a target, everybody firing in concert, they hurled destruction from no less than eighty-four heavy machine guns, ripping their quarry to shreds.

The Pacific war ended, and Deke began the long climb to becoming a test pilot, spending most of the next two years on the ground. He doubled up on his academic and engineering subjects and in two years earned a four-year aeronautical degree from the University of Minnesota. He tried a desk job at Boeing Aircraft before heading back to the military, and Germany, where he cut his teeth on Air Force jet fighters.

It was one jump up the ladder after another. Out of Germany's lush forests and mountains to the California desert and test flying. Deke drove himself relentlessly in test pilot school and found, to his surprise and delight, that "Edwards was where I fit. I had the best job in the Air Force."

That was long ago. *Now* was twelve hundred miles an hour in the F-105 as he approached the test area.

He checked communications with test control. Phil Connelly confirmed telemetry was on the mark. A quiet beginning for what could be a difficult test run in an airplane still full of hidden flaws.

Deke began the run by pushing the F-105 into a tight decelerating turn from maximum speed. The test called for masterful flying that would gauge the turning capabilities of the craft. During that maneuver of clawing around and decelerating, he would wind the fighter down from twelve hundred miles an hour to less than three hundred.

Most pilots would pale at the thought of flying in this manner. Deke knew his capabilities and was confident. He had plenty of air space in which to maneuver, the weather was great, and he'd be sliding down from thin to rich atmosphere where the fighter-bomber could get a better grip of the air.

He sliced the turn closer and closer as the 105's speed fell off sharply. Before him passed the constant blur of earth, horizon, sun, and sky.

That's when it happened. Faster than Deke could react, the F-105 snapped over violently into an inverted spin at high speed.

The world below swam in a maddening whirl. The spin was the worst kind. He had to stop it, now.

Stop the violent rotation. Get the nose down.

He moved the controls precisely. The spin tightened. Blood pounded through his skull. He tried desperately to focus his vision as the 105 spun faster and hurled itself at the earth. All he could see was a blur. He fought to define the horizon and to keep from panicking.

Where the hell was the horizon?

Think!

There was nothing to prepare Deke for the violent spin. No books. No theories worth a damn. No training manuals. What counted now was instinct and a native ability to figure out what was happening—quickly—

and get the hell through the exit door.

Inverted spin? That was it. *Let's try what we know. . . .*

He worked his controls through the spin-recovery techniques that usually worked. No help. This was the meanest, most unpredictable bitch of a situation for a pilot to be in.

He moved his controls through anti-spin motions. His head slammed sideways as the 105 reversed direction.

Nothing to lose. He let loose of the controls. Maybe she'd come out by herself.

Finally he could read his altimeter. He was down to twenty-seven thousand feet. Time mocked him.

He let his senses float free to feel the negative centrifugal force pounding against his body. The negative g-loads were still increasing, forcing his body upward from his safety harness, pushing him away from his seat toward the canopy.

Deke was thinking that he should be getting ready to bail out of what was swiftly turning into a death trap.

He reached down to be sure he could reach the handle of his rocket-propelled ejection seat. First the canopy would go, the seat would be blown up and away, and then he would float free in his parachute.

He couldn't reach the ejection handle.

He was trapped in the airplane.

He forced himself down into his seat as hard as he could. He pulled up on the control stick, reaching, reaching . . . and now he could barely touch the life saving grip.

He knew that he could get out of the spinning fighter. He had a few more precious seconds. He forced himself to wait while his mind worked furiously.

In his mind he saw the 105's speed brakes. They're right there, he told himself. Right there on this plane's ass. Right where they'll—

He hit the control for the speed brakes.

Four huge metal panels splashed outward in the form of a beautiful metal flower that slowed the plane. It felt as if a giant invisible hand had slapped the airplane.

The sudden drag threw Deke against his seat and banged the aircraft's nose toward the desert floor.

The violent oscillations eased and then stopped. The world was no longer racing before his eyes. He smiled. He was back in control.

Well, almost. He was still spinning, but the nose was down and the controls were responding. The 105 kicked out of the spin, reversed direction, and spun the other way. Those speed brakes were working. The next attempt to stop the spin seemed to give him just a tad more control. He neutralized the controls. Rudders even, stick neutral, power off.

The whirling motion of the planet slowed some more.

But he was out of time as the 105 rushed down through the ten thousand-foot mark. He started reaching for the ejection handle because his time was just about all used up.

One more try and—

Whaaammm! She came out of the spin with a crashing motion.

He had control. Good Lord, he really did. He had complete control and he pulled out smoothly, kept the nose down, and lost no time heading for the runway. The F-105 went onto solid ground like a painter's brush.

Deke parked on the flight line and shut the F-105 down. Mechanics swarmed over the airplane. They opened the bomb bay and found a mangled mass of aircraft parts that had been ripped to pieces and hurled from their mounts by the g-forces and vibrations.

Deke stared at the nuts, bolts, rivets, and chunks of metal torn loose. He turned slowly to stare at the big hangar where Republic Aviation maintained its maintenance operation.

Slowly he brought up his right hand, made a fist, leaving his middle finger extended rigidly.

"Up yours, Mister Republic," he said quietly. "Your one-oh-five has been spun, son."

By late 1958 there were other flyers ready to trade the title of test pilot for that of astronaut. One of those ready to answer the call from NASA was a young naval lieutenant named Alan Shepard.

Deke hadn't met Shepard, but he'd heard of him. Alan had already gained a reputation as one helluva pilot, and stories had circulated about how he had been sent by the Navy to Edwards to wring out a souped-up Grumman F11F Tiger.

This winged Tiger had a tough reputation and a nasty habit of biting back.

On the morning of the test, Alan strapped himself into the Tiger, fired up, went through the pre-takeoff drill, and began rolling down an Edwards runway for his day's pay.

The F11F had been experiencing problems with reverse yaw which Grumman had tried to correct. When the experimental engine built up speed to twice that of sound and the pilot entered a turn, the aircraft violently spun around, swapping ends.

Shepard was climbing to seventy-five thousand feet when the engine suddenly flamed out. "I had just enough time to catch my breath before I lost all cabin pressure and the canopy began to frost over," he recalled.

Shepard could barely see through his front windshield, but he had enough visibility to get a visual lock on the horizon to maintain balance and control. His pressure suit sealed his body in a tight grip and continued feeding him oxygen, and a quick scan of the control panel showed he still had the basic instruments with which to maintain directional control of the airplane. For a short while the F11F had been a lump of winged metal, zooming upward without power and losing speed. The still functioning instruments confirmed Shepard was in a slight turn and able to fly the Tiger along a wide arc in a downward glide.

He laughed aloud suddenly. The word glide was a harsh joke. The F11F was gliding all right, but it was falling out of the sky like a Steinway piano in an unstoppable spin.

Alan let the airplane descend for forty thousand feet; then, seven miles above the desert, he went through the engine-starting sequence to try to regain power.

He let the plane fall another ten thousand feet into denser air with more oxygen and again primed the pump. There was no response. Ice coated the canopy, but through the gleaming frost Alan saw mountains in the distance growing larger with unsettling speed.

About this time he began to worry a bit. He was using up the sky in a terrible hurry, and there was now a real chance he might terminate this flight by scattering the plane along the mountainous slopes.

One more shot, he decided, and then it's time to leave this mother. As the F11F fell like a rock through twelve thousand feet, the engine restarted! Immediately Shepard had full control, and as the windshield cleared he set up for a straight-in landing on the desert floor. He rolled to a stop, climbed from the cockpit, gave the F11F the time-honored finger, and met with his test chief to fill out his report.

Then he headed for a bar where Air Force pilots waited with the needle. He didn't care what they said. He needed urgently, as he put it, to regroup all his senses, to recover from the descent, and brace himself with some high-alcohol-content medicinal spirits. Besides, he could give as well as take the ribbing he knew was coming.

The blue-suit boys always told the old story of how they caught a group of Navy pilots flying formation. "They must have been flying formation," the rib went, "because they were all flying in the same direction. Sort of."

Alan didn't mind the rib. In fact, he thought it was a pretty good one. But he, and every other carrier pilot, had one great advantage over the blue-suiters. There wasn't a land pilot alive, no matter how great he was, who didn't hold the carrier jocks in enormous respect for their skill at living, working, and flying to and from a floating runway. There are few things that shake up a land pilot more than the prospect of searching an angry ocean for a bobbing and rolling gray slab that is an aircraft carrier deck.

That's in daylight. At night, in rain and fog and nasty winds, it is like being thrown into a rolling cement mixer. Any time an Air Force jock got his wind up with high-powered bragging, Alan would immediately offer him the job of flying his wing to a carrier deck. Somehow he never got any takers.

Alan Shepard became one of the classics in carrier landings because he found them more interesting and challenging than frightening—or than being plain scared at risking his life with every plunge onto a carrier's deck.

He figures it had to do with his childhood in East Derry, New Hampshire, where his father, a retired Army colonel, created in his son a love of tinkering, building, creating things that flew, rolled, or just plain hummed along. Topping the list were model airplanes, all kinds and sizes. Watching his models fly lit the fire within the youngster, and there was no dousing that flame. Next stop was the local airport for after-school odd jobs, working on machinery that rolled and other machinery with wings that flew. As a high schooler, Shepard was a walking sponge, soaking up information, analyzing, taking things apart and putting them back together, ever questioning, ever seeking and never being satisfied with what he was learning.

Hard work, diligence, and high grades earned an appointment to the U.S. Naval Academy. He could almost feel the gold wings of a naval aviator pinned to his chest, but the Navy does things its own way. Off to the sea for the greenhorn, and only then was he sent off for flight training.

His wings then shone brightly on his uniform, almost as brightly as his Tom Sawyer grin and his disarming charm.

In 1950 Washington tapped Alan for a coveted assignment at the Navy Test Pilot School at Patuxent River Naval Air Base in Maryland. The youngest aviator selected, he went immediately from test pilot school to flight test operations. For the next three years he was immersed in an intensive test program that produced the Navy's fastest and deadliest combat aircraft. After three years Shepard joined the operational ranks as the new operations officer of a night-fighter squadron aboard the USS *Oriskany* in the western Pacific.

One mission especially was memorable. He was strapped tightly within the confines of a dark Banshee jet fighter, preparing for a routine night intercept mission. The launch officer gave the signal, and the Banshee was catapulted into darkness. Carrier combat operations vectored Shepard to the radar-reported positions of bogeys—unidentified aircraft. Moments later Shepard confirmed the bogeys as friendlies—U.S. Air Force planes on their own mission. Everything was going as planned, and Shepard swung about in a wide turn to fly back to his carrier.

It was simple going out for an intercept.

It was a bitch going back.

Weather rolled in fast, and Alan found himself above the overcast. Beneath the cloud deck he descended into heavy rain, flying strictly by instruments in the wet dark, following the homing signal from the carrier.

Suddenly the Banshee's electrical system rolled over and kicked feebly, leaving the airplane and its pilot severely handicapped.

Some of Shepard's radio systems died.

His navigational systems were gone, inert.

His radar beam flickered on and off, an intermittent pulse more maddening than useful.

He was flying blind, and he did not know where he was in relation to the moving aircraft carrier.

As many pilots do when the world starts coming unglued, he spoke loudly to himself. "Your navigational aids have gone south, and it looks like you're gonna have to ditch, old buddy. You're gonna have to set her down in that dark water and take your chances." He paused and added another note of realism, "Your chances stink."

He had company crowding into the Banshee cockpit, an old enemy of pilots in just his circumstances: fear.

Instantly he recognized the intrusion and knew fear could kill him faster than anything else.

"Think, damn you," he ordered himself. He scanned his panel, determined to make the best of whatever equipment was still functioning. It wasn't much, but there was that old rule. You're never out until you quit.

He wasn't quitting.

He keyed his mike, seemingly a useless gesture. His radios were intermittent, so—

"Malta Base, this is *Foxtrot Two*. Do you read? Over."

Astonishingly, through the storm, he received a reply.

"*Foxtrot Two*, this is Malta Base. I just barely read you three-by-five. I say three-by-five but can read. Over."

Alan didn't waste a moment in response.

"This is *Foxtrot Two*, Malta Base. I'm in a little difficulty here. I'm IFR and my nav aids are erratic. Need assistance."

"*Foxtrot Two*, do you wish to declare an emergency? We're not picking up . . . Repeat, we don't hold you on radar."

"No! No damn emergency, Malta Base," he snapped. "I want to try a couple of things. I'll get back to you."

"Roger, *Foxtrot Two*. Malta Base standing by."

The rain pelted the windshield. He was lost over a very dark sea. No navigation aids to steer him to the carrier. Radios cutting in and out that might die for good at any moment. One helluva sticky mess of New Hampshire maple syrup for him to be in, but he rejected punching out, dropping into the stormy sea by parachute. That meant giving real thought to ditching into the dark pit that was the ocean.

Terrific . . .

But declare an emergency?

"Me?" he questioned aloud. "A fighter pilot? A carrier fighter pilot? Declare an emergency? No way!"

It had nothing to do with being macho. Alan Shepard was way beyond that nonsense. Neither was the issue one of inability to live with failure. No ego trip involved here.

To Alan, declaring an emergency the carrier ops man felt was inevitable meant that he had already concluded in his own mind that he had flown himself into a hole from which he couldn't get out.

No way.

Declaring an emergency meant that he couldn't handle his airplane in this situation by himself. That radio call was a blatant admission that he needed help. To admit that, he swore to himself, meant that somehow he had failed, he no longer had that special something to handle his own problems.

Believe in yourself . . .

Damn right. Time to reaffirm in his own mind that his confidence was unshaken, his skill as great as ever in using to the utmost whatever the Banshee had left to give.

He would bring in this machine and its one soul on board despite all the odds.

He felt his old calm settling through his system. He turned on the memory machine and reviewed all the steps he had absorbed in years of training and flying. He went back to the basics. No fancy stuff here. He flew to the point where he believed the aircraft carrier was and he rolled into a simple expanding square search. A basic expanding box. Fly so many minutes on this heading, do a ninety-degree turn, repeat, until you've completed the square. The next turn would be a larger square, because that was the pattern the sides of which represented the visibility limit. Increasing the search pattern this way meant he wouldn't omit anything beneath him.

"Let's try it lower," he told himself. The nose eased down. Payoff! He was now at low altitude, burning fuel at a horrendous rate, but he was below the clouds. He kept searching. One side, two, three, four sides. Still no ship.

He expanded the square and flew lower. He didn't have that much fuel left, and it was going through the Banshee like a fire hose under pressure.

A dim red light just ahead of him.

It had to be a destroyer in portside formation with the carrier! He eased into a turn toward the center of his squares, searching through the rain and scud clouds that had suddenly appeared.

Beautiful, beautiful!

The *Oriskany* grew swiftly in his vision.

He warned himself this party wasn't over yet. He had to make a night landing in foul weather onto a pitching carrier deck, and he had maybe five minutes of fuel remaining. If he was wrong, his aircraft would drop into the sea like a hot rock.

Through the rain the carrier's lights blurred. He saw them as if looking through a shifting veil. No time to worry about it. He rolled the Banshee into a final approach.

Careful, careful now. He trimmed the fighter for the approach. There, the speed is pegged. Okay, down with the gear. Three beautiful green lights on the aircraft's control panel beamed at him, confirming the gear was down and locked. Nothing on any Christmas tree ever looked that good. He didn't know until that moment if he could get the gear down. Not with an electrical failure. Call it luck, fate, miracle. The landing gear was down. So was the hook that would share the deck's lifesaving cable.

Gear down, flaps down, speed pegged. It's one shot down the slot, Alan. There's your ship, and the weather stinks, and the carrier deck is weaving and bobbing, but so what—it's right there in front of you.

He didn't concern himself with going around if he screwed up this approach. He didn't have enough fuel left for that fancy maneuver. He held the Banshee steady, sliding down the invisible line, keeping his approach smack on the numbers.

He smashed with teeth-jarring, bone-shaking impact against the deck.

Instantly he felt the tailhook snag the cable. Smash, hell! That was a "normal carrier landing."

"Normal landing, your ass," he called out to himself.

He laughed. There's no question about coming home in a "controlled crash."

Any carrier landing you walk away from is terrific.

He climbed down and there was a jaunty lift to his walk.

CHAPTER FOUR

THE ASTRONAUTS

"THE TEST PILOTS we are considering must meet certain minimum standards," the NASA announcement of astronaut qualifications read. "Each man must have at least fifteen hundred hours of logged flight time; jet pilot training, experience and full qualifications; at least a bachelor's degree for academic qualifying; and he must pass national security requirements. He may not be taller than five feet eleven inches, he must weigh less than one hundred eighty pounds, and must be under forty years of age."

And, of course, no women, thank you. It was academic; there were no female test pilots in the military services. Of the 508 male test pilots serving the country, 110 met the space agency's standards.

On April 9, 1959, NASA introduced the astronauts to America at a gala press conference held in the ballroom of the historic Dolly Madison House, NASA's temporary headquarters on Lafayette square.

The ballroom went from full to bursting with a steady stream of additional press arrivals, while backstage the seven test pilots were being fussed over by the NASA brass. Reporters yelled, shouted, and cursed one another as they elbowed and jockeyed for the best positions for their cameras and long boom microphones. It was a madhouse of excitement and anticipation. In moments the space agency would present America's first astronauts. Seven brave volunteers who would lead America into space and leave the Russians behind. Men who would dare to ride the rockets that so often exploded in raging fireballs over Cape Canaveral.

None of the Mercury Seven believed a word of the hype. Hell, they were test pilots and that was that. They simply wanted to fly higher and faster, and in the process they'd do anything to establish America as the world's leader.

But they dreaded speaking to the huge and unruly audience out there, because NASA expected them not to say anything that would reflect badly on the agency.

Officials began herding the seven young men toward the stage, announcing them in alphabetical order to the tumultuous crowd.

Alan Shepard took his place near the end of the line and was pleased to see Deke Slayton move up alongside him. They both had made the final cut, and this was the first time Shepard really had a good look at Deke. He was relieved to see that Ironman Slayton was as nervous as a cat dancing on a frying pan. Alan watched Deke's bow tie riding up and down his Adam's apple as he kept clearing his throat.

"Shepard," Deke leaned toward him. "I'm nervous as hell. You ever take part in something like this?"

Alan grinned. "Naw." He raised an eyebrow. "Well, not really. Anyway, I hope it's over in a hurry."

"Uh huh. Me, too," Deke said quickly.

Alan's smile was sugar and honey. "Those, uh, bow ties coming back in style?" he asked Deke. They were already moving in line toward a long felt-covered table on the stage.

"What's wrong with my damn tie?" Deke hissed at him.

"Oh. Well, nothing, really," Alan said nonchalantly. "I doubt if the cameras will pick up that smeared egg or catsup or whatever that guck is on it."

Deke didn't know Alan yet, and he didn't know that he was pulling his leg. Deke looked down and tried to bring his bow tie into his line of sight. He was still fussing as NASA officials seated them at the table and placed an Atlas-Mercury rocket model and a Mercury spacecraft model on the floor before them.

Bright overhead lights illuminated their shining faces, and they blinked from the glare as they stared at the heaving ocean of faces, lenses, and flashbulbs. Reporters stood, sat, balanced, squatted, and even hung from ladders. NASA gofers continued their rounds as they handed out press releases. Slowly the chatter subsided. A NASA official stood before a microphone.

"Gentlemen, these are the astronaut volunteers. Take your pictures."

Cattle time! It was like a moving herd. The disorganized sea of elbows, arms, knees, legs, and waving hands heaved forward. Some people remained upright. Others crawled along the floor for a better angle. The duck-walkers maneuvered painfully through resisting bodies. Men and women growled and snarled and cursed, jabbing with elbows, shoving, even kicking their way into position to stab cameras into the faces of the new astronauts. They seemed to be trying to shove cameras up astronaut noses and down astronaut throats. Shepard nudged Deke Slayton and Wally Schirra on either side of him. "I can't believe this," he announced just above the uproar so they could hear him. "These people are nuts."

Shepard stole a glance at Deke, who was fiercely worrying about his tie and the "guck" he couldn't see. It was ironic that this fighter test pilot, who'd flown combat across both hemispheres of the globe, and who'd faced death eating up the skies over Edwards with the fastest jets the country could build, was as self-conscious and uncomfortable as if he were tied stark naked to a tree, about to be lunch for a very hungry lion.

Deke had different thoughts about Alan Shepard. From where he sat, he marveled at the cool front that Alan was offering to the press. He saw Alan chuckling and laughing at the contortions of the media. Like the asshole crawling on his belly. Jeez! He's got that camera almost against my nose!

"Hey, Slayton! Look over here, willya? Yeah, look right this way! Smile!"

"I'm smiling," Deke muttered under his breath. "Shepard's smiling. Schirra's smiling. All of God's children are smiling. Keep smiling, that's it," he told himself.

"You say something, Slayton?" Shepard queried.

Slayton showed him a mouthful of teeth.

"God, that tie," Shepard said.

"Okay, ladies and gentlemen," a NASA public affairs official barked at the press. They recognized old-timer Walt Bonney. "Please move back," Bonney pressed. "Take your seats. Take your seats! Thank you."

Squirming, crawling photographers and TV cameramen jostled their way backward, falling against and over one another, mouthing threats and profanities, shoving and jabbing for the best possible position for more pictures and film. "It's a worm farm out there," Slayton muttered to Shepard.

With the press corps receding, the astronauts found it possible to relax just a bit. Shepard found himself grinning. Slayton nudged him, "What the hell's so funny?"

"Just thinking. If NASA still had some of those sensors on me to measure tension during those tests we went through, I'd fail this show right now."

Deke ignored him. He was back to his tie and the egg or whatever it was that had been photographed for posterity. Alan gave him a gentle elbow. "There's nothing on your tie, Slayton," he said quietly.

Deke looked startled. "What?"

Alan grinned. "Gotcha!"

"Gotcha? What? What the hell are you talking about? Gotcha what?" Light dawned. "Shepard, you—"

Bonney's voice cut him off with his introduction of NASA administrator T. Keith Glennan. The administrator made a brief greeting and then rolled off the names of the Mercury Seven for the gathering. Malcolm Scott Carpenter, a Navy lieutenant, veteran of the Korean War, and test pilot; Leroy Gordon Cooper Jr., an Air Force captain, test pilot, and fighter jock; John Herschel Glenn Jr., a Marine lieutenant colonel who had flown nearly 150 combat missions in two wars, downing three MiGs; Virgil "Gus" Grissom, an Air Force captain, a veteran of one hundred combat missions against MiGs in Korea; and Walter M. Schirra Jr., a Navy lieutenant commander, also a veteran of Korea. Glennan concluded by introducing Alan Shepard and Deke Slayton.

The uproar prior to this moment was by comparison a subdued rumble. The audience erupted in heartfelt response. They were serious! Those maniacs of the press—maniacs only moments ago—stood and applauded and whistled and cheered, but the best of all were the smiles and brightness of their eyes. It was against journalistic ethics for members of the media to take sides, to show approval or disapproval. It was their job just to report facts as simply and objectively as possible, but on this day the astronauts were overwhelmed by the good wishes from these people.

Deke Slayton knew he'd never forget this moment. He watched the audience before him. They were almost mad with reverence. "My God," he thought. "They're applauding us like we've already done something, like we were heroes or something." Deke nudged Shepard. They both knew that from this day on they were part of something extraordinary.

What they didn't know at the time was that in the coming hours, and for years to follow, their families would find themselves pinned to the same stage. And none of them were even remotely ready for the onslaught.

Edwards Air Force Base was about as remote as any pilot's wife would ever want to find herself and her family. Dry desert, dry-as-a-bone ancient lake beds, dry mountains, desiccated scrub, and dust that blew across roads and into homes. That was Edwards. A wasteland. But after a time, especially with close friends, Edwards became a real home, the solitude acceptable.

After NASA presented the Mercury Seven, reporters, photographers, television crews, and public relations people flooded Edwards. Marge Slayton stared in disbelief at the first waves of the invasion. She wondered if life for her family would ever be normal again as she watched her young toddler son, Kent, charge at cameras and attempt to dismantle every microphone he could reach. She did her best to field the nonstop torrent of questions booming at her.

It quickly became obvious that these news people apparently had no respect for the privacy of the families at Edwards. Marge knew that she must shoulder the same burden that Deke had in the news conference, so she did her best to give straightforward, honest replies to the barrage of questions. She explained she and Deke had married in 1954. Yes, it was in Germany. Yes, Deke was then a fighter pilot on active duty. No, she wasn't German. She was a civilian secretary working for the Air Force.

"Of course, I'm pleased my husband was selected as an astronaut. It's a great honor."

"Do I have any fears about the unknowns of space flight? Do you know of anyone who's going to be boosted out of this world who wouldn't be apprehensive?"

"Yes, I'm familiar with the dangers. Yes, I've lived with them for years through Deke's service as a fighter pilot and a test pilot."

"Yes, I stand by him. Yes, I support Deke all the way. No, I have no reservations about what he has chosen to do."

"Yes, I—"

The telephone rang. She caught her breath and picked up the instrument. "Hi, honey." She wanted to reach out and hug him. Deke was calling from Washington to tell her about the news conference. Before they finished talking, they both understood the publicity wave was gathering power like a rolling sonic boom.

The wave engulfed them the next morning. There, on the front page of newspapers throughout America, was a photograph of Marge on the phone, talking to her husband, with Kent in diapers by her side.

The other astronauts, and their families, were soon overwhelmed as the media pursued the nation's newest celebrities. In Ohio, Gus Grissom's wife, Betty, and sons, Scott and Mark, learned the joys of being inundated by an excited and aggressive NBC News team, while in Virginia Beach, Alan Shepard's wife, Louise, gathered together daughters Laura and Julie and niece Alice and fled their home one jump ahead of the press.

Louise drove herself and the girls to a nearby Virginia Beach. It was April, but the day was cold, the wind strong and uninviting to people expecting warm sun. Despite the cold the girls frolicked at the edge of the surf while Louise walked slowly along the sand and wondered how Alan was handling this new experience called a press conference. He'd been so excited when he left for Washington. She'd wanted so badly to be with him, to share this great moment in his life.

That was one of their secrets of the marriage: sharing. Sharing their thoughts, their plans, dreaming together, reaching out beyond the present into the possibilities of the future. They both loved golf, and so they shared that like the great team they were. This had been the pattern of their life from the day they first met, when Alan was attending the Naval Academy. When life and most of the people about them were bright and young and eager and full of life.

There was reality to be accepted along with love and happiness and promise. Marrying Alan was also signing up for the full duration of his life as a pilot. It wouldn't be all manicured lawn and nine-to-five and planned vacations. He would come home with news that she knew sooner or later would intrude upon their time together. Aircraft carrier duty. Assignment to distant parts of the world—the Mediterranean Sea or Korea or other places totally strange to her—long periods of emptiness that stretched into lonesome months.

She hated that part of her life but accepted that it must be, and no complaint passed her lips. It was simple. He flew for his country, she sustained the home for the family.

"Mrs. Shepard?"

The sudden intrusion of the reporter's question startled her. She was on the edge of acknowledging that whether or not they were pilots, or astronauts, or a mixture of both, they were fiercely competitive. She knew these men. Friends they might be, and they were, but they'd stomp right over one another to get a flight ahead of the rest of the group. She turned to the man speaking to her.

Two men. "We're from *Life* magazine, Mrs. Shepard," said one. Smiling, very smooth. "We'd like to take some pictures."

She locked eyes with the speaker and returned the smile. "Okay," she answered, "but why?"

"Your husband," said the photographer. "You know, of course, that your husband is one of the new Mercury astronauts."

"Yes, I know," she said. "That's about all I know. I'm afraid—"

"We understand," the smiling man broke in quickly. "We're not reporters, ma'am. We take pictures," he offered in explanation. "We don't ask questions."

"That will be all right," she said. She didn't change position. She could see they expected her to slip into a pose for their cameras.

The cameras clicked; the two men worked different angles. She stood patiently for a few moments and then turned toward the girls. "Laura, time to go," she called. The faces of the photographers registered their surprise that she had chosen to leave before they were through with their pictures. Quickly she gathered the three girls and they returned to their car.

Driving home, Louise somehow felt she had made an escape. Then, rounding the corner, her driveway in sight, she knew the photographers on the beach had been just a harbinger of an invasion. She couldn't believe her eyes. Cars and vans filled the street, many had been driven onto lawns and sidewalks. She was reminded of a great cattle drive, and her home was the final destination. People milled about between cars, along her driveway, and on her front lawn. She recognized the faces of several of her neighbors. The others were strangers. And they'd come armed. Men and women with notepads, clipboards, tape recorders, cameras, movie cameras, microphones on booms, bright lights. All eyes were staring at her, the crowd ready to move in for close-ups.

"This can't be," she said quietly. The girls stared in surprise. "Mom! What is all this?"

Louise knew. She had already escaped once to the beach. But this was overwhelming! She felt apprehension building to panic. "Stay close to me, girls," she told them as she edged her car into the single space left in her driveway. She was still moving as the microphones and camera lenses swarmed the car.

She turned off the engine and took a long breath. There was no way out of this. She could have kept driving, but this was *home*. She stepped from the car, and strident voices railed at her from all sides.

"How does it feel to be the wife of an astronaut?"

"Are you worried he'll be killed?"

"How long have you been married?"

"What do your kids think?"

"Do you really want him to go?"

"Are you worried he'll be killed?" That question would be repeated again and again, annoying her like some kind of angry bug she wanted to grind into the earth with her heel.

She'd been brought up to be gracious and understanding, and she ordered herself to be so now. She left the girls in the car and strode into the midst of the crowd of reporters milling on her front lawn, smiling and, when possible, answering questions in a polite and orderly fashion. This was crazy, but at the same time she was aware this was for Alan, and her tremendous pride in her husband helped calm her. She weighed every word she spoke to avoid saying anything that might reflect adversely on him.

A photographer, unusually courteous, requested her to stand by her mailbox. The request seemed harmless enough but proved to be a great error. The address of 580 Brandon Road glared prominently in the picture that appeared in hundreds of newspapers and opened the floodgates for a torrent of mail that made her home the receiving end of a postal blizzard.

There was no way the neighborhood could not see and hear the uproar from the front lawn of the Shepards. Mrs. Clark, whom Louise knew, remained in the safety of her home next door, but sent her son, Sumner, to investigate the commotion and report back.

Sumner walked across to the Shepards, and pushed his way into the crowd of reporters and photographers. His eyes widened with what he heard. It couldn't be! But it was, and the moment he was certain of the news he ran home.

He burst through his front door. "Mom! Mom! You gotta hear this! Mr. Shepard's going to the moon!"

Mrs. Clark gaped, but only for a moment, and then she shook her finger at her son. "Sumner, you are not a bit funny."

CHAPTER FIVE

TRAINING

HEROES?

Deke Slayton shook his head in disbelief. How the hell had seven guys suddenly become heroes?

Oh, they had taken their chances in testing the country's hottest and newest aircraft, but heroes?

"It happened without us doing a damn thing," he laughed. "We seven show up for a news conference, and now we're the bravest men in the country. Talk about crazy!"

But in the meantime, despite the bright lights and the accolades, there was still work to be done, and shortly after the press conference, the astronauts and their families were moved to Langley Air Force Base in Virginia. NASA's new Space Task Group also set up its headquarters there.

For the next two years they would train to be astronauts while engineers worked to develop and perfect the Mercury spacecraft that would fly them out of the world. "And back again safely," they reminded newsmen at every turn.

Deke put it as candidly as he could. "None of us knew a damn thing about being an astronaut," he said. "Space flight was science fiction. We didn't yet grasp the big picture, so it was natural for us to be as friendly and helpful to and with each other as we could. We were treading the new path together. We flew as a unit."

Langley was their home base, but they traveled widely. They studied the Mercury spacecraft then under development at the McDonnell Aircraft plant in St. Louis, where each of them received a visual reminder that, as Deke put it, "the thing ain't got no wings!" Then it was off to Convair at General Dynamics, where the intercontinental ballistic missile Atlas, still tricky and unproven, was being modified—its usual hydrogen bomb warhead replaced with the bell-shaped, blunt-edged Mercury. More important to the astronauts was the fact the Atlas was being man-rated, made as safe and reliable as humanly possible, with three or more redundant systems being built in to do the same job. If one system failed, another would take over and protect the astronaut. Finally, the Mercury Seven flew to the launch complex at Cape Canaveral on the east coast of Florida, where it would all happen.

The Cape Canaveral peninsula had been at the forefront of settlement in Florida in years past. Here, bear shared undeveloped land with alligators and deer, and Indians buried their dead on sacred grounds. Later, other men would come to hack out farmland. But taming the cape was tougher than expected; snakes and an infuriating number of persistent mosquitoes had driven most of the humans away. Now, beneath the palmetto scrub land and sand, the rows of launch towers and blockhouses and hangars and offices were connected by thousands of invisible electrical arteries, a finely woven network of underground cables through which flashed the impulses of energy, vital messages, and electronic commands that would ignite the rockets and launch the Mercury Seven into space.

The astronauts had multiple roles to play. Among them was hand-holding. Members of Congress, government officials, and industry leaders came to Cape Canaveral with little knowledge of the technological challenges at hand, expecting that the dollars they had allocated would guarantee immediate success. Putting a man on the moon wouldn't be quite that easy, and it would take time. The Mercury astronauts mixed with the powerful visitors, who were shocked when the first Atlas, topped with an unmanned Mercury spacecraft, was launched on a tail of fire into the clouds.

Where it promptly blew itself all to hell and beyond.

Congressmen looked at the astronauts as the Atlas tumbled to earth in flaming chunks. "You're going to get on top one of those things?" they asked.

It wasn't just the Atlas that failed to perform as expected. Charlie Chaplin couldn't have provided a better script for one Mercury-Redstone launch when the astronauts and a visiting congressional delegation watched as the rocket struggled to rise from the pad at the moment of ignition before quickly settling back on its pad when an electrical problem shut off the engine. The visitors observed the escape tower rocket atop the Mercury capsule hurl back fire and fling itself over the cape, leaving the Mercury and the Redstone on the pad. The sight of an out-of-control tower belching fire brought loudspeakers blaring warning for everyone to take cover immediately. It was a sight unprecedented in the new space age at Canaveral as astronauts, congressmen, generals, engineers, and reporters dived beneath bleachers and behind anything to gain cover from flames shooting above the launch pad. Then, while the huddled crowds watched, befuddled and amazed, the top of the capsule popped open, and the drogue parachute, the main parachute, then the reserve parachute, all unraveled and spilled down the side of the booster. The escape tower then crashed about four hundred yards from the pad after scooting to a height of four thousand feet.

"That was a hell of a mess," said Flight Director Chris Kraft.

It was also the price tag for development. It has always been that way with fiery machines in need of maturing. All part of the new school for astronauts.

Several months into the tests and training, a potentially sinister medical problem came knocking on Deke Slayton's door.

Deke and the other astronauts were making routine runs in the NASA centrifuge, an enclosed cockpit at the far end of a giant metal arm that rotates fast enough to mash a man's brains down into his throat. The centrifuge generates the forces a pilot feels in high-gravity maneuvers, such as very tight turns or sharp pullouts from a dive, increasing the gravity forces, or g-forces, on the man being slung around at the end of the centrifuge arm. If he's pulling a 5g load, then suddenly he weighs five times his normal weight.

The idea was to see how long a pilot could hang in there before so much blood was drained from his brain that he would pass out. Deke was a whiz at it. Technicians called the centrifuge the "County Fair Killer," and Deke handled g-loads that amazed even those running the machine.

But on the test that Slayton would never forget, the flight surgeon, Air Force Lieutenant Colonel William Douglas, was fastening sensors to Deke's skin.

Suddenly Douglas stopped short.

"What is it, Bill?" Deke asked instantly.

"Not sure. What I think I'm getting is an irregular heartbeat."

"You're kidding."

"Nope. Don't sweat it," Douglas said. "Could be the equipment."

Douglas rechecked the sensors, adjusted the monitoring equipment, and called for a centrifuge run at 3g. Slowly the flight surgeon felt his own heart sink.

Deke's irregular heartbeat wasn't in the equipment. It was in Deke's chest.

Douglas ordered the centrifuge to a halt. One look at the doctor's face and Deke had the bad news. Something was wrong. And Deke knew the score. If the medical tests showed any cardiac irregularity, it could be the end, right then. He would be grounded.

"We'll get back to you, Deke," Douglas told him.

Deke nodded.

A day passed. Then another. Still no answers.

Marge Slayton saw Deke's jaw stiffen as he fought for patience. Immediately he quit smoking. He knocked off the coffee. He ran like a man possessed. Every day he pounded pavement, running along nature trails and dodging trees as he raced through the woods. He was challenging his own heart.

It didn't help. The irregular beat would last for two, sometimes three, days, and then it would vanish from the medical tests. But no one was accepting a sudden cessation. The medical boys had plenty of time. Sure enough, an average of ten days after the irregularity slipped away, it came back.

How was this possible? Deke was in great shape. Physically his body was as close to perfection as the human body could be.

He knew, hating it, that his heart was like a rough-running engine. Sometimes fast, sometimes slow. It wouldn't give in and he wouldn't give in, and now and then, like a dark shadow gaining mass in his chest, he could sense it. Mean, nasty, hateful.

The mystery was that no one knew just what was happening inside Deke's body. Dr. Douglas spared no effort to find out. He called in several of the leading cardiologists in the country to examine Deke. They were baffled.

Deke wanted to chew nails.

The top specialists lacked answers, but their report left no doubt about how NASA should proceed. "Despite the irregularity, it is our opinion that this condition in no way affects the performance of astronaut Slayton. It is also our opinion that he should be accepted for space flight operations."

Bill Douglas gave Deke the word, and the word was good, and Deke let out a war whoop of joy.

NASA accepted the doctors' reports that Deke was healthy, and they subjected him to more rigorous and demanding tests with every passing day. He was slammed into deep pools in a sealed spacecraft so that he could become proficient in unhooking his harnesses, opening the hatch, swimming to the surface, and activating his escape-and-survival systems. All such tests were part of the routine of two years of training. What the astronauts experienced in centrifuge runs and weightless runs on board a specially equipped aircraft called the "vomit comet," all paled alongside what the astronauts felt was the most punishing test of all, a "godawful, unforgivable exercise" in Deke's words.

Or survival training in the wild, as NASA called it. Men preparing for flight at eighteen thousand miles an hour more than a hundred miles above the earth were "dumped" in deserts, atop rugged mountains, in Panamanian jungles and other remote sites, and left to fend for themselves. NASA generously equipped each man with a small portion of water, his survival gear, and his wits. A sense of humor and a strong stomach willing to accept lizards, snakes, insects, and whatever else could be gleaned from the inhospitable surroundings were also essential.

The astronauts almost preferred the wilds to the frenzy of reporters who were after one thing above all: the inside word. The scoop. Newspaper and television editors and directors did everything but arm their media reps to get the one story they all wanted to release.

Who would be the first American to be launched into space? That was the name they sought.

But they weren't sure which source to track for the inside story. So they went after everybody in NASA who might have the answer and who could be persuaded to give them the biggest news break of their careers.

But all their efforts failed. What the press did not know was that the issue of who was to be the first in space would ultimately be settled in a way the astronauts never expected.

It was the old story of the unexpected coming out of left field.

THE SELECTION

DEKE SLAYTON LOOKED DOWN at the aimless scribblings on the sheets of paper in front of him, and he was suddenly angry at having to wait. He crumpled the papers and angrily flung them into the trash can.

It wasn't just the waiting. Good Lord, a life in the military is primarily a life of waiting. It was the tension. It was so damn intolerable it was almost painful, and the seven Mercury astronauts fidgeted at the desks in their Langley office, doodling and making small talk.

Eyes kept returning to the clock on the wall: 5:15. Hell of a day. It was January 19, 1961. Tomorrow Jack Kennedy would be sworn in as president of the United States. But right now Robert Gilruth was more important to these men than a dozen Kennedys.

As chief of the Space Task Group, Gilruth ran Project Mercury. A brilliant engineer and manager, who had once directed Langley's Pilotless Aircraft Program, he was the biggest name in the manned space flight business, the high priest with the final say on who would fly in space.

Gilruth might have been King Arthur about to select one man to be knighted and given the most trusted mission. First to go. That afternoon he'd called the astronauts. "How about hanging in, guys, after quitting time?" More statement than question. "I have something important to tell you."

The king would point his finger at one man, and that one man would be seared into the pages of history like a branding iron on fresh leather—the first man to be sent into space.

Alan Shepard turned suddenly to Slayton. "Deke, what do you think?"

"I think I wish to hell he'd hurry up," Deke said with a half-snarl.

Gus Grissom joined in from the sidelines. "If we wait much longer, I may have to make a speech."

That eased the tension. Gus Grissom making a speech? The original tight mouth stringing words together? Aloud? His colleagues grinned.

Silence followed, tension began building again. The pilots each reviewed where they stood in the program. There'd been a big break in mid-December when a Redstone carried an unmanned Mercury capsule straight and true through its trajectory. That's when Gilruth dropped the first shoe. "Everybody better start thinking about who goes first atop a Redstone."

Okay. They thought. Then Gilruth threw them the other shoe. "I want you guys to take a peer vote. If you couldn't make the first flight, select the man you think should go." He smiled at their discomfort. "Write your choice on a piece of paper and drop it by my office."

Mercury's top man knew how to get their attention. The astronauts couldn't determine whether Gilruth had really put the "who goes first" decision in their hands or if he was playing a clever game. Either way the team knew he could simply select the man he and his superiors wanted, and the astronauts would never be the wiser.

Clever, all right. But it caught the pilots off balance, forcing all of them to make a second choice because each wanted to be first.

Deke Slayton had maintained a tight watch on his competition. He judged it would go the worst for John Glenn in a peer vote. Ever since their first press conference he had been wrapped up in the American flag. NASA's fair-haired boy never let anyone doubt that he was the best pick to lead the way. And his toughest competitors, his six fellow pilots, admitted he'd worked as hard as anyone, that he'd gone the extra mile, and he was a hell of a good man to carry the flag into space.

He'd also polished a lot of apples. Sometimes that backfired, especially when the media harangued incessantly that Glenn was the "father figure" of the Mercury Seven. That was often exacerbated by Glenn preaching to the others about maintaining a clean-cut public image. He held aloft the sacred icon of image above all. The guys should knock off the partying, racing down highways, and the philandering that John condemned as skirt-chasing, all of which he felt could lead to negative publicity that might tarnish the Mercury Seven program. His moralizing led to colorful and heated exchanges among the pilots, and it wasn't pleasant banter.

Gilruth had chosen the right place for these hyped astronauts to wait. No fancy customizing in their office. Bare G.I. steel desks and chairs. Walls cluttered with flight plans, diagrams, and spacecraft.

The astronauts gathered here often to hold what became dreaded among the engineers as the "pilot seances," where they dreamed up innovations and changes that drove the engineers up the wall. Like putting a window in the Mercury capsule and—

Gilruth entered the office. He cut right to the point. "What I have to say to you is confidential. Keep it that way. Each of you has done an outstanding job. We're grateful for your contributions, but you all know only one man can be first in space."

Deke's heart picked this moment to pound in his chest. He judged the others were going through the same cardiac drill. He ignored the pounding. "What I'm about to announce," Gilruth continued, "is the most difficult decision I've ever had to make. It is essential this decision be known to only a small group of people. We'll make it known to the public at the appropriate time."

He hesitated until the long pause threatened to bulge and burst about his captive audience. Gilruth was a master at working the clock. Again, without preamble:

"Alan Shepard will make the first suborbital Redstone flight," he said, with all the emotion of a man driving his car to a gas pump. Every man except Shepard experienced deep shock. The odds were seven to one, but so what? Deke Slayton may have been the only one among the six not selected to be pleased it was Shepard.

Deke was in a quandary of his own making. His private list had his name first, Shepard second. Yet Alan's selection still dealt a blow to Deke's sense of self-worth.

Gilruth's voice continued. "Gus Grissom will follow Alan on the second suborbital flight. John Glenn will be back-up for both missions."

Wham! Deke couldn't believe that he wasn't even among the first three. It didn't matter that another three men were going through the same shock. Deke was at the point where he wondered if he was even included in the future flight program. Was it his heart flutter?

He made no bones about his feelings. "I was shocked, hurt, and downright humiliated," he said later. "I looked at Alan. He was staring at the floor, his face blank. Then he managed a thin smile. In spite of my feelings, I reminded myself that Shepard, other than being Navy, was one hell of a test pilot.

"Then reality walloped me right between the eyes. Of course! Politics! Gilruth, who had to play the strings of the congressional banjo in troubled NASA budget times, now had his Navy man, Shepard. Grissom wore Air Force blue. Capping off the trio was Glenn of the Marines. The Army had no test pilots. So none of the military services could fault Gilruth. Not even the incoming president. No way was it an accident that both Shepard and John Kennedy were Navy. Kennedy, the famed PT boat captain. Kennedy, the naval war hero. All true enough."

Yet, intentionally or not, Gilruth had divided his astronauts. Alan Shepard, Gus Grissom, and John Glenn were the real astronauts in as far as the press would tell the story. The rest of the group had begun their slide into the background.

Wally Schirra, Gordon Cooper, Scott Carpenter, and Deke Slayton. Spear carriers. That's how they described themselves. Soon after, NASA made it starkly plain that the astronauts were all required to go along with the public relations strategy that had been decided upon for the selection announcement.

It stated merely that Shepard, Grissom, and Glenn had been selected as candidates for the first Redstone flight. The three would continue training and, as the launch neared, one of them would be tapped for history. No mention that the pilot for that plum already had been designated. It was a charade.

Gilruth offered a lame justification for his approach. Secrecy about the selection would protect Alan Shepard's privacy. It would prevent the press and public from focusing on one individual, perhaps crushing him under undue pressure. Considering the astronauts had an exclusive contract with *Life* magazine for their personal stories, and *Life* reporters and photographers were all over them, it was an unsatisfactory excuse from the get-go. But that was how Gilruth and his superiors were calling the shots.

It didn't happen right away, but Deke Slayton was finally to discover just how the selection game was being played. All the pilots had assumed that the first guy to boost out of the atmosphere would be selected on his skills as a veteran flyer, the best "stick-and-rudder" jockey. Of that group, the consensus of opinion pointed to Deke.

NASA managers went in a different direction. They concluded that Shepard was the smartest and most articulate of the bunch, the most talented and capable of making instant decisions in difficult situations, and they wanted him on the first suborbital mission so he could tell the scientists what the grinning chimps whom they planned to send up first wouldn't be able to report in words.

Gus Grissom was recognized for his skills in lightning-fast engineering problem-solving. Perfect to follow Shepard, and from the beginning not many NASA bosses felt they would ever need to make more than two to four Redstone suborbital flights.

Glenn's record spoke for itself. He had downed three MiGs in combat, but, more importantly, he had brought back airplanes so badly shot up they were judged unflyable by other pilots. Not only had he brought them back and landed them safely, but when he climbed from the cockpit the maintenance officers marked these planes as junk to be cannibalized for parts.

When NASA decided that two suborbital flights were enough, the charismatic Glenn was assigned the first Mercury orbital mission. Like the other pilots, Deke remained in the dark about his role. He could not know that because of his flying skills, he would be tapped to be the second American in orbit. Specifically, Deke had been listed as the man who would begin the development of orbital flight techniques and, along with Wally Schirra, an equally talented stick-and-rudder man, put the pilot fully back in the control loop.

The key to the selection process was that all seven of these men were solidly proven. They had flown to the edge in testing the nation's newest and fastest jet planes, and they had survived the hazards of that dangerous occupation. They all would contribute their skills to America's space effort.

John Glenn was the first to step forward to shake hands with Shepard. The others surrounded him, offered congratulations, and then quietly left the room. They would be astonished later to realize that nobody had even suggested a round of drinks to toast Alan.

Shepard understood the disappointment and frustration of those who left. He felt sorry for them, but inside he was dancing a heel-clicking jig all the way home. He felt the hurt of the other guys, but they all knew from the opening bell one would go, six would watch.

So he went home like he'd just finished a countdown and burst through the front door with his Tom Sawyer grin. "Louise! Louise, you home?" he shouted.

She came into the living room, and her three words said it all. "You got it!" She threw her arms around him, and he squeezed her until she almost winced with the pain. "You got the first ride!"

He shouted at her, "Lady, you can't tell anyone, but you have your arms around the man who'll be first in space!"

"Who let a Russian in here?" she mocked him.

"Nah. We'll beat those guys."

"Keep thinking that way." She hugged him tighter.

"Thinking, hell. I'm going to push for an early trip. We had a great flight with a Redstone and the capsule last month. They don't need to fly that damn chimp. If we drop that flight, we'll kick-start the program and in a couple months I'll be in space."

In the weeks to come, the heavy layer of silence about the selections conspired against Shepard. A couple of astronauts went all out to over-rule Gilruth's decision. They wanted Glenn to lead the way, and they emphasized the wild antics for which Shepard was justly infamous. The competitors said Shepard was too lighthearted for the job. And don't forget *Life* magazine, they stressed. Alan just didn't have the "perfect image" the magazine was painting of the astronauts.

You can't keep that kind of carping quiet too long, and finally Gilruth stepped in. "I want this backbiting stopped *right now*," he said by way of warning. "Alan Shepard is my choice. That's it."

There's nothing like final judgment to get rid of bickering. Now that the law had been carved on NASA's clay tablets, the astronauts joined ranks and gave Shepard their full support. Once again, and long enough in coming, they marched to the beat of a single drummer.

The reality was simple. If the first flight came off as planned, then all seven men would have their crack at looking down at earth far below. They had no fight with one another. Their struggle was to develop their flight hardware to safe and reliable mission capability. Get off the launch pad and come home alive and well.

They were reminded, often, that they could all wake up one morning and hear Russian being spoken from space.

Next stop: Cape Canaveral. Shepard began his campaign to get the chimp grounded and an American in space.

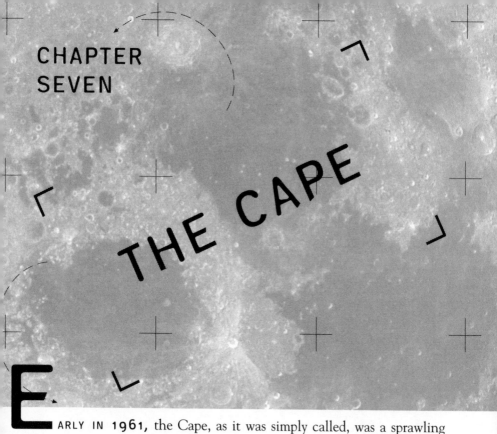

CHAPTER SEVEN

THE CAPE

ARLY IN 1961, the Cape, as it was simply called, was a sprawling gateway to the future. It was the most vital and intensely exciting place in the country, a fifteen-thousand-acre sandspit that had been reshaped into a port of blinding searchlights surrounding active launch pads. It was a place where rows of rocket gantries and block-houses and hangars and office buildings were lined up neatly behind a centuries-old lighthouse, which stood like a defending giant on the Cape's jutting tip.

The lighthouse served as the focal point for countless jokes pulled on the uninitiated. Many times when one of the rockets would ignite and roar from its launch pad, newcomers and visitors were left staring at the lighthouse wondering why it hadn't left the ground. The fact that they'd been told the white-and-black tower was the rocket to be launched made the joke even more hilarious. The stunt was pulled so often that the Air Force made a film with the lighthouse painted over an Atlas roaring into space.

The cape was a place where humor was mixed with the long workdays to keep the assemblage of man's latest high-tech creations working, and the astronauts led the parade of space age pranksters.

The Mercury Seven met and became friendly with Jim Rathmann, a local General Motors dealer and race car driver who had won the Indianapolis 500 the year before. Jim got them all good deals on sports cars of their choice, and then the fun began. The astronauts competed fiercely to see who could get the most speed and performance out of his car and, after a long day of training, they would set up drag races on the long, straight road that stretched past the row of rockets and gantries.

Alan, Gus, and Gordo had bought Corvettes, and at first Alan performed pretty well in the races. They'd line up and roar down the straight Cape road, sending rabbits and deer and wild hogs running through the palmettos and scrub brush. It was a great way to get rid of the tension that built up during the workday.

One day Gordo Cooper left Alan Shepard in the dust at the starting gate. He burned Alan's tail good, and from then on Shepard began losing. He wasn't even in the same race. Alan turned to Gus. "What the hell's going on?" he demanded.

"You lost, Alan," Gus grinned.

"I know I lost, damn it, but why?"

"Guess you lost your touch."

"My touch, my ass," Alan said. "There's something wrong with this car."

"Sure," Gordo laughed, "and you didn't eat your Wheaties today."

"My Wheaties? I'm taking this car in," he fumed, getting into his 'Vette and heading for Rathmann's place.

Jim was still in his office, and Shepard demanded to know what was wrong.

"Leave it with me, Alan," he said. "I'll see what I can do."

Jim Rathmann was in on Gordo's gag though, and when Shepard picked up his 'Vette a couple of days later and challenged the guys again, he lost again. The car wasn't any better. God, it was even worse, and Alan took it in again, and Gus and Gordo had Jim do a special paint job on it.

During World War II fighter pilots painted swastikas or rising sun flags on the side of their cockpits to represent the number of enemy planes they had shot down and, when Shepard's 'Vette was returned, it had four Volkswagens, a couple of other underpowered cars, and four bicycles painted on its side.

Suddenly, Shepard knew he had been had, and he yelled, "That's it. I demand a meeting," and, once he had Jim and Gordo and Gus in the office, the three, laughing like well-fed hyenas, confessed.

"Alan," Rathmann began, "we changed the rear end in your car."

"You what?"

"We changed the rear end ratio," he laughed. "Your car now has more speed."

More speed? It was immediately clear to Shepard. These laughing dogs had set his 'Vette up so that it would do 125 miles an hour all right, but to do this the car's pickup was so slow he needed twenty minutes to get up to speed.

"Gotcha!" Gus laughed, slapping him on the back, and Shepard began to shake his head. He had been had, and he had to admit it was a classic. Suddenly Shepard was laughing as hard as they were, and the one-up gamesmanship soon spilled over into their training on the Redstone launch pad.

The Mercury operations boss, Walt Williams, was a serious man. Everything about Project Mercury was innovative, the latest in engineering theory, and this quiet man moved about the rows of rocket gantries and blockhouses and hangars and office buildings with a never-changing expression of determination. A smile was not a frequent visitor to Williams' face and, this particular day, when the astronauts were working on the Atlas launch pad, he suddenly remembered he had to make a luncheon speech in town.

"Damn," he said. "I don't have a car, and I've gotta be there in twenty minutes."

"Take my Corvette, Walt," Alan Shepard graciously offered. "I'll catch a ride in later with Gus or John."

"Thanks, Alan." Walt finally managed a slight smile as he rushed from the launch pad and got into the hot sports car, where he sat fussing with the unfamiliar controls. He finally figured out how to start the car, put it in gear, and roared off.

Shepard grinned. No sooner had Williams turned onto the main road than he phoned the cops. "This is astronaut Alan Shepard," he shouted. "Some sonofabitch has just stolen my Corvette. He's headed for the south gate."

But Williams had the last laugh. When he drove Alan's Corvette through the security gate, the guard on duty recognized him and waved him through without checking the license plate.

Despite the glowing press reports about how well things were going with the astronauts and the Mercury operations team, the reality was that conflict was a part of every day at Cape Canaveral. Arduous project schedules and the long wait to get up into space made the astronauts feel stifled, and they were always looking for a way to relieve frustration. On such occasions the lure of space flight dimmed, and conversation among the team turned to their exciting and fast-paced days as test pilots.

The irony of playing second fiddle to a chimpanzee was particularly galling to these highly intelligent and skilled men. NASA had decided to send a chimp into space as a precaution before sending Alan Shepard. Shepard was ready to have a chimp barbecue with the hairy visitor, but NASA insisted the little ape go first. The agency meant well. But all Shepard could think about were Russian boosters rolling to their pads for the first manned space flight.

"The only way for us to go," Shepard said, "was to work hard and play hard, and that's what really made the train go. Gus, John, and I dominated the simulators while the others readied themselves for the assignments they'd have on launch day."

But no matter how hard they worked and played, there were disagreements between the astronauts and NASA officials. First, they weren't flying. Anyone who believes pilots can easily shunt aside their time in the skies, no matter what else invites them, simply doesn't understand the nature of a pilot. NASA's schedule did not include proficiency flights in jets. So the astronauts departed Langley or the Cape on the slightest pretext—to check out the spacecraft or the boosters—because no matter where they went, they flew in sleek jets.

There were other frictions, too. At the Cape, Alan spent most of his time in a "procedures trainer." This was a replica of the actual spaceship in which he would be boosted more than a hundred miles into space. It also duplicated the severe semi-supine flight position, with the pilot lying on his back, his legs vertical to the knees and then dropped down so that he was shaped like a squared-off pretzel.

No one liked the trainer. It was like taking a straight-backed chair, placing it on its back, and then "sitting" in it. This is where the astronaut trained to reach all his instruments and controls until he could go through every motion of his scheduled flight with his eyes closed and never miss hitting the right button or lever.

Similarly, "crew quarters in Hangar S were spartan, austere, nondescript, and totally uncomfortable," Shepard remembered. "Our sleeping quarters could be reached only by going down a long, poorly lit hallway, an unpleasant walk during which we were assailed by hoots, screeches, screams, and howls of a small colony of apes housed out back."

The astronauts decided the humiliation of stepping aside for a squat, grinning monkey was bad enough, but they didn't have to live with the howlers and dung flingers. So they all abandoned Hangar S and took up residence at a motel along the beach where they lived like human beings.

It was an enormous psychological break. Alan, Deke, and the others spent hours jogging along the fine, hard sands of Cocoa Beach, drinking in the fresh salt air while formations of pelicans floated overhead. Headquarters and domicile was the Holiday Inn in Cocoa Beach, run by Henri Landwirth, who as a young boy had been confined in one of Hitler's concentration camps. Somehow the Belgian-born youngster survived the horrors of the camp and managed to make it to the United States with a single twenty-dollar bill. He threw himself into becoming an American citizen, settling in Florida, where he became an innkeeper. Before Project Mercury, Henri Landwirth had gained a small amount of fame as host to visiting congressmen, military officers, journalists, and foreign visitors who came to Cape Canaveral.

If the astronauts gave Henri or any of his staff trouble, he would throw them out, his melodious Flemish accent reminding them, "Customers I can always get! Good help is hard to find!"

Henri was one of a kind, and the longer he knew the astronauts, the fonder he became of them and the more he mellowed toward them, muttering under his breath, "Boys will be boys." He supplied them with a wealth of hospitality and food and provided protected areas for them to relax and unwind.

Gordo Cooper rewarded him one night by having the pool filled with fish so he could sit poolside with fishing pole in hand. The rest of the guests weren't pleased with having to swim with fish while dodging Gordo's multi-hook lure.

This fish-in-the-pool stunt held the record as the most outlandish astronaut prank until one night the Mercury-Atlas launch team and the astronauts decided to move their party to Henri's motel from a boat in the nearby Banana River. The waters on the river grew too rough, and the service on the boat wasn't all that good, so the launch team and future spacemen picked up the whole damn boat, carried it by hand across busy streets and planted it in Henri's swimming pool.

They stood there on the boat in the middle of the pool, shouting, "Rum for the crew, wenches for the officers," until some of the wives dumped them overboard and Henri rolled his eyes in dismay.

It also was a grave error to allow Landwirth and Wally Schirra to get together. Wally was a hopeless practical joker, and he elicited from Henri long-suppressed desires to pull a few pranks of his own. One day Henri and Wally emerged from Wally's motel room, Landwirth supporting a grievously wounded astronaut with a bloody towel wrapped about his arm. NASA officials and the reporters who covered the astronauts day and night rushed over to see what had happened.

Wally pointed at a large field with palmetto scrub north of the motel. "We went after something moving in there. I don't know what it is"—he groaned with pain—"but it sure as hell tore up my arm." Eyes stared at the thick, bloodied towel, and the group crowded into Wally's room to see what he had captured and had put in a large box on his bed, covered with a blanket. Wally pointed at the box. "Be careful, dammit. That thing's dangerous. I think it's a mongoose."

"Big mongoose," Henri confirmed.

Doug Dederer from the *New York Times* shook his head. "They don't have mongoose in Florida."

"Maybe it got loose from a zoo," Wally said with disdain. "Who cares where it came from? Dammit, Doug, look for yourself."

Dederer was a six-foot-two, 220-pound Connecticut Yankee. He moved toward the box. "Careful!" Wally warned. "Just pull the blanket back an inch or so."

The big man bent over and slowly peeled away a portion of blanket. *Wham!* A huge spring-loaded hairy thing with long teeth burst upward through the blanket. Six grown men shot through the motel room door. Dederer beat them all back to poolside, leaving Wally convulsed on the floor, choking with laughter, hugging the "jack in the box" monster he had lovingly built and then added fur and teeth to.

The "mongoose" in the months to come sent some of this nation's best fighter pilots jumping through windows and hurtling over beds as it burst through the blanket.

Keep a pilot on the ground long enough, and he'll do just about anything to bust out of his doldrums. The Mercury Seven made close friends with professional comedian Bill Dana, who had developed a routine known as the Cowardly Astronaut. Every chance they had, the astronauts would troop off to where Dana was performing and join him on stage as part of a wild impromptu act. Years later, long after the men of Mercury had retired from NASA, they still played straight men at dinners and parties to the crazy pseudo-Mexican who called himself Jose Jiminez.

By late January 1961, events were coming down to the wire. As flight time neared, the fun and games faded away. A serious tone settled over the launch, support, flight, and recovery teams. Redstone was working well, and Shepard was scheduled to be launched in about six more weeks.

But the charade to mislead the press and the public as to who was going to lead the way continued. The astronauts hated the roles Glenn and Grissom had to play: pretending they were still in the running for the first flight. The media's general consensus was that John Glenn would be chosen. The secrecy surrounding the selection continued right up to launch day, with Bob Gilruth deciding that Shepard's name would be made public only after the Redstone lit off and was flying. "There was even some incredible dumb talk," Shepard said, "about bringing all three of us out to the pad on launch day, all of us dressed to fly, with hoods over our heads! That way not even the launch team would really know who was *numero uno*."

The charade left a sour taste in the astronauts' mouths, but it was minor compared with the knowledge that the chimpanzee was still going before a man could fly. All that chimp would do was go along for the ride and bang levers and push buttons in the capsule, and get jolted with electricity if he didn't perform his tasks properly. Despite the astronauts' protests that the whole concept was ridiculous, the medical teams and psychologists insisted there were too many unknowns about space flight, especially weightlessness and what could be unexpectedly high g-forces, to risk a man's life without first sending up a living, breathing creature as a possible sacrifice.

So the chimps trained to go into space. It was a great circus act. Scientifically it was a huge waste of money, time, and manpower. Rube Goldberg could have designed the machine intended to test simian skills. In the box-shaped gadget, when a certain color light or series of lights flashed, the animal was trained to push either a right- or left-hand lever. If he performed as commanded, he was rewarded with a banana pellet funneled to his mouth through a dispenser. If the chimp failed to push the right lever, he was whacked with a slight electrical jolt to his foot.

This was the culmination of many years of research, the effort of hundreds of engineers, and the expenditure of several million dollars. Banana pellets and electric shocks to be administered by a madcap space-borne slot machine. It was about as close to insane as the astronauts could imagine.

Yet the effort went forward, and NASA selected one ape, crowned him Ham, and on January 31, 1961, the Mercury Seven gathered to watch the momentous liftoff of the slot-happy chimp. The flight turned out to be a bit more interesting than planned. The Redstone had a "hot engine" and consumed all its fuel five seconds ahead of schedule. The automatic control system sensed that something was wrong. Instantly it ignited the escape tower above the Mercury, and it blew the spacecraft away from the rocket with a great shriek of flame that sent the craft much higher and faster than it was intended to go.

The medics figured the ape would be squeezed by eight times gravity, when, in fact, Ham experienced more than twice the original estimate. Teeth bared, he was one unhappy ape when he didn't get his reward. The on-board equipment failed, the electrical system and light tests went haywire and, drifting weightless, Ham was banging on every lever he could. He did everything right, and for his efforts was rewarded at every turn with a nasty shock instead of a banana pellet.

He also sailed 122 miles farther down range than the planned 300 miles. He came down with crushing deceleration, the parachute opening slamming him about in the capsule, and he hit the ocean surface with another ear-clamoring bang. Then began the stomach-churning motion of rolling sea. By the time the recovery choppers showed up to lift the craft out of the waves, it was on its side, filled with so much water they had a sputtering, choking, near-drowned chimp on their hands.

They returned Ham to the Cape. NASA went through some idiotic official greeting, but Ham came out of the capsule biting anything, human or otherwise, that came near him.

Alan Shepard reviewed the telemetry tapes and records of the Great Chimp Adventure, as Ham's flight came to be known. He knew he could have survived that trip, but he also knew immediately his own planned flight was in deep trouble. If only the damn chimp ride had been on the mark, then Alan would launch in March.

But Ham's flight had been miserable, and in Huntsville, Alabama,

Wernher von Braun was showing signs of a new conservatism he had developed, as his responsibility for men's lives factored into his decisions. Von Braun, to the utter dismay of Shepard and the rest of the Mercury pilots, said simply, "We require another unmanned Mercury-Redstone flight."

Working with the engineers, Shepard confirmed that the problem with Ham's Redstone had been nothing more than a minor electrical relay. The fix was quick and easy, and the Redstone was back in perfect shape. "For God's sake, let's fly. Now!" Alan implored NASA officials. "Even von Braun should be satisfied with what we've found!"

He was wrong. Dr. von Braun stood fast. "Another test flight."

Shepard stalked off to the office of Flight Director Chris Kraft. He was steaming. "Look, Chris, we're pilots," he snapped. "When there's a failure, dammit, we fix it."

"I know, Alan," he said. "I know where you're coming from."

"Well, what about it?" Shepard asked. "It's an established fact that the relay was the problem, and it's fixed."

"Right."

"So, why don't we go ahead? Why don't we man the next one?"

"Why waste time, right?" Kraft smiled.

"Right."

"Because when it comes to rockets," the flight director shook his head, "Wernher is king."

"King?"

"King."

"Forget it, right?"

"Right."

Shepard walked away, brooding. The March 24 Redstone flight was an absolute beauty. Shepard could have killed. He knew he should have been on that flight. He could have led the world into space. He should have been floating weightlessly, looking down on a sharply curving horizon, while the Russians were still wrestling with a balky rocket booster. "We had 'em," Shepard recalled. "We had 'em by the short hairs, and we gave it away."

CHAPTER EIGHT

FIRST IN SPACE

FOR A LONG TIME he drifted between sleep and wakefulness. It had been that way throughout the night. Floating between memories as sounds drifted through the walls of his room. Strange, he couldn't tell if the sounds were from past memories or from the present, but he accepted the ones that reminded him of his father. A carpenter, a skilled craftsman who had worked hard to make so special their wooden home in the village of Klushino, near Smolensk of the western Soviet Union.

Harsher memories intruded. Great guns firing, shells exploding. Earth-shaking rumble of German tanks moving through his hometown. He was a wide-eyed boy then, watching his parents' world coming apart as they fell under enemy occupation. They obeyed the Nazis, and whenever possible they foraged food from the fields and the scattered wreckage of their village. Another sound grew louder, and he seized it in his dream state. Airplane engines. At first only German. Then other planes came by with red stars on their wings, and there was terrible fighting and the tanks that pushed into Klushino were Russian. As quickly as that war ended, young Yuri A. Gagarin crammed day and night, in school and at home, so that one day he would qualify to become a pilot in the Red Air Force.

They moved to the larger town of Gzhatsk. Yuri completed school, completed special courses, and in 1955 entered the Air Force. Two years later he won the coveted wings of a jet fighter pilot. He had become an expert parachutist as well. For two years he served in operational units and then, in 1959, he volunteered for an exciting new program.

Cosmonaut! He swept through the rigorous training, excelling in everything he did. On April 8, 1961, only four days before this night of dreams, his commander gave him the news. "You will be the first to travel through space."

Unreal. It all seemed unreal. But it was true. And his close friend, Gherman Stepanovich Titov, would be his backup. Today was the day. His door opened.

Sleep and dreams vanished. He met with Titov, technicians, doctors, engineers, the political commissar. Everything moved smoothly through breakfast, final medical checks. Sensors were attached to his body before he donned the pressure suit and heavy helmet. Fully protected from space, his teammates helped him into the bright orange flight suit that would aid the recovery crews in spotting him after landing.

Sunrise was still to come as he arrived at the launch pad. He stood quietly for several minutes, studying the enormous SS-6 ICBM that would send him into orbit. No warhead atop the big rocket. It had been replaced with the *Swallow*, the Vostok spacecraft of more than five tons.

Gagarin stood on a ramp part way up the stairs to the elevator. He turned to the select group who would witness the moment that would separate the past from the future. He spoke clearly to those men:

"The whole of my life seems to be condensed into this one wonderful moment," he began. His audience stood silently transfixed. "Everything that I have been, everything that I have done, was for this," Gagarin added. Yielding to the emotion of the moment, he lowered his head, regained control.

He looked up again, smiling. "Of course I'm happy," he said, his voice stronger. "Who would not be? To take part in new discoveries, to be the first to enter the cosmos, to engage in a single-handed duel with nature . . . " His smile broadened. "Could anyone dream of more?"

His words spoken, he waved farewell and entered the elevator to the top of the support tower. There he climbed a short ladder to the platform alongside his Vostok spacecraft. Technicians and his close back-up team assisted him through the hatch. They secured his harness to the specially designed seat. Gagarin nodded, signaling he was ready. The hatch closed and was sealed.

Countdown delays were as common in Russia as they were at the Cape. Gagarin received word that a faulty valve had been detected. "But it will soon be fixed. Be patient, Comrade Gagarin."

Soon came the countdown phase he had been waiting for.

"*Gotovnosty dyesyat minut.*"

Two minutes more . . .

He felt motors whining. Excellent. He knew what the sounds meant. Moments later he ordered himself to relax in his contoured seat. The gantry with the service level was pulling away. Tall towers leaned backward and away so that the launch pad took on the appearance of a huge steel-cabled flower petal. He felt the bumps and thuds of power cables ejected from their slots in the rocket. Now the powerful SS-6 drew power from its own internal systems.

The final seconds rushed away; a voice cried, "*Zazhiganiye!*"

Gagarin needed no words to tell him he had ignition as twenty powerful main thrust chambers and a dozen vernier control engines ignited in an explosive fury of nine hundred thousand pounds of thrust. Thirty-two engines strained, explosive hold-down bolts fired, and the first man to leave earth was on his way.

Moscow time: 9:07 A.M. America slept, unaware of Gagarin's jubilant cry from his ascending fire machine. "Off we go!" he cried aloud, bringing smiles and grins to the crews in the launch control center. With the SS-6 well clear of the launch pad, many of the men whose duties were through rushed outside to see the rocket accelerating faster and faster. Binoculars showed them the liquid-propellant strap-on boosters ejecting from the rocket's main core and a single, dazzling ball of flame rising with increasing speed. In just those first few minutes of ascent, Yuri Gagarin was traveling faster than any man in history. Then, the booster was bending far above and away over the distant horizon, leaving behind a twisting trail of smoke as a signature of its passage.

Through the increasing g-loads Gagarin maintained steady reports. He was young and muscular, and he absorbed the acceleration punishment easily. With the strap-on boosters jettisoned and the main core engines burning, the acceleration generated a force of six times normal gravity.

Gagarin heard and felt a sudden loud report, then a series of bumps and bangs as the protective shroud covering Vostok was hurled away by small rockets. Now he could see clearly through his portholes a brilliant earth horizon and a universe of blackness above. Finally the central core exhausted its fuel, and explosive bolts fired to release the final "half stage" rocket to complete the burn to orbital height and speed.

At last the final stage fell silent. More booms and thuds as the spent rocket was discarded.

The miracle was at hand. A human being was falling around the earth at 17,500 miles an hour. Gagarin in *Swallow* had entered orbit with a low point above earth of 112.4 miles, soaring as high as 203 miles before starting down again.

Those on the ground listened in wonder at Gagarin's smooth control, his reports of what he was feeling, how his equipment was working. Then he went silent for several moments as a never-before-known sensation enveloped his body and his mind.

He felt as if he were a stranger in his own body. He was not sitting or lying down. Up and down no longer existed. He was suspended in physical limbo, kept from floating about loosely only by the harness strapping him to his contoured couch. About him the magic of weightlessness appeared in the form of papers, a pencil, his notebook, and other objects drifting, responding to the gentle tugs of air from his life-support system fans.

He forced himself back to his schedule, reporting the readings of his instruments. As critical as were those reports, there was even greater interest in what Gagarin felt and saw. He told ground controllers weightlessness was "relaxing." He took precious moments as he orbited the earth, covering five miles every second, to report, "The sky looks very, very dark and the earth is bluish." He waxed enthusiastic about the startling brightness of the sunlit side of the earth. He raced through a sunset and a sunrise and, almost before he realized the passage of time, he was nearing the end of man's first orbital flight.

He would use his manual controls only in an emergency. Now he remained both physically relaxed and mentally vigilant as he monitored the automatic systems turning *Swallow* about for retrofire.

Rockets blazed. The sudden deceleration rammed him hard into his couch. He smiled with the full-body blow; everything was working perfectly. He watched the timer. There! Another series of sharp explosive reports as *Swallow* and its electronics pack separated from the equipment module.

He had circled the globe once in eighty-nine minutes.

He had plunged across East Africa and now began his return to earth, flying backward.

He became weightless again. Not for long this time, as *Swallow* arced downward into thickening atmosphere. He felt the first caress of weight from deceleration. Now he was a passenger within a blazing sphere. Through the spaceship's portholes he saw flames, at first filmy, then thickening and becoming intense blazing fire as friction from the atmosphere heated the ablative covering of *Swallow* to thousands of degrees. The protective coating burned away with increasing fury. He was in the center of a manmade comet streaking toward the flattening horizon. Though inside a fireball, he could not feel the heat.

Then he was through reentry burn. The *Swallow* had slowed to subsonic speed. Twenty-three thousand feet above the ground the escape hatch blew away from the Vostok. Gagarin saw blue sky, a flash of white clouds. Small rockets within the spacecraft fired, sending the cosmonaut and his contour couch flying away from the Vostok.

Gagarin watched a stabilization chute billow upward from the seat. Everything worked perfectly. For ten thousand feet he rode downward in the seat. In the near distance he saw the village of Smelovaka.

Thirteen thousand feet above the ground he separated from the ejection seat and deployed his personal parachute. He breathed in deeply the fresh spring air. What a marvelous ride down!

On the ground, two startled peasants and their cow working in a field watched as a man wearing a bright orange suit, topped with a white helmet, drifted out of the sky. The man hit the ground running. He tumbled, rolled over, and immediately regained his feet to gather his parachute. Gagarin unhooked the parachute harness and looked up to see a woman and a girl staring at him.

"Have you come from outer space?" asked the astonished woman.

"Yes, yes, would you believe it?" Gagarin answered with a wide grin. "I certainly have."

The shrill ringing penetrated the fog of his sleep—annoying, persistent. For a moment Alan Shepard remained confused, extricating himself from deep slumber. Only for the moment. Then he reached for the clamoring telephone in his motel room.

"What?" he barked into the phone.

The voice at the other end of the call was soft, polite. Considerate. "Commander Shepard?"

"Uh-huh. Yeah, this is Shepard."

"Have you heard?"

He was sharply attentive now. He didn't like those words. "Heard what?" he asked cautiously.

"The Russians have put a man in orbit."

Shepard blinked. He sat straight up, rubbing his eyes. "They what?"

"They've put a man in orbit."

The phone almost slipped from Shepard's hand. He sat quietly another few moments to brush away the last fog of sleep from his mind. Then his disbelief found voice. "You've got to be kidding."

The caller was a NASA engineer who knew Shepard. "I wouldn't do that, Commander Shepard." He appeared apologetic for being the messenger of shocking news. "They've done it. They really put a man in orbit."

Shepard managed a courteous response, thanked the man, and replaced the phone in its cradle. A single phrase kept repeating itself over and over again in his mind:

I could have been up there three weeks ago . . .

He turned on his bedside radio. Excited voices spoke of Vostok and orbit and Yuri Gagarin. Alan called the other Mercury troops. They'd heard. They were all glum. There wasn't a race anymore.

It was bad enough for the astronauts to get the news in the middle of the night by telephone. It was worse for the United States to be seen falling behind the Soviet Union. It was worse that the public relations mouthpiece for the Mercury Seven, Colonel John "Shorty" Powers, didn't know how to keep his mouth shut at a tense moment.

A reporter woke Powers and started asking questions. Groggy, befuddled, and angry, Shorty snarled into the phone, "If you want anything from us, you jerk, the answer is we're all asleep!" He slammed down the phone. The moment was heaven-sent for a newsman out for the perfect cutline, and by morning the reporter's newspaper headline shouted:

SOVIETS SEND MAN INTO SPACE;
SPOKESMAN SAYS U.S. ASLEEP

NASA officials were all over the seven astronauts before dawn. "We need a clear-cut statement for the press."

The astronauts allowed they were disappointed but made certain to offer sincere congratulations to the Soviets for a terrific technical feat. Once again John Glenn galloped to the rescue. He had a secret. Be blunt and truthful.

"They just beat the pants off us, that's all," he told the press. "There's no use kidding ourselves about that. But now that the space age has begun, there's going to be plenty of work for everybody."

John went over like rich cream. The press lapped it up. His candor only helped convince the remaining holdouts that John Glenn had been picked to be the first American in space.

The charade continued.

In the long run, the question of who was the first astronaut in space would never mean that much. When the first man flew, the guesswork would vanish and that would be the end of it. But would he fly? Given Gagarin's flight and the overwhelming power of the Soviet boosters, there were suggestions in Washington that the U.S. man-in-space program be canceled. The dark mood had as its banner the feeling that the United States could never catch the Russians now, so why waste the time and effort and money to run second-best?

In the White House, the country's young new president was bedeviled with the reality of Soviet superiority in powerful rockets. At first he seemed reconciled that space belonged to the Russians. In fact, John F. Kennedy emphatically told a news conference after the Gagarin flight that the nation would not try to match the Soviet achievements in space, choosing instead "other areas where we can be first and which will bring more long-range benefits to mankind."

But Kennedy was not comfortable with that stand. Only a few short months before, he'd pledged to "get this country moving again." In his election campaign against Richard Nixon, he had made major issues of the missile and space gaps between the U.S. and USSR. He had campaigned eloquently that America belonged in space because "space is our great new frontier."

How could he allow the United States to simply quit? No matter that for the Russians space was most certainly their great new frontier. America had come from behind before, and it must do it again.

Kennedy didn't have any help from the head of his Science Advisory Committee, Jerome B. Wiesner of the Massachusetts Institute of

Technology. Wiesner wanted to gut the whole national space program. Rip NASA down to bare bones and reorganize from the ground up. Concentrate on aeronautics and yield the space race to the Russians. End the Mercury program quickly with Kennedy's signature on a cancellation document, because if Mercury failed, if lives were lost, the country would blame Kennedy for the failure. Wiesner suggested the U.S. should expand its leadership in science, communications, and military satellites. "Go with the winners," was his forceful recommendation.

Kennedy knew the American people better than that. He rejected that first sign of quitting. He called the vice president into his office and told Lyndon B. Johnson that from that moment on, Johnson, a strong backer of the space program, would run the National Space Council and get things moving. Then he called James E. Webb, a tough-willed North Carolina attorney who knew the routes through government, industry, executive, and political pastures. Kennedy got right to the point. "Jim, I want you to run NASA."

These were forceful moves, and yet not until Yuri Gagarin sped about the earth did Kennedy move in the direction of a powerful national space advocacy. Until then he had been preoccupied with a flagging economy, communist incursions throughout the world, and the takeover of Cuba by Fidel Castro.

Gagarin changed all that. Two days after the cosmonaut finished his flight, Kennedy had Johnson, Webb, and Wiesner in his office. The president wanted answers to how the U.S. could catch up with the Russians. Johnson and Webb both shared the same view: "Catch up, hell. Let's pass 'em."

With the advent of the Mercury program, there had been discussions among scientists, explorers, and in the press that man might one day be rocketed to the moon. "What about the moon?" Kennedy asked. "Can we do it? And can we beat the Russians to it? Find out and get back to me." Johnson and Webb were gung ho for the idea. Wiesner mumbled and fumbled that all was lost, space belonged to the Russians, and the U.S. might as well become accustomed to it.

Gagarin's solo orbit of the earth capped a series of events that rocked the new administration. The Russians were now the pioneers of successful manned flight around the earth and challenged America to catch up. It would be like the race between the tortoise and the hare. If one nation made a misstep or faltered along the way, its opponent would catch it and race on ahead. Not a shred of doubt existed that this new race for space would be a high-flying competition, played out in full view of the world.

Far more disturbing than the Soviet achievement in space was the simmering cauldron of communism in neighboring Cuba, where Fidel Castro's tyranny spawned a growing anger among Cuban exiles in the United States and in Cuba's neighboring countries—exiles who were determined to overthrow him by force of arms. Kennedy was in the thick of a complicated political situation not of his own making but inherited from the Eisenhower administration. President Eisenhower approved a CIA-sponsored covert operation in Cuba that would train, finance, supply, and equip an invasion force to assault Cuba. The operation was a covert one as Kennedy and his predecessors avoided the appearance of interfering in the political processes of Latin countries.

In truth, Kennedy wanted nothing to do with such a politically hazardous and morally questionable invasion. His aides, and the exile leaders, swore to him that the invasion would spur the Cuban people to an uprising against Castro. The Cuban exiles' army was ready. All that would be required for a successful invasion, they assured him, would be the support of American air power to crush the communist jets and other fighters that would be sent to intercept the attacking bombers.

Guatemalan airfields echoed with the thunder of twin-engine B-26 Invader bombers whose crews had been rehearsing for months. They would strike strategic targets throughout Cuba. A thousand miles away, the swampy environs of southern Florida bristled with scores of Cubans training for invasion landings and hill-fighting in their native land. They were ready. If the invasion did not proceed now, as planned, the opportunity to overthrow Fidel Castro might be lost forever.

The Cuban exiles begged for air cover. Kennedy assented and promised that an American aircraft carrier would be stationed off the Bay of Pigs to provide air support for the invasion. The invading army of Free Cuba insurgents would swarm ashore behind the devastating cannon, rockets, and bombs of American jets.

The fanfare surrounding Gagarin's flight faded in the tense hours before the predawn invasion on April 17. While sixteen B-26 bombers readied to attack airfields at San Antonio de los Banos and Santiago, John Kennedy fidgeted with indecision. Making a decision his closest aides and advisors would never understand, Kennedy abruptly ordered that the force of sixteen bombers be reduced to eight. Aware that their lives were seriously imperiled, Free Cuba's finest pilots commenced their strike against the communist airfields.

Cuban fighter planes counterattacked, reducing the insurgents' planes to broken, flaming wreckage. Screaming with their last breath, the pilots begged for the American air cover that had been promised them by Kennedy.

On the ground, the invading army troops slogging ashore at the Bay of Pigs were cut to bloody shreds by machine-gun crossfire from Castro's defensive positions.

On the off-shore American aircraft carrier, planes loaded with bombs, rockets, and cannon were poised for takeoff. Engines running, they were only moments away from smashing Castro's defending army. The orders came forth from the White House.

"Stand down your aircraft. Do not attack. Repeat. Do not attack."

The few who miraculously emerged alive from the slaughter of thousands of men at the hands of Castro's guns hurled hateful invectives at the American president. Kennedy was denounced as a coward, a liar, a man whose word was nothing, and a man whose lack of nerve and leadership ability had cost the lives of thousands who died needlessly.

The press had their own field day with the ghastly photographs, television footage and eye witness accounts from those who had survived the broken promises of John F. Kennedy. The young president's standing with the American people would never be lower.

If ever Kennedy needed a bold new step, now was the time. He rested his case solidly on the future of the U.S. in space. LBJ answered his summons and heard a no-nonsense president talking. "I want you to tell me where we stand in space. Do we have a chance of beating the Soviets by putting up a laboratory in space? Or by a trip around the moon, or by a rocket to go to the moon and back with a man? Is there any other space program that promises dramatic results in which we could win?"

Kennedy and Johnson were striking out for the future but, elsewhere in Washington, Wiesner and his cronies wanted to put on the brakes. The first American manned space shot was now on the calendar for May 2. A failure at this point could be devastating for national morale and prestige. On April 25 the American space program started to unravel. An Atlas booster lifting an unmanned Mercury capsule drifted off course, was blown up, and returned to earth in a huge gout of flaming debris. Three days later a Little Joe rocket boosting a Mercury on a test of its emergency escape system spun out of control and was destroyed.

At a meeting the next day in the Oval Office, Wiesner told the president he must order a delay in the first manned liftoff. He recommended further that if Kennedy insisted on going ahead with it, the flight should be carried out later and in secrecy to avoid an overpublicized fiasco should the mission fizzle.

The small group of officials who had convened wondered how NASA could carry out such a flight in secrecy, with hordes of newsmen aiming cameras at the Cape, looking down from airplanes and helicopters, and using their own special techniques for tapping into NASA control and communications lines.

Kennedy was resolved to make the United States a true spacefaring nation. There would be no secret launches of the manned civilian space program, he ordered. We will act in the open, for the public.

He received support from Johnson, who told the group the Atlas and Little Joe failures had no bearing on the reliability of the Redstone and Mercury spacecraft. The executive director of the Space Council, Edward Welsh, also stood up to be counted. He rattled off a list of Redstone reliability figures. "The flight will go," he said. "And it won't fail. The risks of failure are no greater than our crashing in an airliner between here and Los Angeles just because the weather isn't perfect. So why postpone a success?"

Kennedy nodded. They had a GO.

The astronauts felt they had been sealed within a giant pressure cooker. The word was out that if the U.S. was to have a civilian manned space program, the first shot must succeed.

The entire team—astronauts, engineers, scientists, technicians, *everybody*—went on twenty-four-hour availability. Glenn and Grissom became Shepard's alter egos, his shadows, always there to support him in every way.

They were still plagued by questions from the media about who would be first to go. They had enough to do without having to deal with that issue, but with Alan spending most of the time in the Mercury simulator, you didn't need a crystal ball to judge that most likely he had the nod.

Three hours before scheduled launch on May 2, the name of Alan Shepard floated to the surface of the national media. The Associated Press even went with a confirmation of Shepard.

Alan Shepard regarded the news with relief. They could get that monkey off their backs and concentrate on the flight.

Alan was ready, the rocket was ready, and the range was ready. But Shepard did not lift off that day. Low cloud cover rolled in, and Walt Williams scrubbed the launch. The flight operations director was right. He wanted a clear view of that Redstone and its precious human cargo all the way through fuel burnout.

Alan Shepard moaned and groaned. "I guess I'm destined to stay forever on this planet."

His flight surgeon, Bill Douglas, grinned at Alan's discomfiture. "Not hardly, Al. That's a departure we'll all make someday. The difference is you want to leave and come back."

Shepard offered a patient smile and went back to the simulator for more make-believe space flying.

Three days passed. Shepard felt the delay actually eased the tension that had been growing within him. Before the May 2 launch he'd been plagued with visions of rockets tumbling out of control or blowing up violently in the air. Understandable; he'd seen those actual sights.

He solved his problem in typical Shepard fashion. "I backed off and regrouped and hit it again," he said. He recognized he was experiencing normal apprehension and not fear. The entire reasoning process was old hat to the test pilot. He knew how to turn off whatever gnawed at him.

He'd been through the drills and was amazingly calm as the new launch date of May 5 neared.

The night of May 4, however, the other astronauts and support teams brought their own tension onto the scene. Everyone except Shepard walked on eggshells. Despite the strong feelings about weather, rocket reliability, the escape system, anything and everything, no one dared to broach those subjects. The air became so thick Shepard left for his bedroom. He had some feelings of his own he wanted to share. He phoned his family in Virginia Beach.

On the first ring he heard her voice. "Hello."

"Louise!"

"Hi, Alan." Pause. "I was hoping it was you."

"Yeah, it's me. Tomorrow looks promising. I think we've got a go."

"That's wonderful." Alan could feel her smile reaching him all the way down that telephone line.

"Weather's supposed to be good," he said.

"It will be." Not a doubt in the world. "It'll be your good-luck day."

"I do feel good about it. Everything okay at home?"

"Everything is just great, Alan. The folks are here."

"I figured. I'd like to talk with them and then the girls."

Several minutes later, Louise came back on the line. "We'll be watching you on TV. Be sure to wave when you lift off."

"Right," he laughed. "I'll open the hatch and stick out my arm."

She chuckled, warm and loving. "You do that. Take care of yourself, sweetheart. Hurry home."

"I will."

"I love you."

"Me, too, Louise."

That was pure tonic, wonderful and soothing.

Flight surgeon Bill Douglas woke Alan early. Just like another day at the office, Shepard grinned to himself as he shaved and showered, and then polished off a breakfast of filet mignon, eggs, orange juice, and tea. He left the breakfast table to place himself at the tender mercies of the doctors, who did their usual poking, prodding, and measuring, and then they attached a battery of medical sensors to him. Shortly after 4:00 A.M., suited up, he departed Hangar S with Douglas and Gus Grissom for the launch pad. They rode in a transfer van, which Shepard likened to a "cramped cattle car."

Shepard turned suddenly to Grissom. "Hey, Gus, do you know what it really takes to be an astronaut?" He asked the question in a parody of Bill Dana's frightened astronaut Jose Jiminez accent.

"No, Jose," Gus replied. "Tell me."

"You should have courage and the right blood pressure and four legs."

"Why four legs, Jose?"

"Because they really wanted to send a dog, but they decided that would be too cruel."

The van stopped at the launch pad. Instantly everything changed. Alan Shepard stepped out into an alien world of glaring floodlights, banshee wails from a breeze blowing across super-cold fuel lines. He was a creature from another planet in his silvery space suit.

He looked up, for the moment overwhelmed by the gleaming blue-white lights. Then he began the final walk toward the gantry elevator. "Up" was six stories above him.

Abruptly he stopped, almost mesmerized by another long look at what he considered the "awesome beauty" of the Redstone booster with its

Mercury spacecraft. They had, he told himself, "an air of expectancy about them." He watched plumes of vapor venting from the liquid oxygen tank. "I never again will see this rocket," he said confidently.

He tilted his helmeted head up and back. He wanted to burn the scene permanently into his memory.

He moved into the elevator. The door closing behind him was an unplanned signal for applause and cheering by the men and women who'd worked day and night, always under the shadow of the great Russian boosters, to start America's own high road into space. Alan turned, waved to his launch team.

He had the strangest feeling he was taking them along with him.

Alan started to call out to them. The words choked in his throat. He felt that if he tried to speak, his emotions would not let . . . he forced his attention to Bill Douglas.

"This is for you," his flight surgeon said, handing him a box of children's crayons.

"Just so you'll have something to do up there," Douglas said.

Then he was in the sterile White Room that surrounded the capsule that would take him out of the world. John Glenn greeted him with word that everything was ready. The two gripped hands firmly, and Alan began the squeeze into his spacecraft.

The name *Freedom Seven* had been painted on the capsule's side. Alan's choice. *Freedom* because it was patriotic. *Seven* because it was the seventh Mercury capsule produced. It also represented the seven Mercury astronauts.

No streamlined space liner this. Mercury was a truncated cone ten feet high and just over six feet wide at the base of the ablative heat shield. Once Alan was shoehorned into his space chariot and all suit connections were completed, he could move his eyeballs and not much more. For this flight NASA had placed a parachute chest pack on a small ledge inside the capsule. The only time Shepard could use it was after his main parachute opened and then something went wrong that required an emergency exit. Then he'd have to clip on the chest pack, open the hatch, and wriggle his way out. Anyone who knew the tight quarters of the *Freedom Seven* knew the chute was along so that if the flight came apart, no one could say that the pilot never had a means of personal escape.

Settled in, Alan saw a small notice attached to the instrument panel.

NO HANDBALL PLAYING IN HERE.

Glenn looked through the hatch, grinning at Alan reading the sign. Alan returned the notice to him.

They readied the hatch. Alan held up a gloved thumb to Glenn. "See you soon," he said. Behind Glenn the gantry crew shouted their farewell. "Happy landings, Commander!"

The hatch closed.

Alan Shepard was alone, squeezed into a spacecraft with less room than a telephone booth. He felt butterflies trying to squeeze in with him, reminding him he was sitting atop a great tube of a bomb. He pushed aside the thought and went to work, running down his check lists, testing the radios, all switch settings.

The butterflies went elsewhere.

Shepard looked through the periscope viewer in the center of the instrument panel. He saw the crew still at work outside the spacecraft. Then they began to diminish in size as the gantry rolled back on steel rails from the Redstone.

The periscope gave him a view of clouds lit by the morning sun. Far below he watched the launch crew finishing last-minute details at the base of the rocket.

He glanced at his capsule timer. Only fifteen minutes to go in the countdown. The view outside dimmed. Cloud cover rolling in! Damn!

The countdown clock stopped. Everybody sat on tenterhooks, waiting for the sky to clear. Everybody hated countdown delays. It just allowed more time for something to go wrong.

It did. The launch director ordered the gantry rolled back to the rocket. A small electrical part had suffered a glitch. Not much, but it had to be fixed, resulting in an hour-and-twenty-six-minute delay.

The countdown hold was long enough for Gordo Cooper, principal prelaunch communicator in the blockhouse. He had to hunt for things to say. Bill Douglas fretted like a mother hen. Wernher von Braun came on the communications line, as unhappy as Shepard with the delay.

"Gordo?"

"Yeah, Alan."

"Tell Shorty Powers to call Louise. I want her to hear from us that I'm fine, and explain that I'm going nowhere fast."

"Got it."

The delay was taking its toll in a physiological manner as well. Shepard had been seated atop the Redstone for so long that the need to urinate was becoming urgent. And that was a problem. The suborbital flight was scheduled to last only fifteen minutes. No one thought it necessary to equip Shepard or the Mercury with a urine collection system.

"Gordo!"

"Go, Alan."

"Man, I got to pee."

"You what?"

"You heard me. I've got to pee. I've been in here forever. The gantry is still right here, so why don't you guys let me out of here for a quick stretch?"

"Hold on," Gordo told him. He came back a few minutes later. "No way, Alan. Wernher says we don't have the time to reassemble the White Room. He says you're in there to stay."

"Gordo, I've got to relieve myself!" Shepard shouted. "I could be in here a couple more hours, and by that time my bladder's gonna burst!"

"Wernher says no."

Alan's temper was soaring. "Well, shit, Gordo," he said, "we've got to do something. Dammit, tell 'em I'm going to let it go in my suit."

"No! No, good God, you can't do that," Gordo shouted back. "The medics say you'll short-circuit all their medical leads!"

"Tell 'em to turn the power off!" Alan snapped.

The solution was that simple. Gordo had a chuckle in his voice. "Okay, Alan. Power's off. Go to it."

No science fiction writer had ever penned this scenario. Shepard simply couldn't hold back the urine flow any longer. But since he was in that reclining semi-supine position the liquid pooled in the small of his back. It was as if they'd designed the suit for such an urgency. His heavy undergarment soaked up the urine, and with 100 percent oxygen flowing through the suit he was soon dry.

The countdown resumed.

The gantry was gone.

Alan watched the waves breaking on the beach. Just what the doctor ordered. Calming and soothing.

Five minutes.

Three minutes.

Two minutes and forty seconds and counting.

Shepard heard the dreaded word, "Hold."

Gordo was on the line immediately. "Alan, uh, we're gonna hold here at this time. We've, ah, got a little computer problem here—"

"Shit!" Alan yelled. "I've been in here more than three hours. I'm a hell of a lot cooler than you guys. Why don't you just fix your little problem and light this candle?"

They fixed the problem; the count resumed. At T-minus 2 minutes Alan heard Deke's voice. Pure comfort to hear that man. From this point on Deke Slayton, the man who had that unspoken deep understanding with Shepard, would be his main contact for the mission.

Gordo was in the blockhouse just a stone's throw from the pad. Deke sat before his console in Mercury Control two miles away. He shared the control room with fifteen men who sat behind three banks of consoles to measure every moment of the flight.

Deke's voice became a professional monotone as he counted off the final seconds.

Just before liftoff, Shepard had a final message he spoke only to himself. "Deke and the man upstairs will watch over me. So don't screw up, Shepard. Don't screw up. Your ass is hauling what's left of your country's man-in-space program."

Vibration rattled the capsule as the Redstone's internal pumps came alive.

CHAPTER NINE

FREEDOM SEVEN

THE INVASION FORCE gathered outside the home of Louise and Alan Shepard was on its own "hold," sipping coffee and consuming doughnuts brought to them by the Shepards' neighbors. Photographers, television camera crews, reporters and broadcasters, playing the waiting game, hoping Louise Shepard would emerge from her home to talk with them, tell them how she felt, what were her emotions, everything from pride to fear of—

No, she would not permit contemplation of something going wrong, even to herself, let alone to the wolves at the door. The moment was familiar except for the siege outside. She'd waited before when Alan had flown tests closer to the earth. She knew the clammy feelings when he was late, but that was a straight road to a nervous breakdown, and she had pushed all that away from her life before now. Test plane or tall rocket. It didn't matter.

If there was a danger that was real, unexpected, Alan would have told her. They lived by that agreement. No false bravado. Just the truth, plain and simple.

She had unquestioned faith and confidence in her husband. If the metal parts held together and the flame burned bright and true, and success hinged on the performance of Commander Alan Shepard, then her confidence in the successful outcome of this flight would be the same as it had always been.

Now, there was something new for her, her children, their family. They could do more than listen. The new idol of the space age was that large, squarish box that snatched pictures magically out of thin air and displayed them on the television screen. Now they could hear and see what was happening.

She understood the pressure on the media to ask her questions and share her thoughts and feelings with readers and viewers and listeners throughout the world. In many ways she spoke for them all. They could transfer their own empathy for whatever it was they thought she was enduring. It was a tug of war between what people wanted to see, hear, and feel, and the intensity of her own desire to retain the integrity and privacy not only of her family but of Alan himself.

The newsmen and women waiting outside were made up of both compassionate beings and story-hungry flacks with no concern for the feelings of others. They represented the broad spectrum of a nation eager for news. But she was Mrs. Alan Shepard, and they would respect that, period. Through the long night she had heard footsteps coming up on her front porch, each time followed by a pause and the sound of retreat as the news people read the note she had left on her door:

THERE ARE NO REPORTERS INSIDE. I WILL HAVE A
STATEMENT FOR THE PRESS AFTER THE FLIGHT.

She was grateful they had chosen to respect her wishes, to accept her word there were no reporters in her home. A rumor had circulated among the gathered press that *Life* magazine had a reporter and photographer inside.

Louise watched the crowd outside, then turned from the window, lifting the small transistor radio she'd carried all morning to her ear. The station was carrying the Cape Canaveral broadcast live. She didn't want to miss a beat.

"Louise!" Her father called from where the rest of the family was before the TV. "Better get in here! They've picked up the countdown!"

She caught the sudden change of voice from the radio. "This should be it," the broadcaster said quietly. "Everything looks good. The weather is go, and Mercury Control says Alan Shepard and his *Freedom Seven* are go . . ."

She joined the family, staring at the slender rocket standing alone. It looked like a marble pillar or a stone from some ancient Greek painting. A painting alive with a white cloud of cold vapor drifting from the rocket in answer to the wind. She knelt before the TV.

Instinctively she reached forward to touch the live picture of the spacecraft, to touch her husband.

T*-minus seven.*

Alan drew strength from Deke's firm voice.

Six.

He spoke in his mind to his friend. "Hang in there with me, Deke . . ."

Five.

He pushed his feet firmly against the capsule, bracing himself.

Four.

Alan had his hand up near the stopwatch on the panel. He must initiate the timer at the moment of liftoff in case the automatic clock failed.

Three.

Left hand on the abort handle. The escape tower was loaded.

Two.

He was talking aloud to himself in the tradition of pilots about to enter the unknown. "Okay, buster, you volunteered for this thing. Now make it work!"

One.

"Don't screw up, Shepard . . ."

Zero.

Deke's voice rose in pitch as he sang out, "*Ignition!*"

Rumbling far beneath Shepard. Pumps spinning at full speed. Fuel gushing through lines, joining in the combustion chamber to be born anew in fire. He tensed his body.

Before he could think about what came next, a dull roar boomed through the Redstone, rushed up into the spacecraft, shook the world with a surprisingly gentle touch. Thunder grew, louder and louder.

"*Liftoff!*" Deke called.

Alan felt movement. Again he readied himself for vibration and shock. In anticipation he had already turned up the volume of his headphones. He didn't want to miss a word from Deke because of the still-increasing thunder.

Freedom Seven swayed slightly.

His heart pounded.

He had first motion.

"You're on your way, Jose!" Deke shouted, jubilation in his voice.

"Roger, liftoff, and the clock has started," Alan called out. Now he felt the power beneath him. The sound and vibration surprised him. Redstone came to life gently, a slumbering giant greeting the sky with a yawn and a stretch. But by God he was on his way . . .

"This is *Freedom Seven*. Fuel is go. Oxygen is go. Cabin holding at 5.5 PSI." The hard data came from Alan like a ticker tape.

"I understand, cabin holding at five-point-five," Deke responded.

How incredible. It seemed the two calmest, most assured people along the entire space coast this day were Alan Shepard and Deke Slayton.

Even before the first swaying movement of *Freedom Seven*, other machines were out in force in preparation for Alan Shepard's liftoff. Military helicopters with rescue teams moved to the west of the launch pad, over the scrub brush and palmettos of Merritt Island, and to the east, skimming the ocean offshore. Streaking toward the pad in F-106 jets were astronauts Wally Schirra and Scott Carpenter, primed to chase and observe the Redstone as long as they could before it sped from sight. Tracking and search planes cruised from low-level to stratospheric heights, and the sea was dotted with swift crash boats and Navy ships, all coiled to spring to *Freedom Seven* in the event the unlikely, the unthinkable, might happen.

Every road and pathway leading to and from the launch pad showed the flashing lights of fire trucks, ambulances, crash trucks, security teams, communications teams, and whatever might be needed to back up that one man already slicing into high flight.

At the center of Cape Canaveral's fifteen thousand acres was a press site thrown together of trailers, television trucks, prefab offices, bleachers, high viewing stands, camera mounts, a blizzard of antennae, and a snake forest of cabling along the ground. Tension on the site was as strained as anywhere else, for the fourth estate was hooked up to receiving facilities not only in the United States but throughout the world. Of the more than three hundred newsmen and women who'd sweated out this first manned launch, working down to split-second timing, proud of their self-discipline in telling the world Alan Shepard was on his way, many simply and plainly blew their cool.

They were screaming, "Go! Go! Go!" without regard for timing or microphones or anything save watching the Redstone lift off from the ground. Tough and grizzled news veterans lifted faces unashamedly showing tears as they pounded fists on wooden railings, against their equipment, against defenseless backs of their compatriots.

Beyond the Cape, down along the causeways, on the beaches, and lining the roads and highways, a great army had assembled to witness an epochal moment in history. Five hundred thousand men, women, and children, in cars, trucks, motorcycles, trailers, motor homes, anything that would roll and move, had gathered, nudged, pushed, shoved, and squeezed as close as they could get to the security perimeter of the Cape to watch and shout encouragement.

They went mad at the sight of the Redstone breaking above the tree-line; their combined chorus of hope and prayer was almost as mighty as the roar of the rocket.

This was pure, naked, uninhibited emotion. It gathered substance over the ocean surface, along the beaches, in the palmetto scrub, from every point in the compass beyond this space community.

In Cocoa Beach, people left their homes to stand outside and look toward the Cape. They went to balconies and front lawns and back lawns. They stood atop cars and trucks and rooftops. They left their morning coffee and bacon and eggs in restaurants to walk outside on the street or on the sands of the beach. They left beauty parlors and barber shops with sheets around their bodies. Policemen stopped their cars and stood outside, the better to see and hear. Along the water, the surfers ceased their pursuit of the waves and stood, transfixed, swept up in the fleeting moments.

It was a moment in a town when time stood still.

Fire was born, the dragon howled, and the Redstone levitated with its precious human cargo.

That was but the beginning. When the bright flame came into view, even before the deep pure sound washed across the town, something happened.

Something . . . wonderful.

Men and women sank slowly to their knees. Praying.

Others stood praying.

Crying.

There was no holding back.

All that moved in Cocoa Beach were beating hearts and falling tears.

Flame lifted *Freedom Seven* higher, faster.

Not bad at all! Damn, Shepard, this is smoother than anything you ever expected. Hang in, guy. It's going beautifully.

Shepard spoke to Mercury Control. "This is *Freedom Seven*. Two-point-five-g. Cabin five-point-five. Oxygen is go. The main buss is twenty-four, and the isolated battery is twenty-nine."

A comfortable, assured "Roger" came back from Deke.

Shepard was at two and a half times his normal weight. So far the flight had been a piece of cake. Flame beneath the Redstone lengthened as the outside air grew thinner.

Shepard was through the smoothest part of powered ascent, and then he reached the rutted road, the barrier he must defeat before he would leave the atmosphere behind.

Redstone was pushing, pulsing, hammering at shock waves gathering stubbornly before its passage. Alan was slicing from below the speed of sound through the barrier to supersonic flight straight up. Now he was in the reefs of Max Q, the zone of maximum dynamic pressure where the forces of flight and ascent challenged the booster rocket.

Buffeting began, an upward dash over invisible deep and jagged pot-holes. His helmet slammed against his contour couch. He had the mental picture of a terrier shaking a rat in its jaws, and the rat was called Shepard.

Eighteen inches before him the instrument panel became a blur, almost impossible to read.

A thousand pounds of pressure for every square foot of *Freedom Seven* was trying to crack the capsule like a brittle walnut.

Hang in there. His own voice talking to him.

He started to call Deke far below, changed his mind. No matter that he was being rattled violently like a steel ball in a cage, he was thinking. That's why they gave him flight pay. He must keep his mouth shut until he was past the foaming rapids pounding the capsule.

A garbled transmission at this point, he thought, could send Mercury Control into a wide-eyed flap. It might even trigger an abort by someone overzealously guarding his safety.

No calls, no mistaken abort. "Shut up. If I need an abort, I'll tell Deke."

As if in answer to a silent plea to the flame beneath him, the Redstone slipped through the hammering blows into smoothness beyond. Out of Max Q, Shepard grinned and keyed his mike.

"Okay, it's a lot smoother now. A lot smoother."

If nothing else, Deke was the original laconic man. "Roger," he said calmly.

Louise Shepard stared at her television screen, watching the rocket lifting magically from its launch pad. On the screen the flame seemed as tiny as it was bright. She tried desperately to listen to the words being exchanged between Mercury Control and her husband. She would have been grateful to hear the Redstone's roar, but the girls in the Shepard household were ecstatic, excited, wild, cheering and shrieking at the top of their lungs. Louise didn't even think of trying to quiet down the children.

That was their father in that rocket. This was their moment, too.

And hers. She smiled to herself, brought a hand up to her lips. "Go, Alan," she said quietly, unheard in the din of the room. "Go, sweetheart."

Mercury Control called out the time hack. "Plus two minutes . . . "

Alan Shepard was now twenty-five miles high and accelerating through twenty-seven hundred miles an hour.

Increasing g-forces mashed him down into his couch. It hurt and it felt terrific.

What a ride!

"All systems are go," he called down to Deke.

Every moment of prelaunch and ascent was prime time for news coverage of the flight of Freedom Seven. Merrill Mueller of NBC was broadcasting across the length and breadth of America and through a far-flung international radio network. Mueller was a veteran, the voice of confidence, unflappable, unshakable. He'd done his newscasts through raging battles of war, and he'd been the voice that issued forth from the deck of the USS Missouri when the Japanese surrendered in Tokyo Bay in 1945. He never lost his cool, he was magnificently composed, and now he was describing to the world the launch of America's first man to hurtle into space.

He had a thousand things to say about the astronaut, his family, the mission, the Redstone, the oddly shaped cone in which Alan Shepard rode. He could do play-by-play on a live broadcast as though he'd rehearsed it for a week.

But he'd never seen a man disappearing in the bright sunlit sky as a single source point of silvery flame as he left the earth.

The master of the newscast felt his voice fading. He tried desperately to regain control. Finally the dean of description swallowed hard. He could think of only one thing to say to his audience.

"He looks so lonely up there . . ."

Then he fell silent.

Redstone increased Shepard's weight to a thousand pounds as he called out the force of six times gravity to Deke. He found it difficult to talk as the g-forces squeezed his throat and vocal cords. He drew on the techniques of fighting these loads he'd perfected in test flying. They heard him clearly in Mission Control.

Another moment of truth was at hand.

Cutoff!

That was the instant when the Redstone engine shut down, when the booster became an inert empty tube from which Shepard had to separate.

Above the astronaut's head a single, large, solid-propellant rocket blazed to life, spewing back flame from three canted nozzles. These broke connecting links to yank the tall escape tower, now no longer needed, away from the *Freedom Seven* and send it racing along a safe departure angle.

Systems functioned precisely and on a rapid schedule.

Three small separation rockets at the base of the capsule ignited. *Freedom Seven* pulled away from the Redstone.

A new light flashed on the instrument panel before Shepard.

"This is *Seven*. Cap sep is green."

Shepard was on his own, slicing high above earth along a great ballistic arc.

"Roger," Deke confirmed. He had little to say. It was up to Alan to talk. He had something to talk about.

Alan called out the programmed reports. Then he took moments to drink in the sensations of being separated from his world.

Moments before he had weighed a thousand pounds. Now a feather on the surface of the earth had more weight than this man—who had none.

"I'm free!" he exulted.

Being weightless was . . . wonderful, marvelous, incredible. A miracle in comfort.

The tiny capsule seemed to expand magically as pressure points vanished. Were he not strapped in and down to the couch, he would have floated about in total relaxation. No up, no down, no lying or sitting or standing.

A missing washer and bits of dust drifted before his eyes. He laughed at the sight.

He had expected silence at this point. Atmosphere was something far below him. No rush of wind despite so many thousands of miles an hour. No friction. No turbulence. It should have been the silence of ghosts.

But these ghosts were real, and they made their own sounds. It was the murmur of *Freedom Seven*, as though a brook were running mechanically through its structure. Inverters moaned, gyroscopes whirred, cooling fans had their own sound, cameras hummed, the radios crackled and emitted their tones before and after conversational exchanges. The sounds flowed together, some dulled, others sharper—a miniature mechanical orchestral chorus. Shepard smiled. He was hearing the Concert of the Spacecraft, a strange and unexpected company to remain with him as he hurtled through the soundlessness of space. Those sounds are most welcome, he told himself. They mean things are working, doing, pushing, and repeating. They were the new age sounds of life.

Weightlessness was still new, refreshing, exciting, but this was a romance kick-started by a great rocket. Shepard took to zero-g with fierce pleasure, bonding not only naturally but eagerly with this new world without weight.

He felt *Freedom Seven* initiate its slow turnaround. Other sounds! Of course, the control thrusters firing in vacuum. But within the hull of the capsule they exerted pressure, and that pressure came to him as thudding sounds, dimmed bangs carrying wonderful satisfaction. His ship was obeying its autopilot-commanded flight plan, turning on schedule, rotating into a position that would assure the blunt end of the capsule facing in the direction of reentry.

Reality reached in through the capsule's titanium shell, an invisible hand to tap him on the shoulder. Shepard grinned with the realization that this sense of comfort and freedom, the humming sounds of the spacecraft, had blanked out the fact that he was zinging along high above the planet's atmosphere at better than five thousand miles an hour.

But there was nothing by which to judge speed. You need relative comparison for that—a tree, a building, a passing spacecraft. His view of the outside universe was restricted to what he could see through the capsule's two small portholes, and through those he saw that very deep blue, almost jet black sky. Only one reference point was available to him. He had to look at the earth below. Otherwise every sense he had told him he wasn't moving.

He hesitated. That look at earth would have to wait. The mission check list came first. "Got to go flying, guy," he said. Until now *Freedom Seven* had flown its profile on autopilot. The only aspect of the flight different for Shepard than for the chimp that preceded him was that he could give a verbal report of events.

Now he wrapped the gloved fingers of his right hand about the three-axis control stick. He reached out to switch from autopilot mode to manual control. One axis at a time, he warned himself.

"Switching to manual pitch," he announced to Deke.

Major Laconic was right there. "Roger."

He squeezed the stick to one side. Tiny jets of hydrogen peroxide gas spat into space from exterior ports on the spacecraft. Instantly he felt the reaction as *Seven's* blunt end raised and lowered in response to his hand commands. He couldn't believe the incredibly smooth movements of his small spaceship. It was doing precisely what he demanded.

"Pitch is okay," he said briefly. "Switching to manual yaw."

"Roger. Manual yaw."

Alan fed in reaction thrust to the yaw axis, and again *Seven* danced slickly to his tune, shifting left or right.

"Yaw is okay. Shifting to manual roll."

"Roger. Roll."

Again *Seven* moved on invisible silken threads. Shepard was elated. "Finally," he shouted within his mind, "we're doing something in space on our own. We're first with it! Manual control of a spaceship. Dyn-o-mite!"

Not for an instant would Shepard even think of degrading the sensational orbital flight of Yuri Gagarin in that heavy Vostok he called *Swallow*. He'd gone higher and faster and had raced all the way around the planet, but the Russians had played it very tight against the vest with a supercautious approach, and Gagarin had been a fascinated passenger.

The *Swallow* had flown its entire mission on autopilot. Gagarin had gone along for the ride. Not that the Vostok hadn't had a manual back-up system in case the automatics failed. The Russians didn't want to risk using the system and having it fail on them.

Shepard's flight advanced confidence in the onboard steering systems of Mercury. As Shepard saw it, the *Freedom Seven* had proved itself a good little flying ship, and he had become the first pilot to maneuver a space vehicle.

He felt damn good about it.

He smiled as he reported, "Roll is okay."

Deke almost made a speech as he responded. "Roll is okay," he confirmed from his console. "It looked good here."

Alan Shepard took a deep breath. Now, he told himself. Now you may fly part of this mission for yourself. Go ahead, man. Take a good look at the earth.

His portholes still looked outward, toward the blackness. So he moved his head to look downward through the periscope, and cursed himself. While still on the pad, looking through the scope, he'd stared into a bright sun. Immediately he had moved in filters to cut down the glare. He'd forgotten to remove those filters and now, looking through the scope, instead of a brilliant blue earth with a curving horizon, he saw his planet only in shades of gray.

He reached for the filter knob and, as he did so, the pressure gauge on his left wrist bumped against the abort handle. He stopped that movement real quick. Sure, the escape tower was gone, and hitting the abort handle might not have caused any great bother, but this was still a test flight and he wasn't about to play guessing games.

He looked again through the periscope. Even through the gray the sight was breathtaking. The sun's reflection from the world below was strong enough to give him a picture.

"On the periscope," he radioed. Then, with great excitement, "What a beautiful view!"

"Roger."

"Cloud cover over Florida, three- to four-tenths on the eastern coast, obscured up through Hatteras."

He drank it all in, amazed to look down on the world from his seat with the gods. He watched the curving edge of the planet fall away below the southeastern United States.

Clouds obscured the Florida coastline south to Fort Lauderdale, then yielded to sunshine and the rich green of Lake Okeechobee's shores and down to the spindly curve of the Florida Keys. He shifted slightly to see the Florida panhandle extending west, saw Pensacola clearly. The horizon arced away to offer a tantalizingly bare glimpse of Mobile, beyond which, just out of visual reach, lay New Orleans.

He looked northward across Georgia, at the Carolinas, and saw the coastline of Cape Hatteras and beyond.

He looked down, beneath the tight little craft, studied Andros Island and Bimini and saw other Bahamian islands through broken cloud cover. "What I'd give," he said, "to have that filter out of there so I could see the beautiful waters and coral formations around those islands."

He was now at his highest point, 116 miles. Back to the mission profile. *Freedom Seven*, obeying the intractable laws of celestial mechanics, was swinging into its downward curve, calculated to carry Shepard directly to the Navy recovery teams waiting for him in the waters near Grand Bahama Island, some three hundred miles southeast of the Cape.

He worked the controls to the proper angle to test-fire the three retrorockets. They weren't necessary for descent on this suborbital, up-and-down mission, but they had to be proven for orbital flights to follow, when they would be critical to decelerate Mercury spaceships from orbital speed to initiate their return to earth.

Deke remained with him every second of the way and began the countdown from Mission Control. "Five, four, three, two, one, retro angle," Deke confirmed.

Retro sequence was set. "In retro attitude," Shepard announced. "All green."

"Roger."

"Control is smooth," came the words from space.

"Roger, understand. All going smooth."

"Retro one," Alan sang out. The first rocket fired and shoved him back against his couch. "Very smooth," he added.

"Roger, roger."

"Retro two." Another blast of fire, another invisible hand shoving him backward.

"Retro three. All three retros have fired."

"All fired on the button," Deke said with satisfaction.

The weightless wonderland vanished like swiftly dissipating fog.

Gravity was back, and the build-up came swiftly as *Freedom Seven* plunged into the atmosphere. Alan switched to manual control to get as much flying in as he could while under g-forces. He worked the controls until the small thrusters could no longer counter the massive pull of high g-loads. Rightfully, he switched to automatic mode for the rest of his trip downward.

Deke was on the horn. "Do you see the booster?" There was a touch of concern in his voice.

Before launch, engineers had voiced some concern that when Shepard fired the retros, his speed would be slowed enough so that the empty Redstone, following its own ballistic arc, might catch up to and interfere with the flight of the Mercury.

Shepard judged the issue as trying to fix something that wasn't broken. Old maid hand-wringing. Even though the Mercury and the Redstone had boosted out of the atmosphere, there was still some drag associated with upward lofting after burnout of fuel. The Redstone was so much bigger and its mass so much greater that even remnants of atmosphere would to some degree interfere with its ballistic arc and keep it well below the *Freedom Seven*. The way Alan judged the ballistic profiles, the Redstone would be considerably lower than his position and would soon slam back into denser atmosphere to begin its self-destruction as it plunged toward the ocean.

Shepard's calculations proved to be right on the money. Well below *Freedom Seven*, the Redstone was tumbling wildly out of control, increasing its drag and imposing terrible forces on the rocket structure. The Redstone was like a helpless, frightened whale attacked by invisible sharks of reentry, pieces being torn from the body of the rocket, chunks hurled away, and the metal structure heating up swiftly from friction. Behind the Redstone, the first sputum of its destruction fled backward in the ionized trail it created through thickening air.

Below, a freighter plowed northward through calm seas on an uneventful journey. Until the first American manned space flight entered the scene. Both the Mercury and the Redstone were whipping huge sonic waves through the atmosphere. Without warning, shock waves ripped downward through the sky to smash against the vessel. The windows on the bridge rattled and flexed wildly from the sonic boom howling across the water. A shock wave cutting the sky at the speed of sound is a fearsome thing to meet, especially when you have no idea the sky's about to fall on you.

The terrified crew thought their ship had exploded. The captain jerked his head around. "Damn, what the hell was that?" he shouted.

Before anyone could answer, a whirlwind of ear-piercing whistling and howling force tore across the ship. Someone pointed up and screamed, "Omigod! Look!"

A white and black shape, the charred and still burning Redstone, tumbling crazily, crashed out of the sky, sailed high over the ship's deck, and smashed into the Atlantic several miles east of the freighter, sending multiple water geysers into the air. The Redstone didn't die easily and seemed determined to end its time with a spectacular water ballet. It skipped across the water's surface, twice parting the sea, sending up two towering walls of water. Then a final careening plunge into the waves, and it settled forever into the ocean.

The crew stared, stunned. They stood frozen in place until they heard the shouts of their radio operator. "Hey, on deck! Everybody! Listen!"

He placed the microphone by his radio, calling in the blind for anyone who might answer, who could tell them what insanity was going on. NBC radio engineer Joe Sturniola was at his short-wave gear on Grand Bahama Island and picked up the freighter's call. He heard clearly the operator saying, "It couldn't have been an airplane. Not from what we saw. We don't know what it was!"

Sturniola answered immediately. "You people have just been missed by the rocket that carried the first American astronaut into space."

"Rocket?"

"Affirmative."

"Astronaut? Into space? American?"

Sturniola grinned. "That's right. This is Grand Bahama Island, and astronaut Alan Shepard will be arriving here shortly."

Okay. This is *Freedom Seven* . . . my g-build-up is three . . . six . . ." His voice faltered as a great invisible hand squeezed him with brutal force.

"Nine . . . " he grunted, using the proven system of body-tightening and muscle rigidity to force the words through a tortured throat. Words still spoken under control. Grunt talk that worked.

"Rog."

Deke didn't want to miss a word from the plunging spacecraft.

"Okay . . . Okay . . . " Alan's voice rose as the intensity of the struggle increased. Eleven times the normal force of gravity, getting close to "weighing" a full earth ton. But he had pulled eleven g-loads in the centrifuge, and he knew he could keep right on working now.

He did.

Deke stabbed into the silence after his last call. "Coming through loud and clear, *Seven*."

"Okay," came the grunted oath from what was now high atmosphere instead of space. No matter how severe the punishment, so long as he kept repeating at least that one word they knew he was on top of the situation.

"Okay . . . " They noticed the change in his voice. Less quaver. Lower pitch. They tracked him on radar, knew his changing altitude, but hearing the man's voice was what really counted. The capsule slowed rapidly, and the g-loads were fading away.

"Okay . . . this is *Seven*, okay. Forty-five thousand feet. Uh, now forty thousand feet."

Shepard was through the gauntlet of punishing g-forces and deceleration and blazing heat of reentry. He felt great. Right behind his back the temperature had soared to 1,230 degrees, a critical test for the spacecraft and no small feat for the man inside. Under the worst-case conditions of the scorching dive, his cabin temperature hit a peak of 102 degrees. Inside his suit the reading topped at 85 degrees. Not at all bad, he grinned. Just nice and toasty.

But it was still an E-ticket ride on a roller coaster that had some bumps and grinds to go.

His altimeter showed thirty-one thousand feet when Deke's voice reached him again. "*Seven*, your impact will be right on the button."

Great news. Flight computations were as close to perfect as could be, and so were the performances of the Redstone and the spacecraft. The Mercury was arrowing directly for the center of the Atlantic recovery area close to Grand Bahama Island. The Cape lay three hundred miles to the northwest, and with Alan's loss of altitude it soon would be out of radio contact. No time for long good-byes as he signed off with Deke, telling him he was going to the new frequency.

"Roger, *Seven*, read you switching to GBI."

Deke was eager to cut the hell out of Mercury Control Center as fast as he could go. Shepard almost laughed aloud. He knew Gus would be right there with Deke, and the two of them would clamber into a NASA jet and burn sky, blazing their way to GBI so they could be on the ground waiting for him when Shepard was delivered by helicopter from the recovery vessel.

"*Seven*, do you read?" came a new voice, using the GBI line.

"I read," Alan called back, starting to look for the armada of ships in the recovery fleet waiting for him.

But this game wasn't quite over. He still had to reach that fleet and in good shape. That meant the parachute system had to work.

Perfectly.

Or all that had gone so beautifully up to this moment would mean nothing.

He stared through the periscope. Above him, panel covers snapped away in the wind as the spacecraft fell. "I watched gratefully," he said afterward. A simple and total statement referring to his sight of the small drogue chute whipping out first to stabilize the craft.

"The drogue is green at twenty-one, and the periscope is out."

Down went *Freedom Seven* and Alan Shepard, tugging and snapping the small but vital drogue parachute. The altimeter unwound and aimed for ten thousand feet where the main chute was to open. If it failed, well, he already had a finger on the "pull-like-hell ring" in the tiny cabin, which would haul a second, reserve chute out of its pack.

"Standing by for main."

Freedom Seven continued like the champ it had proven to be. Through the periscope he saw "the most beautiful sight of the mission. That big orange-and-white monster blossomed above me so beautifully. It told me I was safe, all was well, I had done it, all of us had done it. I was home free."

"Main on green," he reported. "Main chute is reefed, and it looks good."

Freedom Seven swayed back and forth as it dropped lower. The main chute unreefed and blossomed into a magnificent orange-and-white paneled flower. In contrast to moments in his immediate past, Alan Shepard tiptoed gently toward the ocean.

He opened his helmet face plate. Quickly he disconnected life-support hoses to his suit and then released the straps that had kept him properly snared within the cabin. He wanted nothing to impede a hasty exit, just in case the last few minutes of the mission held surprises for him.

From a thousand feet up he saw the water clearly below him. The heat shield had dropped four feet as intended, to deploy the perforated-skirt landing bag, which would act as an air cushion when the Mercury and the ocean met.

He braced himself for—

Splashdown! "Into the water we went with a good pop! Abrupt, but not bad. No worse than the kick in the butt when I'm catapulted off a carrier deck. This is home plate!"

The spacecraft tipped on its side, bringing water over the right port-hole. He smacked the switch to release the reserve parachute that kept the capsule top-heavy. While he waited for the shifted balance to right his small spacecraft (and lousy boat), he kept in mind the chimp's near disappearance beneath the ocean and checked the cabin for leaks, ready to punch out.

He stayed dry. Shifting the center of gravity worked, and the capsule came back upright.

Planes roared overhead. "Cardfile Two Three," he called. "This is *Freedom Seven*. Would you please relay all is okay?"

"This is Two Three. Roger that."

"This is *Seven*. Dye marker is out. Everything is okay. Ready for recovery."

Green dye spread brilliantly across the ocean surface from the capsule.

"*Seven*, this is Two Three. Rescue One will be at your location momentarily."

It went like another practice run. Within minutes Rescue One, a powerful helicopter, was overhead. Alan opened the hatch, clutched a harness dropped from the chopper and was hoisted aboard.

Rescue One zeroed in on a waiting aircraft carrier, the USS *Lake Champlain*. Sailors lined the deck, cheering and waving wildly. "This is one of the best carrier landings I've ever made," he told a chopper crewman.

Until this moment, when he stepped out on the deck of the aircraft carrier festooned everywhere with red, white, and blue decorations, "I had not realized the intensity of the emotions and feelings that so many people had for me, the other astronauts, the whole damned manned space program. This was the first sense of adulation, a sense of public response, a sense of public expression of thanks for what we were doing. I got all choked up."

With moisture in his eyes, he thought it's no longer just our fight to get "out there." The struggle belongs to everyone in America.

That was the best of all.

From now on there was no turning back.

CHAPTER TEN

NASA IS MADE

LOUISE SHEPARD'S hand flew to her mouth. She felt her heart stop as ahead of the helicopter a flock of pigeons filled her view. Their closing speed made the birds appear to explode around them. Just as quickly they were gone, beneath the big Marine helicopter lifted instinctively by the pilot.

Alan Shepard grinned broadly at his wife. "Gets interesting sometimes, doesn't it?"

She squeezed his hand. "Wow! I can do without the heart-stoppers, thank you."

He laughed. "The birds wouldn't dare. Not while we're on the way to see the president."

Washington, D.C., rolled beneath them. Alan glanced behind his chopper at the other two helicopters carrying the rest of the Mercury astronauts and their wives. It felt strange to be in a military machine and wearing civilian clothes. At least the women looked marvelous, all of them wearing white orchids.

Alan watched in silence as they flew past the Lincoln, Jefferson, and Washington monuments. To his right the Capitol dome reflected bright sunlight. He turned to the White House directly before them as the helicopter began a gentle descent. He had always been impressed with the great sandstone building that was home to the leaders of this nation.

They touched down in a flawless landing. Leaving the machine, Alan turned to the pilot. "Beautiful," he said. Nothing could have pleased that Marine more than that one word.

The Shepards stood on the White House grounds facing the president of the United States.

John F. Kennedy extended his right hand. "Welcome, Commander, and congratulations," he said warmly. Alan and Louise exchanged greetings with Jackie Kennedy. By now the remainder of the astronaut entourage was at hand, and they all moved toward the Rose Garden.

"I almost came to a dead stop when I saw the huge delegation waiting for my appearance," Shepard recalled. "Cabinet members, congressional leaders, and government officials were gathered on the porch outside the White House. The White House staff crowded along a covered walkway."

Not bad for a farmboy from New Hampshire, he thought as he followed Kennedy to a small wooden platform. "Ladies and gentlemen," the president began briskly, "I want to express on behalf of us all the great pleasure we have in welcoming Commander Shepard and Mrs. Shepard here today. I think they know as citizens of this great country how proud we are of him, what satisfaction we have in his accomplishment, what a service he has rendered our country.

"And we are also very proud of Mrs. Shepard," continued Kennedy. Alan winked at his wife; she managed a fleeting smile back for him. Then Kennedy offered a verbal bow to the other six astronauts.

"Our pride in them is equal," he said. A glint seemed to appear in Kennedy's eye, and he eased humor to the fore. "They," he emphasized, "are the tanned and healthy ones." He gestured to the astronauts with officialdom standing behind them. "The others are Washington employees."

Kennedy completed his remarks and turned to present Shepard with the NASA Distinguished Service Medal. The commendation slipped from his hand and fell to the platform floor, treating those watching to the unexpected sight of the president and the astronaut nearly bumping heads as they both bent over to recover the medal.

Kennedy got there first, and he was just as quick with his wit. "This decoration," he said, "has gone from the ground up." The recovery was great and brought applause and laughter from the crowd. He just handed the medal to Shepard, but the first lady leaned forward. "Jack, *pin* it on him."

"Let me pin it on," Kennedy told Shepard. "I'll do my duty."

Shepard didn't miss the unspoken message that came with the medal and felt a special bond with the president, who had also worn the three bars of a Navy commander. Shepard was moved. "Today even eclipses last Friday," he said of his Mercury flight in accepting the medal. "As a matter of fact, I got far less sleep last night than I did the night before the flight. Seriously," he added, "I appreciate this honor. But the accolades of today should really go to the NASA teams who made it possible."

The official ceremonies wound down, and Jackie Kennedy left with the astronaut wives for a private tour of the White House. The president's mood was changing even as he led the astronauts, Lyndon Johnson, several staff members, and picked NASA officials into the Oval Office. Kennedy eased into his rocking chair, asked a few questions about Shepard's flight, then cut right to the heart of the matter. He wanted to know what NASA was doing. Not planning. Doing. "I want a briefing," he said without further preamble.

The astronauts exchanged quick, meaningful glances as they listened to the course the conversation was taking. James Webb and Bob Gilruth of NASA were thinking far beyond Mercury. "My God," Shepard thought, "these guys are thinking about sending a man to the moon!"

It also was obvious to Alan and Deke that NASA officials had gone over this subject before with the president.

Webb told Kennedy that he felt the country could send a man to the moon within just a few years. Other opinions were expressed. Kennedy listened carefully and turned to the astronauts.

"We're not about to put you guys on a rocket and send you to the moon," he said disarmingly. "We're just thinking about it."

Shepard stepped in immediately. "I'm ready," he said. Six astronauts echoed his sentiments.

Kennedy didn't respond. He was reflecting on the heavy pressure on his administration. The Bay of Pigs debacle had tarnished his "new frontier" policy, and powerful communist excursions throughout the world seemed excessive. The Russians were prodding and probing and waving their big rockets as evidence of the superiority of their political system.

No one doubted that Kennedy, with his extensive combat experience and thorough knowledge of technical matters, was fully aware of the limited significance of Shepard's suborbital mission, especially compared with the extent of Yuri Gagarin's orbit around the world. Yet Kennedy had been willing to stand strong behind Shepard, for the Navy pilot had borne up with true grit as the underdog in his flight even before he left the ground. And he'd done it before the world, cameras staring almost down his throat, everything out in the open rather than concealed behind the security cloak the Soviets had maintained.

Had Gagarin's flight failed, the world likely would never have known a man was involved. Had Shepard's mission failed, there was no way the world could have missed watching a man go to his death.

John Fitzgerald Kennedy was determined to exploit to the hilt what these seven young men offered. What the "new frontier" demanded. The power of courage and honor.

On this day a future was being shaped in the mind of the man who led his nation. Kennedy judged that his leadership would be graded in terms of what he did. These seven test pilots with him carried no banners for special favors or political handouts. They flew for their country, and they would do anything their country asked.

Kennedy reasoned that he had been put into the office of president to lead. With every passing moment, seven young men gazing steadily at him, John F. Kennedy was discounting the advice he had been receiving from the advisers and hand-wringers who felt America should step back from the challenge offered by the Russians.

He remembered Jerry Wiesner's insistence, then demands, to kill Shepard's flight. Give space to the Russians, his science adviser had said. Let them put up the manned ships. Let them have the moon. *Give up!*

Kennedy had other ideas and had already seen what his astronauts were capable of. He'd listened to his people, heard their shouts and roars of approval of the man who had flown atop the candlestick and performed superbly.

Kennedy was now convinced that the entire world judged America in terms of how well it performed, now and in the future, in the new arena of leadership. If he declined to challenge Russia, history would pass him by.

He knew also, having thought long and hard of the paths that had brought him to this moment, that he might set a goal far enough away, far enough in the future, so that America could come from behind to win. He was painfully aware the Russian boosters were massive compared to the best rockets America yet could fly. It was still a case of the United States orbiting cupcakes while Russia orbited tractors. Those were the facts.

Whatever happened, America wouldn't win the race in the future if it was restricted to earth orbit. The Russians were far out front in that realm. So the key was to bide time while the U.S. jump-started a new program.

Right then and there, it is judged, President Kennedy made his decision to go for something the Russians were no more prepared to do than the United States.

The moon.

Kennedy rose to his feet and motioned to Shepard. "What's on tap for you, Commander?"

"They've got a parade scheduled, Mr. President, and then we're to meet some members of Congress. After that there's a luncheon—"

"They'll wait," Kennedy broke in. "Right now you come with me. You and Newt and your wife and Lyndon are coming with me to the National Association of Broadcasters convention for a quick appearance. Then you can make that parade." Shepard nodded. He knew that Newt was Newton Minnow, the White House communications chief.

Shepard did not like what was happening. His patience was evaporating swiftly. He disliked, intensely, being used. Walking in on the broadcasters' convention with the president would be showing off a war trophy named Shepard, and it smelled. He mollified himself somewhat by remembering that no matter who else he was, Kennedy was also his commander in chief, and you can excuse almost anything if you're obeying orders. The fact that he and Louise received a standing ovation did diminish his objections to some degree.

While Kennedy attended to the broadcasters, Lyndon Johnson enveloped Alan and Louise Shepard and installed Alan as the star attraction in the parade up Pennsylvania Avenue from the White House to the Capitol. It was embarrassing for Alan that the other Mercury six astronauts, and their wives, ended up trailing his motorcade.

Lyndon Johnson made certain the Shepards sat with him on the open back of the parade limousine, and Alan feared the vice president was going to wrench his arm from its socket with his nonstop waving to the crowd. But it was to Shepard that the tens of thousands of onlookers were waving, and that meant, Alan knew, that enormous public support was building with every passing moment for financing a major manned space flight program. If this wave of enthusiasm and hero worship was played to the maximum, the astronauts could regard it as the beginning of the real countdown to the moon.

Shepard's mood alternated between enthusiasm for the possibilities of going to the moon and growing anger and distaste for the manner in which his fellow astronauts were being treated.

"I hated their having to ride along in my wake, staring at the back of my head," he said. "I was shocked at one point to look behind, and there were Deke Slayton and John Glenn trying to hitch a ride in a news media pool car.

"Deke and John had stepped down from their car at the Capitol building and in the uproar they somehow lost sight of their vehicle. They were trying to get a ride any way they could so they could stay with the motorcade.

"The media guys didn't recognize them, and they were shoved physically away from the press car. 'Beat it, you guys,' I heard a photographer yell. 'We don't need any gatecrashers with us!'

"I heard Deke try to explain that he and John were astronauts. To which another photographer yelled, 'And my grandmother is the queen of England! Get lost!'

"I was livid," Shepard said. "I turned to Lyndon to tell him to stop the car. I was going to run back there and help Deke and John. But about that time a couple NASA guys ran over to them, explained to the press who they were, and they got their ride.

"At that moment I swore myself to an oath," Alan said. "No more 'I' did this or 'I' did that. From then on, whenever I spoke, it was 'we astronauts,' and 'we did that' and 'we're going to do this.' "

The day finally ended. Much of it had been great, much of it left a bad taste in their mouths, and most of them would have been happy never to see Washington, D.C., again. Yet their visit haunted them. Not the crowds or the parade or the shrieking ovations. None of that.

A quiet conversation. A man speaking from a rocking chair. Was Kennedy really serious about going to the moon?

Twenty days after Shepard flew the *Freedom Seven* into space, Kennedy settled the questions in his mind. He had judged the comparative merits and capabilities of rocket boosters of both the United States and the Soviet Union. He accepted the present overwhelming superiority of the Russians. He was determined to take the long-range gamble that American science, technology, and industry would persevere, pull itself together, and with clearly defined goals be able to surpass the Soviets.

Kennedy had confidence in the American people, in America's scientific-industrial machine.

In ringing tones he told Congress it was time for the nation to take longer strides, to become the leading player in space, and to lead both America and the world into a better future for the entire planet.

With stirring candor Kennedy pointed out that the Russians had immensely powerful rockets. He stressed that while America had waffled on the future, the Russians had thrown themselves into building powerful missiles that also had proved brilliantly adaptable to missions in space.

Kennedy laid it on the line. The Russians were leading the way. He could not promise the Congress or the American people that America would catch up and overtake the Soviets. There were no guarantees.

But he would guarantee that if the nation failed to make the effort, then assuredly America would not only be last, but also would remain last for uncounted years to come. Alan Shepard had demonstrated to all Americans, however, that if the United States committed to new future goals, then our stature must be advanced in full view of all the world.

There were several moments of silence as Kennedy paused to catch his breath, and a feeling of expectancy, a touch of magic, fell across the great hall of Congress.

"I believe this nation should commit itself to achieving the goal, before the decade is out, of landing a man on the moon and returning him safely to earth. No single space project in this period will be more impressive to mankind, or more important for the long-range exploration of space, and," he paused, "none will be so difficult or expensive to accomplish."

No sooner had he spoken these words than the Congress leaped to its feet with a roar of approval and thundering waves of applause. Kennedy knew his timing had been perfect, his message what the nation wanted to hear. Now he had thrown the gauntlet down at the feet of the Russians. If this reaction was any sign of the future, then his "new frontier" had absorbed new life and vitality. It was on the way back. He had absolute confidence that this was a gamble his administration could not lose. Even if reelected, he would be out of the White House before the decade was over.

If the program worked as planned, it would be another president in Washington who would talk to the first Americans on the moon.

In the astronaut office complex at Langley Air Force Base, Alan Shepard listened to Kennedy's words on a radio by his desk. His face reflected changing emotions: disbelief, hope, wonder. Finally he could no longer contain himself.

He turned to the others. "Did I hear what I think I heard?"

Deke Slayton nodded. "You heard right," he confirmed. "The man wants to send us to land on the moon. *Land*, not just fly around the moon," he emphasized.

"Sure, but don't you remember what he said in Washington?" Alan retorted. "He was going for a flight that would loop around the moon. Take the safest shot."

Gordo Cooper looked up, infinitely patient. "It just ain't possible," he joined in. "First, we don't have the rockets. Second, we don't have the spacecraft, and third, by God, we don't even know how to navigate our way out there and back."

Deke looked at Gordo and laughed at his fellow Air Force pilot. "Hey, Gordo, don't forget Kennedy's a Navy man like Shepard. We've got only fifteen minutes of flight time in space, and now he figures we'll just get us a road map, or we can follow the railroad tracks all the way to the moon."

"Yeah, that's the way a Navy pilot would get us there," Cooper chuckled, "following the railroad tracks."

Alan Shepard showed them a mouthful of teeth and shot the middle finger of his right hand straight into the air.

"Kennedy is nuts," came a comment from the back of the room.

"Nuts?" Deke laughed. "Could the eminent Captain Gus Grissom be calling our president nuts?"

Gus grinned and John Glenn gestured for attention. His face was serious as he spoke.

"This thing may not be so crazy after all, guys," he said slowly. "I know we've got boosters that like to blow up."

"Yeah," came an acid comment. "Like that overblown balloon they call the Atlas. Can't get out of its own way."

"They'll get it right," Glenn said with confidence. "But that's just a whistle stop for us. Titan's coming on line, we all know that. You guys seen the advanced Army projects Wernher's working on?"

Heads turned. "Spell it out, John."

"DARPA," he said. They knew what he meant. That high office in the Pentagon. The Defense Advanced Research Projects Agency. "The Army wants to put up a really humongous communications satellite system. It wants instant worldwide communications. So it's turned Wernher loose on some monster rocket they call Saturn. Eight engines. Something over a million pounds of thrust for liftoff. We could put up a Mack truck with that thing."

Astronauts were coming to their seats. "They're working on that now?" came the question.

Glenn nodded. "The engineers say it's a winner. It's not the thin edge, either. Its whole key is power and reliability. I admit bringing everything on line in eight years is a hell of a job, but—" He grinned. "Any of you guys old enough to remember how down in the dumps we were the day after the Japanese tore us up at Pearl Harbor? Less than four years later Japan was an ash heap."

"Eight years, huh?" Deke chimed in. He winked at his fellow astronauts. "Kennedy never would have said what he did if he didn't believe we could do it. Besides, think of what it means to us. Eight years from now we'll all still be young."

He slapped his knee and chuckled. "In eight years we could all go to the moon. Think about that."

Deke stood up. "There's just one thing I think we ought to remember."

"What's that?"

"If it wasn't for the Russians, we wouldn't be going anywhere."

No one argued the point.

Congress acted as if it had been given a shot of super-vitamins. On the first fiscal go-around it gave NASA a check for $1.7 billion to kick the new program into action. NASA administrator Jim Webb confirmed Kennedy's confidence in him when he spread a great net across the U.S. and began to haul in top talent from leading industrial and science centers from every state in the union. Then he began assigning contracts as much on the scattering of congressional districts as any other factor. He was assuring that the nation's most ambitious technological effort wouldn't be burdened by governors and congressmen banging on his door because their states had been left out. This tactic gave the program wide political support.

The sweeping national effort to place Americans on the moon transformed the face, industry, science, and lifestyle of the nation. More than twenty thousand industrial contractors and four hundred thousand technicians, engineers, managers, plus those with administrative and other skills were swept up by NASA's huge job vacuum cleaner.

The new missile and space programs culled their names from Greek and Roman and Norse legends and mythology until it became commonplace to talk of Thor, Jupiter, Juno, Atlas, Titan, Saturn, Mercury, Gemini, Pegasus, Orion, Polaris, Poseidon, and others.

Now there was Apollo on whose wings men would fly to the moon.

So went the plan.

But this was 1961. Never mind the miles; the moon was still an impossibly long way off. The difficulties of boosting, navigating, and then the thorny problems to of traversing down to its lifeless surface seemed insurmountable.

In the meantime, it was a matter of taking one step at a time. The nation had flown only one space capsule and it had been weightless for just five minutes.

The second step was to repeat the first, to gather more data and confirm the lessons of the first. That's how building blocks are made.

Gus Grissom would ride the second manned Redstone. Even as engineers and the astronauts readied the slim booster for its suborbital mission, eyes were turning to the launch pads closer to the beach.

The Atlas complex of four launch towers. A rocket with a skin so thin and sensitive it had to be pressurized internally to keep its balloon-like structure from collapsing.

A real confidence-builder, that.

The astronauts knew the Atlas was their ride into orbit. But at the moment there were no volunteers to ride the tempermental booster.

The odds on every Atlas flight were terrible. No one was running to the pad to sit atop 360,000 pounds of flaming thrust that had the nasty habit of exploding into a fireball.

Kennedy's speech had been grand, utterly marvelous.

But the road to the moon still waited to be paved.

CHAPTER ELEVEN

MERCURY: MESSENGERS FROM SPACE

T WAS EARLY EVENING, the time of day when the road leading to the launch center of Cape Canaveral was abandoned to creatures of the night, the floods of skittering insects, and the furry animals some-times large enough to cave in the whole front of a speeding car. The road was made to delight those who liked to drive fast. First a stretch running to the west, a wide turn, and then a long straight stretch that eased to the right in a great sweeping curve that ended in a long straightaway, dead east, toward the Cape's main gate.

Jim Rathmann shifted expertly as he hammered the Corvette through the 100 mph mark on the speedometer. The speed was a gentle load for Rathmann, who only the year before had claimed the winner's position in the grueling Indianapolis 500. He loved the 'Vette as did the man in the right seat, the car's owner. Gus Grissom loved speed just as much as his buddy Rathmann, although Gus usually did his speeding about forty thousand feet above sea level.

Rathmann boomed across a bump in the road, the 'Vette easing into the air with all four wheels spinning, then landing with a satis-fying *whoomp*. It was like driving through falling snow; the lights reflected hordes of insects that splattered in a steady rain against the windshield.

"Whatta you usually take this turn at, Gus?" Rathmann asked.

"About a hundred and five," came the answer.

"Whoo. Pretty good. But tonight we've got those Daytona racing tires under us," Jim reminded him. "Slick as a baby's ass."

"Uh-huh," Gus grunted. "Don't matter, I can make it same speed with them."

"That so? You really could make it with the slicks?"

Gus grinned. "Piece of cake."

"Yeah? Well, hold on to your hat, pard," Rathmann growled. He kept heavy pressure on the pedal, the speed still coming up, and he gripped the wheel tight, cutting inside of the turn, the wheel firm in his expert hands.

Centrifugal force pulled Gus from his seat as they entered the turn, and he locked his fingers around the door handle. At the same moment the slick tires gave up the ghost, all traction vanished, and the Corvette came unglued from the road surface. In an instant the heavy car was in a wild flat spin, swapping ends as it headed directly for the salt marsh off the road. The Corvette, slung low and built like a small tank, crashed, plowing its way to a stop, its hubcaps sunk in muck and squashed bugs.

Gus glared at his friend. "Dammit, Jim, you scared the crap out of me!"

"Hell, we're just fine, pardner," Rathmann laughed. "It's like I said. Those slick tires wouldn't hold the road."

"No shit, Jose," Gus cursed. He looked around. "Dammit, I've gotta fly in a couple of days!"

"Really?" Rathmann growled, as if he didn't know.

"If Walt Williams or some other NASA brass sees my car here like this, I'm dead."

"No sweat. I'll have us outta here in no time."

"Out of here?" Gus shook his head in dismay. "It better be less than no time. The cops could be along any minute."

"I told you, don't sweat it," Rathmann laughed. "They all drive my cars. They won't see a thing if I ask them to look the other way."

"That so?"

"That's so. Just stay low, cowboy," Rathmann said, climbing from the car. "I'll be back with the Lone Ranger."

He disappeared into the night. Gus slouched in the car, knowing his hide was on the line. If Williams got hold of this, he could easily ground Gus. No Redstone mission. Williams would replace him with Glenn, and space flight would be out for Gus Grissom.

"Stupid!" Gus railed against himself. "Plain stupid!"

He was right. NASA was tightening up on the antics of its astronauts. One man hurt or killed shortly before a flight could mess up carefully planned schedules. The minutes passed, and Gus really began to worry. He saw lights coming closer, around the turn.

He couldn't believe it. Rathmann with two wreckers! It took some time and skill, but soon the heavy car was out of the muck and hauled off to be hidden—and repaired—in Rathmann's garage.

Again and again Gus Grissom berated himself for risking both his neck and his space flight. He knew he shouldn't take any more unnecessary risks, but the urge to get even with Rathmann was too great. A few days later, without authorization, Gus took Rathmann, a civilian, to Patrick Air Force Base, breaking all variety of federal regulations. The two climbed into a supersonic jet fighter, fully suited up with oxygen masks and helmets.

Moments later the jet streaked down the runway. Grissom lifted off the field in a steady, fast high climb and bored for high blue. When he reached eight miles above the earth, Rathmann was startled to see the sky suddenly vanish. Instantly it was replaced by a spinning vertical line, which Rathmann realized was the earth's horizon. The jet rolled around the inside of a great invisible barrel, snapped out in wicked high-speed vertical rolls. The rolls stopped with a bone-jarring snap. Blue sky was below, ocean above, and for a while it was impossible to tell which was which. Gus did things with the jet Rathmann found impossible either to describe or even understand. His stomach screamed at him, his head was mashed into his neck and alternately shoved the other way with eyes and ears popping. He was on the edge of yielding his breakfast to the inside of his oxygen mask when Gus decided he'd paid back Rathmann for skidding off the road at Canaveral.

On the ground, the two buddies laughed at each other, pounded each other's shoulders. "Gus, you scared the crap out of me!"

Gus let out a boom of laughter. "Now, where have I heard that before?"

While the launch team readied his Redstone and spacecraft, Gus was pretty much laid-back, letting his crew do their jobs without interference.

But the old rule is that everybody isn't equal on a team, and Gus demanded the best, and he had an eagle eye for missed detail and a short fuse for slop. He could handle honest mistakes just as long as they didn't grow but, when they did, his irritation took dead aim on the blockheads responsible, whom he wanted off his team right then and there. Petty excuses were to him utter bullshit, and woe to the incompetent he found working on the ship his life depended on.

Engineers working on the spacecraft until now had labored on unmanned robots. The machines didn't see, speak, suffer, or lose their cool. When it came to engineering, though, Grissom held his own with the best of them. And when he spotted someone slacking instead of doing serious work, Gus would give the man his choice of taking the stairs down or getting thrown off the gantry. The last-moment tinkerers trying to impress their boss with their knowledge often felt his wrath.

When he ran into an engineering stumbling block, he could really go through the roof. The one problem that drove him to distraction was the same one detected by former President Harry Truman when he was first shown the Mercury. "How in the hell do these guys take a leak?" asked Truman. There was a lot of feet shuffling before someone finally responded, "Uh, sir, they don't." Truman walked away convinced the country's space engineers were loonies.

After Alan's experience soaking up the contents of his bladder with his heavy underwear, Gus was after an answer to Truman's query.

Flight Surgeon Bill Douglas knew better than to put off Grissom, and he went after an immediate solution. Douglas sent the astronauts' nurse, Dee O'Hara, into Cocoa Beach to buy a panty girdle. He reasoned the snug garment would serve quite well as a liquid container. Gus, who first looked aghast at the "medical solution," finally murmured, "Oh, what the hell, I've dealt with worse make-do crap before," and when he climbed into his Mercury capsule, he became the only male astronaut prepared to risk space flight in women's lingerie.

Gus Grissom's Mercury-Redstone lifted from its launch pad on July 21, 1961, and he flew an almost exact duplication of Alan Shepard's suborbital flight—115 miles up, 300 miles downrange. The parallels were stunning, and Gus boomed out of the sky with a perfect splashdown, but that's where any similarities to Shepard's mission ended.

Gus went through the drill of readying the capsule, which he had named *Liberty Bell Seven*, for helicopter recovery, reviewing his recovery check list while waiting for the hookup. He was lying back when an explosion blasted one side of his spacecraft. The hatch, modified to use an explosive primer cord instead of the mechanical locks of Shepard's capsule, had inexplicably ignited and blown away the emergency exit hatch.

Gus saw the waves coming in and had to scamble out of the capsule and swim for his life as he watched the three-thousand-pound spacecraft sink in fifteen thousand feet of ocean.

Engineers puzzled over the detonation. Some asserted that the design of the capsule made an accidental explosion impossible and theorized Grissom may have inadvertently hit an emergency plunger that triggered the hatch release. Grissom denied the charge, repeatedly insisting, "The damn thing just blew." His fellow astronauts backed him all the way, and an accident review board cleared Gus of any wrongdoing in connection with the incident. Unanswered questions about the hatch explosion continued to dog Grissom. The favorite theories about what happened included speculation that an exterior lanyard connected to the igniter entangled with the landing gear straps, that a ring seal might have been omitted on the detonation plunger, or that static electricity generated by the helicopter might have fired the hatch cover. With the evidence on the ocean floor, it was impossible to determine a cause.

In Washington, a few hours after the Grissom scare, President Kennedy signed a bill allocating NASA $1.8 billion in additional funding, including money to start the Apollo program. Congress had given him every penny he asked for.

If the loss of the *Liberty Bell Seven* wiped some of the shine off an otherwise perfect flight, the rest of the luster vanished sixteen days after Gus's splashdown when another powerful SS-6 booster thundered skyward from its launch pad in Russia with Vostok II—*Eagle*—carrying Major Gherman S. Titov, who had been back-up pilot for Yuri Gagarin. Titov went aloft in a ship weighing nearly eleven thousand pounds and stayed in orbit for a full day.

NASA managers and the seven astronauts could only shake their heads. They looked at one another and agreed that the Redstone had done its job. It was time to get on with launching an American into orbit, and suddenly America's most urgent goal was to send a man circling around the earth before the end of 1961. That's the year the Soviets did it, not once but twice, and they did it first. But history could record that both nations did it in the same year.

If. And it was a big if.

Of all the available rockets in the American arsenal, only the Atlas ICBM was capable of lofting the Mercury capsule into orbit. The Atlas worked well as an intercontinental range war rocket, but its thin skin often collapsed under the heavy burden of the Mercury spacecraft and all the heavy features built in to protect human life. As a space booster to kick Mercury into orbit, it was unproven and full of risk.

Three times Atlas rockets had blasted away from the launch pad with an unmanned Mercury on top, and on two of those flights the big rocket had exploded, spewing blazing wreckage of booster and spacecraft into the sea.

The pressure mounted on Mercury Operations Director Walt Williams. The White House wanted an American in space, and they wanted him to go as high and as fast as any of the Russians had. Williams had to balance the angry impatience of the White House against the realities of a non-performing booster. Finally Williams went through the roof. He was sick of listening to the Convair officials, manufacturers of the Atlas, and the miserable litany of nonstop excuses they offered when flights ended in explosions after repeated assurances that the rocket was capable. He was ready to dump Convair and Atlas and tell the president that Atlas was a piece of junk. He wanted to recommend that NASA wait out the development of the Air Force's new Titan ICBM, now undergoing testing. That would slow down the astronaut orbital program, but at least the pilots would have better than a one-in-three chance of getting into orbit.

Williams went after Convair's Cape manager, B.G. MacNabb, and laid it on the line. "Fix it or get the hell out." MacNabb screwed up one eye. "We'll fix it."

MacNabb brought Convair's toughest test conductor to his office. He passed on Williams's ravings and told Tom O'Malley that they were about to get bumped for the Titan missile. O'Malley moved from an Atlas ICBM test pad to the Mercury-Atlas pad, called the launch team together, and gave them the new word. "The next son of a bitch who says no sweat, who tells me or anybody else we don't have a problem, will ride the toe of my boot out the door."

T.J., as he was known, took no lip from anyone. What he got for his hard-boiled attitude was a crew with a new dedication, anxious to work day and night and turn the Atlas into a fine piece of reliable machinery.

Many Atlas systems were modified. The fragile skin was strengthened in the most vulnerable area, where rocket joined with spacecraft, and on September 13, five months after the last blow-up, NASA and Convair were ready to try again. This time the Atlas worked like a precision watch and drilled an unmanned Mercury capsule into orbit. After circling the globe once, ground signals triggered braking rockets and the spacecraft returned safely to earth.

Elation and confidence swept through the manned space program.

John Glenn stepped to the plate.

From the beginning he had lobbied to be the first to get into the Mercury and initiate American manned space flight. The selection by Bob Gilruth gave Shepard and Grissom first nod, after which the remaining five astronauts would fly the suborbital lobs. All the team would then have a baseline of experience.

Vostok I and Vostok II scrambled the Mercury schedule. No more suborbital flights. Mission No. 3 would go for orbit, and Glenn, as back-up to both Shepard and Grissom, was in a perfect position to snag the assignment.

Except that first he had to endure the same humiliation that had bedeviled Alan Shepard. On November 29, 1961, Glenn stepped aside while NASA bioengineers loaded another chimpanzee, Enos, into a Mercury capsule. Again the Atlas performed flawlessly and thrust the ape into orbit. The public said, "Ho, hum," and the rest of the world laughed. But the chimp came home in reasonably good shape.

NASA then tapped John Glenn. His flight was scheduled for December 20. A Christmas present for America. But things just didn't work out as planned. From the day Glenn's booster and spacecraft went on the launch pad, there began a series of frustrating delays, mechanical problems, breakdowns, poor weather conditions, and maddening troubles that would have broken the spirit of a lesser man. Finally, on the morning of February 20, 1962, after eighty-two days of delays, John Glenn was strapped into the Mercury spacecraft *Friendship Seven*. Lady luck smiled on this tenth scheduled launch date and, as the countdown nudged its way to 9:47 A.M., the long-frustrated launch team ticked off the final seconds.

With fingers crossed, observers prayed for success as the Atlas waited on its stand, billowing ice-cold vapors before the wind.

The timers went down to "forty seconds and counting."

The launch team became a human machine.

"Status check," Test conductor T. J. O'Malley barked into his lip mike, which carried to the headset of every member of his team. "Pressurization?"

"GO."

"Lox tanking?"

"Have a blinking high-level light."

O'Malley knew that, despite a false reading of the liquid oxygen level of the Atlas tank, the booster was loaded with more than sufficient lox to keep the fuel burning all the way into orbit. "You are GO," O'Malley snapped. He hesitated a second before the next call.

"Umbilical retract now."

It was "GO."

"Range operations?" O'Malley sang out.

"GO for launch." The entire string of stations along the global tracking network was in the green.

"Mercury capsule?"

"GO."

"All prestart panels are correct," O'Malley called. "The ready light is on. Eject Mercury umbilical."

"Mercury umbilical clear."

"All recorders to fast," T. J. ordered. "T-minus 18 seconds and counting. Engine start!"

"You have a firing signal," astronaut Scott Carpenter told his friend John Glenn from his capsule communicator's position in the blockhouse.

B. G. MacNabb got onto the line to speak directly to O'Malley. "May the wee ones be with you, Thomas." O'Malley was glad to hear the words. He'd take the luck of the wee ones with gratitude. Personally he'd been praying all the way through the countdown. Now he took a deep breath, and the tough old Irish Catholic crossed himself. "Good Lord, ride all the way," he said quietly.

"Godspeed, John Glenn!" The call boomed from Carpenter as he racked down the final seconds of the count. "Three seconds . . . two . . . one . . . zero!"

The blockhouse fell silent, only the sound of instruments breaking the suddenly frigid air. The Atlas was ablaze, flame pouring from its three powerful main engines and two shrieking verniers, the vibration making John Glenn's voice tremble.

"Uh . . . rog-ger . . . the clock is operating . . . We're underway . . ."

And he was. The Atlas became a monolith of intense fire and gleaming silver, the dark Mercury capsule resting atop a tube covered with ice, the orange escape tower lifting above it all.

Nearly 360,000 pounds of flaming thrust. The autopilot ticked away in slick fashion, and the rocket obeyed by gimbaling its engines and working its verniers, and Glenn called out, "We're programming in roll okay."

The huge assembly of people on highways, beaches, atop buildings, and on streets went mad. Atlas crashed upward with a thunderous roar, amid the screaming cries of hundreds of thousands of people.

Max-Q. Tremendous air pressure squeezed the Atlas, buffeted the big rocket, hammered against its reinforced sides, rattled and shook the machine. The Marine along for the ride called down, "It's a little bumpy along here."

Out of Max-Q. Climbing at supersonic speed and accelerating rapidly as the engines increased in thrust and power, the Atlas lightened as fuel burned away. Two of the main engines fell away, reducing the weight even more, and the remaining engine and the two small verniers pushed the Atlas faster and faster.

Glenn was over a hundred miles high.

Engine cutoff. The escape tower was gone, the separation rockets spurted, the *Friendship Seven* pushed away from the now inert booster.

"Roger, zero-g, and I feel fine," Glenn reported. "Capsule is turning around. Oh! That view is tremendous!"

Glenn and America were in orbit. Cheers and tears erupted across the land.

The astronaut was on his way to three trips around the earth, and he reminded himself he had a debt to pay to millions of people who were anxious, almost desperate, to hear from the first American orbiting their world just what it felt and looked like. So Glenn grinned and, when time permitted, he became the narrator for his voyage.

Over the Indian Ocean on his first orbit, Glenn became the first American to witness a sunset from more than a hundred miles above the earth. "This moment of twilight is simply beautiful," he reported. "The sky in space is very black, with a thin band of blue along the horizon. The sun went down fast but not quite as quickly as I expected. For five or six

MERCURY: MESSENGERS FROM SPACE

minutes there was a slow but continuous reduction in light intensity, and brilliant orange-and-blue layers spread out forty-five to sixty degrees on either side of the sun, tapering gradually toward the horizon."

He offered other glowing descriptions of the planet sliding by beneath him, spoke of the snow-white mantle covering high mountains, the rich deep green of Bahamian waters, the sculptured sands of the deserts. He peered down at volcanoes and saw avalanches, told of sun reflecting off the clouds and the highest spires of great cities.

In the blackness each massive thunderstorm became a giant light bulb spitting and snarling with electrical fire. Observing a string of thunderstorm, flattening out along the horizon, it seemed like approaching a battlefield with guns flashing and rockets firing and bombs bursting. It was incredible.

Suspended in the blackest velvet of night, only the motors and instruments of *Friendship Seven* offered any sounds. The remainder of the universe had gone mute. His eyes became acclimated to the darkness, and he turned down the lights of his instrument panel.

Moving through the velvet night, Glenn began to see the stars. They appeared first as a filmy haze, became defined as a blanket, and then he was staring at the brightest, most clearly defined celestial engines he had ever seen—the stars in a glory until so very recently never seen by the eyes of humans.

As he quickly completed his first run through the night, flying backward, he could see the thinnest crease in the darkness behind him, a fairyland breath of slivered light. The breath became a whisper and, swiftly growing to a riotous shout of color, the horizon was transformed magically into a vivid, glowing crescent that separated night from day.

As the sun stabbed across half of the capsule structure, the other half lay in shadow and the dim reflected light from the planet below.

Suddenly John Glenn was no longer alone.

Surrounding the *Friendship Seven*, like tiny light motes from some fable of fairyland, were thousands of tiny creatures. Some came right to his window, and he stared in wonder at the tiny specks. Then he saw they were frost and ice, some shaped like curlicues, others spangled and starry like snowflakes sailing and dancing and swirling in an incredible swarm about the spacecraft.

Glenn was beside himself with awe and curiosity and fascination. He had no idea where this stunning phenomenon had originated. "I'll try to describe what I'm in here," he radioed the ground. Every person in the tracking station below, on Canton Island in the Pacific, snapped to, eyes wide, intense curiosity stamped on their features.

"I'm in a big mass of thousands of very small particles that are brilliantly lit up like they're luminescent," John went on. "They are bright yellowish-green. About the size and intensity of a firefly on a real dark night. I've never seen anything like it."

"Roger, *Friendship Seven*, this is Canton CapCom," came an immediate answering voice. "Can you hear any impact with the capsule? Over."

"Negative, negative. They're very slow," Glenn responded. "They're not going away from me at more than maybe three or four miles an hour."

The ground team was awestruck, hardly more so than Glenn, as they both did their best to determine what was going on "up and out there."

Just as suddenly as they appeared, the shining specks vanished as *Friendship Seven* sped over the Pacific expanse into brighter sunlight. But they were back on the next sunrise, and the next, illuminated by the rays of that rising ball of fire.

While puzzled over the "fireflies," the interest of ground controllers shifted dramatically to a matter of utmost urgency.

Unknown to Glenn, the flight of *Friendship Seven* had become a full-scale emergency. Flight Director Chris Kraft and his controllers felt John Glenn was threatened by a problem totally unexpected, and so critical that his life was in jeopardy.

The monitoring panels in Mercury Control lit up under an item marked Segment 51. That warning light sounded a gut-twisting alarm and indicated that the heat shield of *Friendship Seven* might have come loose from its securing connections.

Without the shield locked into position during reentry, John Glenn would be cremated by temperatures that would build up as great as four thousand degrees during the hypersonic plunge into the atmosphere.

The control team concentrated everything it had on the problem. If the heat shield was loose, what could they do about it to save the astronaut and his ship? They studied a host of ideas, the most experienced engineers throwing their concepts into the pot. One proposal quickly emerged.

Survival might well lie with the retro-pack strapped to the outside of the heat shield. This was a package of six small rockets. Three small separation rockets had fired after the Atlas engine shutdown to push the

Mercury away from the spent booster. Three larger rockets remained to decelerate the Mercury by five hundred feet per second from its orbital speed so it could begin reentry.

The flight program was specific. The retros fired, the capsule slowed only slightly but enough to start its downward slide into the atmosphere. An onboard electrical signal was then sent to break the metal straps. This action separated the retro-pack to send it away from the spacecraft.

Chris Kraft and Operations Director Walt Williams met with the flight team in Mercury Control. If they left the retro-pack in place after the firing, they surmised, then the straps should be strong enough to hold the heat shield in place long enough for the Mercury to descend to an altitude where building air pressure would keep the heat shield in place after the retro-pack burned up.

If, and these were the biggest ifs, the heat shield was loose, and if the retro-pack straps could not hold the shield in place, then the first American to orbit the earth would return as ashes.

Kraft and Williams decided not to alarm Glenn by telling him about the potential problem. Glenn became aware that something was wrong near the end of his second orbit, however, when he was told by the Canton Island station that he was to leave the pack in place after the retro-rockets fired.

"Why?" he asked. Canton CapCom told him he would be fully briefed when he made contact with the Texas station. Glenn liked none of it. No one knew the effect of the retro-pack burning from reentry friction. It could start to bang around and damage the heat shield. It could cause uneven heat distribution about the shield and even damage the shield before John came out of the fiery temperatures of reentry.

In an automated and meticulously programmed mission, any departure from established protocols could have serious repercussions, like one domino toppling against another and taking down a long row. The capsule autopilot was set to follow a particular sequence. If the retro-pack didn't get dumped, then the periscope would not retract, and the periscope doors would not be closed against the heat. To stay alive, John would now have to break to his own flight plan and start doing things manually, taking over from the autopilot.

When the autopilot circuits detected a deceleration force of .05g, the capsule's electronic brain would initiate automatic sequences to begin maneuvers that would safely land the capsule on the ocean surface.

Keeping the retro-pack on meant that John Glenn would be a one-armed man trying to wallpaper a room that wouldn't stay still. If he missed a single critical beat, the whole place could go up in flames all about him.

Glenn was ready to do whatever was necessary, of course, but with the decision to keep the retro-pack on he had changed from being a passenger riding a fiery chariot to earth to a very busy pilot in full, immediate manual control of his machine.

There was something else. They had perfected a schedule. Now, in the most critical time of flight, the schedule was being changed. New factors were being introduced into the situation, and the old safe-and-sane rule of flight, especially test flying, is stick to procedures. You deviate at your own peril.

Glenn wanted to know why the plan had changed and waited for word to come through Texas CapCom.

His heart had picked up its beat. It held now at 96 beats a minute. The ace test pilot and combat veteran had slipped into his familiar role. For the past two orbits his heart had held even at 86 beats. The rise was minor and normal.

Texas CapCom confirmed Glenn was to leave the retro-pack on through reentry. Exactly at 4 hours 43 minutes 53 seconds into the flight he must manually override that .05g switch and retract the periscope and seal its outer doors. He passed out of radio range before he got further information.

Soon he had Mercury Control on the line from the Cape. Finally Alan Shepard gave him the reason for retaining the retro-pack. John was angry he had not been informed earlier, but he concurred with the decision.

Four hours into the flight, off the coast of California, the three retro-rockets fired at five-second intervals. Glenn felt a triple thud at the base of the craft. "I feel like I'm going back to Hawaii," he exclaimed.

The *Friendship Seven* edged into the atmosphere. Heat built up. The capsule swayed slightly from side to side. A sudden *bang* behind Glenn held his attention. He felt it was at least part of the retro-pack breaking away. Glenn called Texas CapCom. He couldn't get through. A layer of ionized air enveloped the spacecraft and isolated it from communications.

Glenn plunged deeper into the searing heat.

His heartbeat increased to 109 beats a minute.

As best they could on the ground, they monitored Glenn's progress. Alan Shepard called with an urgent message for John to get rid of the pack the moment the capsule built up to 1g or greater. The message failed to get through the ionization layer.

Enveloped by a shrieking fireball, Glenn dived earthward. He was completely alone.

Through his window he saw a scene that belonged in a nightmare. A strap from the retro-pack had broken or burned free and was hammering against his window. It burst into fire and flashed away.

Big, flaming chunks of metal whirled and pounded past the window. They were fire devils, which bumped and banged against the spacecraft as they flew past before being swept into the ionized tunnel he was carving out of the atmosphere.

Never before had a man raced earthward at the core of a blazing metal meteor, hand-flying a wingless capsule.

"It was a bad moment," he reflected later. He judged that if everything did not come unglued, he would be okay. If things fell apart, then "it would be all over shortly and, period, there wasn't a thing I could do about it."

Two things dominated Glenn's thinking as the heat built up and *Friendship Seven* rocked from side to side as he fought to keep the little ship steady. He kept waiting for a shield failure that would transform him into a human meteor.

John Glenn's heartbeat held steady at 109.

Despite his apprehension, he kept his cool, flew with consummate skill as he watched the brilliant orange blaze outside his window, the burning chunks flying by. He felt the g forces building. He could have hugged them. It meant he was slowing steadily. He called CapCom; no way to get through the ionized teardrop yet.

The heat shield at his back was at three thousand degrees. No heat pulse on his back to indicate shield failure.

They felt helpless in Mercury Control. No one could help John. All they could do was wait for the communications blackout to end.

It was the longest four minutes and twenty seconds they had ever known.

An engineer stood behind Alan Shepard, part of a group collecting there. "Keep talking, Alan," begged one man. Shepard called again.

"*Friendship Seven*, this is Mercury Control. How do you read? Over."

The words penetrated the *Friendship Seven* like the song of an angel.

Glenn's reply was simple. "Loud and clear. How me?"

A grin split Shepard's face. "Roger," he said. "Reading you loud and clear. How are you doing?"

"Oh, pretty good," Glenn said, "but that was a real fireball, boy!"

Near pandemonium in Mercury Control.

Glenn had reached a deceleration of 7g on reentry. The capsule kept slowing. Now it was oscillating strongly from side to side, rocking badly enough for John to feed corrections with his thrusters. They weren't much good anymore. The atmosphere thickened quickly.

At fifty-five thousand feet he decided to override his automatics and deploy the drogue chute early to gain some stabilization. He was reaching for the override switch.

Glenn's heartbeat was now at 134 beats a minute.

The drogue chute came out. The auto sequences followed like clockwork, and *Friendship Seven* dropped into the water near the recovery destroyer *Noa*.

John Glenn returned to Washington a hero of Charles Lindbergh's stature. The nation flipped over the first American to race into earth orbit. The Russian lead was judged to have been sliced down to something we could manage. The White House extended its invitation to John, and a quarter million people braved heavy rain to watch the astronaut pass. He was then hustled off to New York City, where four million screaming, cheering people greeted him with a tumultuous ovation and a ticker tape parade.

The distance to the moon was starting to lessen.

At NASA headquarters the word went out. Keep the astronauts flying. Get the next Mercury mission into orbit as quickly as possible.

Deke Slayton stepped forward to take the reins for orbital flight Number Two and went to work, with Wally Schirra as his back-up.

He was completely unaware of the black clouds gathering over his head.

Jerome Wiesner cornered James Webb in the NASA chief's office. "Listen to me, Jim," Wiesner said. "Sending this fellow Slayton into orbit could be a terrible mistake. Suppose, just suppose, we run into a failure. When it's all over and we're wearing black armbands, the word gets out that the astronaut flying the ship had an erratic heart condition. Who do you think they are going to blame? It wouldn't matter if his heart had nothing to do with the failure. They're going to take dead aim at you and the administration."

Webb shifted in his seat. He hated this. But the truth had a nasty habit of stinging. He stared at the president's science adviser. "I get your point," Webb said unhappily.

"Take him off the flight, dammit," Wiesner repeated. "It's a risk we can't afford. Get your emotions out of this, Jim."

Webb recognized that Wiesner was still uptight over losing his argument that the president cancel the whole manned space program. He had been overruled in his attempt to quash Shepard's flight. He didn't want anything to go wrong now that would tarnish the president.

If Slayton flew and there was a failure, John Kennedy would end up taking the heat. Wiesner ended his pitch with an unassailable argument: "Why take the risk with the unknown when you have astronauts in perfect physical shape ready to go?"

Webb yielded to the point after asking the Air Force surgeon general to convene a panel of flight surgeons to review the case.

Deke had idiopathic paroxysmal atrial fibrillation, a disturbance of the rhythm in the muscle fibers in the upper chambers of his heart. The cause was unknown.

The disturbance showed up about once every two weeks. To Deke it was as serious as having one blue and one brown eye. He'd been flying punishing machines without anyone ever knowing the condition existed.

Dr. Bill Douglas and the other flight surgeons who examined Deke earlier had agreed the condition was not life-threatening and retained him on flight status. The new Air Force panel requested by Webb came to the same conclusion, judged Slayton to be "fully qualified as an Air Force officer and a pilot."

Still, Webb was worried, and he referred the matter to three eminent cardiologists. They examined Deke, and their consensus was that, while they were unable to state conclusively that the condition would jeopardize Slayton's performance, there were enough unknowns about it that NASA should not take the chance. Deke should be grounded.

"Goddamn it, Bill, those sons-a-bitches can't do this to me," Deke shouted at Bill Douglas when he got the word. "I've worked too long, too hard, for this."

To make matters worse, almost as soon as he learned, Deke had to attend a news conference in Washington about his grounding. "This medical guy and I were to explain all of this, and I was supposed to act like a fuckin' human being," he said. "I could have killed everyone in that room."

But in an incredible moment of self-control, Deke told the press he was grounded. "I am disappointed, of course, but I will step aside as a flying astronaut and support the program in any way I can."

There was more bad news. The rules called for the back-up pilot to slip into the seat of an astronaut unable to make a mission. Which meant Wally Schirra was next to go. But Bob Gilruth decided that Scott Carpenter, John Glenn's back-up during his long flight delays, had more simulator time than Schirra, and Scott drew the assignment.

Deke despised it. He wanted Wally to go, and he wanted to make waves about it, but he also did not want to anger NASA officials. The agency had put him on standby flight status for the future, and he did not want to jeopardize any chance he had, no matter how slim, to fly a space mission. Deke played the game.

NASA also got Deke the hell out of the way for Carpenter's mission. When the *Aurora Seven* lifted off on May 24, 1962, Deke Slayton was at a remote tracking station in Australia.

It was a space flight no one had expected.

When Scott Carpenter rocketed into space, there were no longer any monumental questions about exposure to weightlessness or the dangers of reentry. He had the benefit of the space flight experience of Gagarin, Shepard, Grissom, Titov, and Glenn, and Scott made the most of what he had. On his first two orbits he drank more, ate more, and by wringing out the Mercury spacecraft's control thrusters in one maneuver after another, he virtually depleted the fuel available for altitude maneuvering. He took a basketful of photographs, ran through his scientific program check list with skilled aplomb, which included releasing a balloon from the capsule, and spent as much time as he could catching the spectacular sunrises and sunsets. He was having a ball.

After those first two orbits, which consumed so much fuel, Mercury Control gave serious consideration to bringing him home an orbit early.

But they made a last-moment decision to let Carpenter stay in orbit if he would go into a "drifting mode," which would conserve his remaining fuel. Scott judged it a good idea, and he floated in weightlessness, enjoying every moment. Entering a sunrise, he banged his hand against the inside wall of the capsule.

It was a fortuitous blow with unexpected payoff. The moment he struck the wall, he was flying through a swarm of John Glenn's "fireflies." Again he struck the capsule bulkhead, and more fireflies showered into view. Despite his low thruster fuel, he fired the jets, swung around the capsule, and proved beyond a question the mysterious fireflies originated from vapor vented from the spacecraft. Vapor produced by the human body.

The astronaut's body generated moisture. As he perspired, and especially when he exhaled, body moisture and gases were removed from the spacecraft, dumped through an external vent in the side of the capsule. The instant this moisture entered the cold and vacuum of space, it froze into frost and ice particles. Some swarmed about the capsule or floated away; others clung to the spacecraft side, to be knocked off when Scott thumped the wall. When the sun angle was just right, at sunrise or sunset, these particles became the famed "celestial fireflies."

Scratch one mystery.

Carpenter somehow began to fall behind in the check lists. As the time came for him to fire his retro-rockets, a controller at the Hawaii tracking station reported, "We had the impression he was very confused about what was going on. But it was very difficult to say whether he was confused or preoccupied."

For undetermined reasons Carpenter fired the retros three seconds late, resulting in a twenty-five-degree error in *Aurora Seven*'s yaw position.

"My fuel, I hope it holds out," Carpenter said as the descent began. It didn't, and he had to release his drogue and main chutes early to stabilize a badly oscillating spacecraft.

A few seconds delay at orbital speed translates into big mileage errors. Because of the late retro-firing and the misalignment, *Aurora Seven* whooshed 250 miles beyond its intended landing point in the Atlantic. Carpenter was isolated, beyond radio range. For thirty minutes he was lost to a frantic Mercury Control and to a stunned worldwide audience, which had been following the flight on radio and TV.

A recovery aircraft finally picked up his radio beacon and homed in on the missing astronaut.

The aircraft crew found a bemused Carpenter floating in a life raft attached to his bobbing spacecraft.

He was eating a candy bar.

Behind the public scenes, Deke Slayton was waging his own fierce struggle to return to flight status. Deke went as high as he could on medical levels and finally met with Dr. Paul Dudley White, generally considered to be America's most eminent cardiologist. Dr. White examined Deke with meticulous care and told him: "You're going to live to a very ripe old age, young man. But I must agree with the people at NASA. They don't want to take any chance that's not necessary. It's difficult to fault them for staying on the safe side."

Deke returned home frustrated and depressed. The other astronauts discussed the devastating effect on him, and John Glenn laid it on the line. "We're a team. We've got to pull for Deke and help him."

Shepard nodded. "Gotta give Deke back his pride. We've gotta find a way."

So they came up with the idea of making Deke their boss. Give him the power. His own title, office, whatever he wanted. Gordo Cooper said they'd have to work fast. "There's some hogwash about bringing in an Air Force general, or maybe some admiral, to fill a new job as chief of the Astronaut Office. We need that like, well, hell, we don't need some outside weenie coming in here and telling us what to do."

NASA knew the Mercury Seven could not fill all the space flight slots in an accelerating program, and it was recruiting new pilots for the astronaut corps. It needed someone to manage this new enterprise, to select flight crews, make assignments, plan and schedule training time, be a link between the astronauts and management. In short, be a mother hen to this elite corps.

The Mercury astronauts made a persuasive case to NASA managers, and Deke became Coordinator of Astronaut Activities. Outsiders may have judged the appointment as a sop for a crestfallen astronaut, but that attitude didn't linger. Deke took absolute charge. Many stunned NASA staffers came to regard him as the Ironman. That pleased him. In short order Deke's office was a real power to be reckoned with. The new levels of respect carried over to the entire astronaut team. Everybody stepped back when it came to selection of astronauts for flights. Deke carried the ball and he ran with it, and on October 3, 1962, the Atlas, which was becoming known as "old reliable," belted Wally Schirra and his *Sigma Seven* spacecraft into orbit. He proved his skills, as Deke knew he would. He stayed up for six orbits—nine hours. He had been launched with the same fuel quantity as Glenn and Carpenter, but he conserved fuel in a way that amazed Mission Control. In the process he went through his scientific and engineering check list with an efficiency that would have turned a robot green with envy.

It was just what NASA and the nation had been waiting for. A perfect orbital flight with a perfect ending, *Sigma Seven* splashing down less than four miles from the main recovery carrier near Midway Island in the Pacific. NASA gave the mission the rubber stamp that read TEXTBOOK FLIGHT.

Schirra's flight held the rapt attention of nine test pilots who'd just been given the title of astronaut by NASA—the new crowd who would enter the Gemini two-man spacecraft program and would fly atop the powerful Titan II rocket, modified from a second-generation intercontinental-range war missile. Some would become legends, others would give their lives to the space effort—Neil Armstrong, Frank Borman, Charles "Pete" Conrad, James Lovell, James McDivitt, Elliott See, Tom Stafford, Ed White, and John Young.

Slayton now had fifteen astronauts under his wing. He set the newcomers up for indoctrination and training, and figured the more they saw of the remaining days of Project Mercury, the better prepared they'd be for flying the heavier, larger, tremendously advanced Gemini. With the new spacecraft, Slayton would be starting the test maneuvers and procedures that would lead to Apollo and the moon.

Deke helped draw up a flight plan that would put the Mercury in a category with Russia's heavy spacecraft, in time spent in space. No three- or six-orbit flight for the fourth and final Mercury orbital mission. This would be a shot for twenty-two orbits about the earth, a full day and a half that would tax the endurance of man and machine.

Walt Williams met privately with Deke. "Look, I know that besides yourself, Gordo Cooper is the only Mercury guy who hasn't flown. But maybe it's a good idea to consider moving Al Shepard into this last Mercury flight." Then Williams got out from under. "Of course, it's your call, Deke."

Williams didn't fool Slayton for a second. The issue at hand was that Gordon Cooper was too much of a maverick for some in the space agency hierarchy. His hotshot jet flying and his tendency to bend the rules—like racing cars in the streets of Cocoa Beach or at the Daytona International Speedway—did not sit well with them.

But Deke judged Gordo as nothing less than a terrific pilot who had come up through the ranks flying everything from J-3 cubs to F-106s, and he belonged in that spacecraft. If anyone knew how it felt to have an earned scheduled flight yanked from under his feet, sure as hell Deke Slayton was that man, and he was not about to have Gordo get the shaft.

There was another issue, beneath the surface. Gordo's deep Oklahoma "twang" irritated some of NASA's public affairs officers. To them, pilot qualifications had nothing to do with who should fly the mission. Their reasoning was simple. They did not want a redneck in orbit.

President Kennedy got a smell of this and would have none of it. Even if there had been some objection from on top, Deke would have rejected it. He'd made his decision. The flight was Gordo's.

There remained one obstacle: Gordon Cooper himself. For it seemed the undisciplined, "hot-dogging" pilot could not quite get the hang of putting away his hot-rock jet jockey image and start playing the role of well-behaved astronaut. Just two days prior to his scheduled launch Gordo tested the good-will and patience of NASA's top flight management.

The launch team was on an around-the-clock readiness schedule with Cooper's Atlas and spacecraft, everyone hard at work, when suddenly a tremendous *boom* ripped through the launch pad and gantry. Nobody saw flames they were certain would be leaping in the air, no smoke was rising, no buildings were collapsing, but the blow had been so powerful that—

Then, they saw it. The sleek jet howling away from the Cape in a dizzying climb after laying down a burning, supersonic thunderbolt across the launch center.

Gordo's way of saying, "Mornin', ya'll!"

His blasting arrival was hardly new for fighter pilots, and it was as much an "age-old" ceremonial rite for the astronauts now as it had been before they'd heard about going into space. But this close to launch, with rocket and spacecraft undergoing final critical checks? To Gordo it was a friendly way of announcing his arrival. Made sense; nobody would doubt that he'd be on time for his mission.

Walt Williams was standing in his office when Cooper and the F-102 came by at window level. The sonic boom shook the building, made him drop the papers he was holding, and sent his hands to stop his heart from leaping out of his throat. The Mercury operations director stood there quivering like a large walrus in a fit of uncontrolled rage. He'd spun around fast enough to see the blue blur hurl its supersonic message into the astronaut quarters.

Williams went to the quarters with a black scowl and stomped into the facilities with knitted brow and the scowl cemented to his features. He marched up to Deke Slayton.

"Does Al Shepard's spacesuit fit?" he bellowed at Slayton.

"What?"

"Simple damned question, Slayton," he growled. "I want a simple damned answer! Does his spacesuit fit?"

Deke deferred an answer with a question. "Why?"

"Because he's Cooper's back-up, and I want to know if he's ready to step in for that hot-dog, that's why!"

Deke had heard the sonic boom, and by a valiant effort he suppressed a grin. After all, he was hardly innocent of such greetings himself. He'd shook a few windows in the area, and he took the diplomatic route. He managed to calm down the irate, foot-stomping operations director and, after dinner and a few drinks, got him to allow the mission to continue as planned, with Gordo in the capsule. But it was a tight, tight shave.

Gordo wasn't through.

The next night, on the eve of Cooper's launch, innkeeper Henri Landwirth planned an informal dinner for the men of Mercury and for the Gemini Nine, as the second round of astronauts had become known.

Henri had his chef prepare a magnificent meal, advertised as breaded veal with varieties of potato, a superb salad, and the finest imported wines.

The astronauts murmured their surprise and their thanks to Henri. They all sat down to the great banquet and fell immediately to required toasting and bestowing of good wishes and fortune on one another. It was comradeship at its finest, a measure of friendship to warm hearts and minds. With the general high praise from each group quaffed by the wine, sixteen astronauts sat down to enjoy the gastronomical repast.

Waiters served it on silver trays. Silence descended with a crash upon the table. Henri had prepared a sumptuous feast of fried cardboard, uncooked potatoes, and a bellied-up salad that had been frying in the hot Florida sun for days.

Gordon Cooper had been here before. Indeed, he had his own reputation at air bases throughout the world built on such moments.

To the utter astonishment of the astronauts, and refusing to admit a truly classic "Gotcha!" Gordo chowed down.

The next morning, suited up and strapped into his *Faith Seven* Mercury spacecraft atop the last manned Atlas booster, waiting out a countdown delay, Gordon Cooper fell fast asleep on the launch pad.

Many a pilot and astronaut had done their best to rattle the cage of this man, and just as many had failed, and on that morning of May 15, 1963, Gordo rode *Faith Seven* into orbit where, this time on a preplanned schedule, he became the first American to zonk off into deep sleep while in orbit.

He flew a picture-perfect mission right on through his nineteenth orbit, setting a new American space endurance record with every sweep around the globe. Suddenly everyone in Mercury Control came instantly alert. A green light flashed on a monitoring panel. "Holy Jesus," a controller murmured. "He's on the way back to earth!" The light was the ".05g" signal, scheduled to shine when a capsule began descent into the atmosphere. CapCom made an immediate call to the spacecraft to confirm the unplanned, unexpected departure from orbit and asked Gordo if he was in reentry.

The response was typical Cooper. "Like hell we are," he told the ground. Then he left them alone to figure out what had gone wrong with the sensors to produce this false signal.

This type of glitch meant something was messed up with the electrical wiring, and when there's trouble with the wiring, the glitch is almost certain to grow. It did. Before the next orbit was completed, electrical surges knocked out the navigational instruments that kept Cooper informed of the over-earth position of the spacecraft. Things heated up on orbit twenty-one when the automatic control system rolled over and died.

That meant that Cooper, like Scott Carpenter, would have to fire his retro-rockets manually and guide the capsule through searing reentry heat to a landing.

But to Gordo Cooper, trouble in flight was what they paid him to handle. "Well," he called the flight controllers in his unmistakable twang, "it looks like we've got a few little washouts here. Looks like we'll have to fly this thing ourselves."

Deke smiled. He knew the flying skills of Gordo, and now this was the hoped-for payoff in the face of the NASA image makers who didn't want that redneck tooling about in space.

With just an hour to go in the flight, the control center worked out procedures and maneuvers on a precise timetable, and John Glenn, stationed on a tracking ship south of Japan, radioed them to Cooper.

After a day and a half in zero-g, Gordo fired his three retro-rockets.

Glenn reported: "He held it close, very tight. They were right on time on our marks here. They looked good, sounded good, and were good."

Gordo had threaded the needle perfectly.

Faith Seven came out of space, rolling steadily, the Oklahoma boy flying with a precision that controllers mumbled was tighter than the autopilot had ever delivered.

So tight that Cooper plopped into the sea just four miles from the recovery carrier.

Deke Slayton told him he'd done the best stick-and-rudder job ever. That he'd justified everything the astronauts had ever claimed, fulfilled every promise, and met every goal.

"What a precision ending to Project Mercury, Gordo." Deke smiled, grabbing his hand.

"Aw, shucks, Deke." Gordo grinned. "We aim to please."

CHAPTER TWELVE

HOUSTON

THE SEARCH FOR A NEW HOME for NASA's Manned Spacecraft Center was carried out with all the subtlety of a political cavalry charge. Jeb Stuart could not have done better as Washington insiders manipulated their way through political thickets to lock up a site twenty-five miles south of Houston.

Bob Gilruth, as director of the old Space Task Group, would head the new center, and wanted nothing to do with the Houston site, or any other for that matter. He wanted everything to stay put at Langley, where there was no finer place on earth to sail than the Chesapeake Bay. He and most of the people in his group had been at Langley Field since the days of the old NACA, and he felt there was no logical reason to uproot all those engineers and technicians and move them halfway across the country.

But there just was not enough room at Langley to accommodate the massive expansion needed for Apollo—the ambitious program intended to land Americans on the moon. No room for the growing astronaut corps, for their training devices, for a huge control center to monitor humans en route to another world, for the laboratories that would examine the precious rocks to be gathered on the lunar surface, for the thousands of civil service and contractor engineers, technicians, managers, and other personnel who would toil at the new center to make the dream come true.

"Tell me, Bob," NASA administrator James Webb asked Gilruth, "what has Harry Byrd ever done for you?"

Webb's question was right on the mark. Virginia's Senator Byrd had never done much but criticize the space program. Webb was thinking of a site in California or Texas, two states with strong political figures who were strong supporters of the program.

So Gilruth dutifully set up a committee that scoured the country for a new site, settling on a few locations in both California and Texas, all of which were forwarded to Webb for final selection. Webb chose Mare Island in California and sent the recommendation on to Vice President Lyndon Johnson.

Webb hadn't figured on the clout of two powerful Texas political figures. Albert Thomas of Houston was a key member of the House Appropriations subcommittee. As vice president of the United States, Johnson was chairman of President Kennedy's National Space Council.

Thomas, who had that barren cow pasture in Houston in mind all the while, went to Johnson. They talked, and the vice president overruled Webb and said the site should be Houston, which had been on Gilruth's short list.

Those supporting Houston argued that the area was perfect for NASA, since it had advantages of highway, rail, water, and air transportation. So did many other sites, and the argument of "land availability" faded miserably in the face of the then recent NASA purchase of eighty-eight thousand acres of land immediately adjacent to Cape Canaveral. Indeed, this area of Merritt Island—mostly orange groves—was to become a huge industrial facility and the actual moonport from which humans would leave the earth on journeys deep into space.

Why not, then, consolidate everything there? It not only had everything Houston claimed as its advantages, but it would bestride an already existing support, launch, and tracking network of unparalleled skills.

Lyndon Johnson suggested to his subordinates that they understood little or nothing about the true art of astronautics, or the logistical foundation on which it must rest. Consider Cape Canaveral, he said with well-aimed political darts. It is a military facility. Its main purpose is to perfect the mightiest weapons of war. It is a forest of vertical giants that spell doom and horror for all the world with Atlas and Titan and Polaris, with Thor and Jupiter, and the other secret weapons too horrible to reveal to the public. Missile testing is expanding swiftly at the Cape, he said. The range is crowded, the tracking network overburdened, the communications net and radio traffic so heavy that it would be difficult, perhaps impossible, for NASA to sustain clear communications channels during a long-duration manned space mission.

Launch NASA rockets from the Cape, but communicate with them from afar, preferably from a control center in Houston. A new center, Johnson's allies concluded, must be at least 125 miles of clear and uncluttered distance from Cape Canaveral.

Well, let us see! The Florida contingent, no stranger to pork-barrel politics, roundly condemned Lyndon Johnson as a purveyor of untruths. If a new center must be established 125 miles from the Cape, why, they had the perfect site for NASA. MacDill Air Force Base in Tampa was a superb facility, already built by taxpayer dollars and supported by citywide industrial, business, highway, rail, shipping, airport, and other necessities for any new location. It had large runways, communications networks linking its own command centers to the rest of the world, and it presented an ease of transport and communication with the launch center on the Florida east coast, just 150 miles away. Furthermore, the Air Force was planning to downsize its operations at MacDill severely, which opened, in effect, a huge, government dollar savings welcome for the new NASA group.

Lyndon Johnson knew better than to argue against saving the taxpayers a buck. He assembled his site selection team, explained the situation, and ordered them to come up with a bona fide reason why MacDill simply would not do.

Site requirements changed before a new dawn arrived. A Johnson committee submitted new guidelines that required the second NASA command center to be at least 250 miles from Cape Canaveral.

Johnson smiled. Scratch MacDill, and ring the cash register bells for Houston. At least he heard them ringing in his head. He was fiercely determined to land this plum in his home state. By the time congressional committees had narrowed down the possibilities to ten sites, seven were in Texas. Now the odds leaned in favor of the Stetson Twins. Congressman Albert Thomas galvanized his own forces and arranged the facts neatly for a predictable vote. Thomas just happened to be chairman of the House Independent Office Appropriations Committee, which controlled the funding for future NASA projects.

Charles Donlon, deputy director of the Space Task Group, shook his head in wonderment as he admitted to an interviewer, "It's as though you went through a maze, knowing all the time what door you were going to come out."

NASA insisted that at least a thousand acres be made available, free of restrictions, and preferably donated to NASA to control costs to the taxpayers. That requirement was parlayed with all the skill of a pool hustler cleaning up from the locals in a small Texas town.

LBJ's friends at Humble Oil owned many thousands of acres of pasture land and offered the thousand acres. The company had little use for the land, which had proved worthless as an oil producer. Humble donated the thousand acres not to NASA but to Houston's Rice University, with the proviso that the school must transfer the land to the space agency on a ninety-nine-year lease.

Before long, a new sound could be heard in the area of Clear Lake, the nearest town, and this was the brisk rubbing together of palms in eager anticipation of a fiscal windfall. With the Manned Spacecraft Center (MSC) as anchor, the surrounding acreage soon resounded with the clanking of bulldozers and road graders and cement mixers and truck convoys rolling in and out of the area. The keys were the supporting businesses for the mass of people NASA was moving to MSC. The MSC staff, as well as supporting businesses, new medical facilities, schools, utilities, and the general fabric of a new community, created in astonishingly quick time huge flatlands of thriving suburbs with literally tens of thousands of people clamoring to buy new homes and sign leases on new apartments.

Many friends of Lyndon Johnson and Albert Thomas made a hell of a lot of money when Houston was selected. Those already wealthy smiled as their coffers bulged with new funds, and no one was really surprised when very close friends of LBJ at Brown & Root speared the huge contract to build the Manned Spacecraft Center.

All other factors notwithstanding, Houston was it. Seven hundred engineers and their families loaded up boxes, baggage, and their vehicles for the long transport from the favorable environs of Virginia and the Chesapeake Bay to the flat Texas ranch land, to be the epicenter of the new space effort.

NASA had boosted its image by stretching the last Mercury flight to its limit in May 1963, when Gordon Cooper sped around the planet for more than thirty-four hours. But no amount of praise for Gordo's superb performance could persuade the public that America was catching up. The numbers spoke for themselves and, while NASA was busy building the Manned Spacecraft Center, the Russians were hammering into space and demonstrating they were the faster learners in the public relations business.

On June 14, just one month after Cooper emerged from his harrowing reentry, the Russians hurled Vostok V—*Hawk*—into orbit with Lieutenant Colonel Valery F. Bykovsky. The cosmonaut would remain in orbit just over 119 hours, a number that staggered the space-gazing, awed public.

But even Bykovsky's flight was pushed off the headlines two days later when Vostok VI—*Sea Gull*—ripped away from its Baikonur pad. Bykovsky, still in orbit, was looking down at the launch complex when the Soviet booster sent the *Sea Gull* on its way into orbit *with a woman on board.*

Valentina V. Tereshkova was the first woman to enter this previously all-male domain, and she remained in space nearly seventy-one hours.

With the Russians darting and speeding about the heavens, the word in Washington went urgently to President Kennedy: "For God's sake, *do something!*"

November 16, 1963, John Kennedy followed the requests of his staff and appeared at Cape Canaveral for a firsthand look at the Air Force launch center, as well as at the huge moonport under construction at adjacent Merritt Island.

Kennedy questioned Dr. Wernher von Braun, who showed with obvious pride the huge Saturn I rocket booster being readied for its first all-up test flight. First kicked off its pad in October 1961 for initial launch, Saturn I, with 1.3 million pounds of thrust at liftoff, would be capable of hurling thirty-eight-thousand-pound payloads into earth orbit.

That would even the boost capability between the United States and the Soviet Union.

Kennedy left von Braun and climbed into a helicopter with astronauts Gus Grissom and Gordon Cooper, who with unabashed excitement and pride pointed out the key features of the growing moonport, where one day a monster called Saturn V would stand on its launch pad. Here the name Apollo was gaining substance with every passing day.

John F. Kennedy did not live to see Americans sail across space to the moon. Six days after he viewed the launch areas for Project Apollo, the president of the United States fell to an assassin's bullet during a Dallas motorcade.

Shocked and stunned, America slowed to a stop. With the nation wracked by emotional loss, workers at Cape Canaveral joined in mourning for the passing of a man who had placed America on its course to the moon.

By the end of 1963, a shroud of uncertainty seemed draped over American space projects. NASA's key supporter was dead. A strange lassitude infected both the American and Russian manned projects. With the last flights of Mercury and Vostok, human orbital flights would not take place again for more than a year. It was time for review, study, reorganization, and striking new commitments for the future.

Lyndon Johnson and Jim Webb refused to drink from the poisoned well of inactivity and remained committed to America's progress in space.

The moon! The moon was the goal. Easier said than done, of course, but at least now, with hard experience behind them, NASA leaders and engineers were defining the critical technical problems to be addressed and overcome.

All future spaceships must be able to do more than roll around a point on reentry and waggle small and large blunt ends while sailing along an unchangeable orbital trajectory. That was fine for Project Mercury in its limited role of determining if humans could live and work in space. But now it was painfully clear that *maneuverability* in space was the new holy grail to be achieved.

Thus was born Project Gemini, with a four-ton, two-man spacecraft that must move through space with unexcelled freedom and reliability so that it might deliberately plunge into the problems the future Apollo would face and provide the answers for the manned moon vessels still taking shape on the drawing boards.

Specifically, the astronaut twins of Project Gemini were to perfect all the key techniques for reaching the moon—rendezvous and docking with another spacecraft, long-duration flight, and walking in space.

Two unmanned Gemini spacecraft boosted into orbit atop their Titan-

II boosters and flew very successful missions. NASA considered the new ship ready for its maiden voyage with a manned crew and scheduled it for early spring in 1965.

Deke Slayton considered who would be the best man for the critical, first "all-up" Gemini test flight, and without hesitation he selected Alan Shepard once more to command and lead the way. His copilot would be Tom Stafford, who thus became the first man of the second group of astronauts to receive a flight assignment. Deke followed the same pattern for the back-up crew, selecting Gus Grissom as command pilot and John Young, from Stafford's group, as his copilot. It was time to work the new kids on the block into the flight scheme.

Preparations went smoothly for six weeks, when Alan Shepard awoke one morning feeling nauseated.

Alan met with Deke Slayton to report on what he encountered. He laid it out straight. "All of a sudden, Deke, I fell. I was so dizzy! The room was spinning around and suddenly I'm on the floor. I got up holding onto the wall, and right away I got so sick I vomited. I thought, Jesus, what the hell did I have to drink last night? It must have been one hell of a hoorah, but, well, that just wasn't the case."

For several days Shepard felt fine and worked with precision, his episode apparently well behind him. Until the fifth day, when he again experienced head-spinning and vomiting.

Once again the room whirled madly. He heard and felt an awful ringing in his left ear. It went away.

Then it came back, several times. Alan knew something was terribly, dangerously wrong. He checked in with the flight surgeons.

"You've got a serious problem with your left inner ear," a doctor told him after a thorough examination. "You have what is called Ménière's syndrome."

"What the hell is that?" Alan shot back.

"Certain people who are Class A-type individuals, who are hyper, driven, motivated, highly competitive, will occasionally develop this problem," the doctor responded. "Fluid pressure builds up in your inner ear, and it makes the semicircular canals, the motion detectors, extremely sensitive. This results in disorientation, dizziness, and nausea. You've been experiencing all of this. You also have glaucoma, an elevated pressure in your eyeballs. That's just another indication that as an individual you're highly hyper."

Alan listened patiently to the diagnosis and suddenly found his patience wanting. "Don't give me that crap!" he said angrily. "I'm just a plain fighter pilot. Jesus . . . " He lapsed into an angrier silence, aware immediately that what he had just heard could lose him his Gemini command.

He was right. A panel of NASA medics yanked his wings, and suddenly he found himself in the bleachers with Deke instead of out there on the playing field.

"Look, Alan," a doctor said warmly, "there is no known operation for this condition, but there's a chance, about twenty percent, that what you've got will dissipate, you know, cure itself. *Without* treatment. And with the medication we've prescribed for you, the odds get better."

"How much better?"

"Some."

"How much, dammit!"

"Slightly better."

A dispirited Alan sat down with Deke for a heart-to-heart. "Deke, what the hell do I do? Should I just hang it up? Quit? I've had my moment of glory, if you want to call it that, but it's a hell of a long way from being enough. Something burns inside me, deep down and bright, and it tells me *I've got to fly again*, to keep flying, and one day *go all the way to the moon*. Man, you've been grounded now for more than two years. What are your odds for a flight someday? What are mine?"

Deke didn't answer immediately. The man before him was his friend, his fellow pilot—dammit, the first American in space—and he knew the pain coursing through Alan Shepard. Finally, Deke spoke slowly: "Listen to me, Al. I've talked with the medics. They say you've got a chance, a real chance, to come back. To get back into the cockpit. They tell me your odds are a hell of a lot better than mine. So until you get your medical situation straightened out, *stick around*. I've got a job for you.

"You can keep up with your medication while you back me up. I'm moving up to a new job as chief of flight crew operations. That job includes oversight of the Astronaut Office. I need a man to directly supervise that office, to replace me and work with me. That keeps you right in the heart of the operation and gives me the best man there is to support me. How about it?"

Shepard chewed on his lower lip. It sounded great, but—

"Christ, Deke, I don't know how good I'd be changing diapers and feeding astronauts. You've handled it like a pro. But me? I'm a different story. It would be awfully tough to take a crew down to the Cape, pat them on the ass, and watch them fly. It's the sort of thing that can tear a guy up inside."

"So? So it'll be tough. But, dammit, you're tough, too," Deke growled. "We're both tough, and I don't think we're through with being grounded. Hell, I know I'm not. I intend to make it into space yet, Al. I'm keeping my foot in the door, and yours should be right there next to mine."

Shepard grinned. "I want to fly again so damned bad. And if this keeps that door open—yep, I'll do it," he said, shoving his foot alongside Deke's.

The two feet side by side brought an old grin back to Slayton's face. "Hang in there with me, Navy," he winked. "We'll figure out a way to get our asses into space."

"Okay, Air Force," Alan laughed, "you've got a partner."

It would take years. Many years.

But Alan and Deke would do it.

CHAPTER THIRTEEN

SPACE WALK

LEONOV NODDED TO BELYAEV. They checked their timers and instruments. They had been breathing pure oxygen. "We are free of nitrogen now," Pavel I. Belyaev, commander of the *Diamond* (Voshkod II) told his fellow cosmonaut. "Ready," Alexei A. Leonov nodded.

Moments later their suits inflated to full pressure as the two men prepared for any emergencies.

Leonov released his restraint harness, turned for Belyaev to check his backpack. Leonov received all his power from a battery pack and oxygen bottles integrated within his suit. A large airlock extended from the side of the spacecraft. Leonov floated toward the airlock and sealed the hatch behind him. Now he depressurized the airlock itself, secured a safety tether to the airlock interior. When the gauges read zero pressure, he opened the upper hatch to—the void of space.

Leonov floated, then pushed himself gently through the hatch. "I'm getting out," he radioed to Belyaev. His excited commander radioed the mission control center outside Moscow. "At this moment," he called with unrestrained elation, "a man is floating free in space!"

The first human satellite drifted away from Voshkod, turning and tumbling slowly, weightless, stunned with the sight of the earth below. He rolled about to look into a "flow of blindingly bright sunlight, like an arc of electric welding." He pulled gently on his lifeline and hurtled toward Voshkod's hull. He quickly put out a hand to check himself and bounced into a spin. He had learned something about the fickle ways of force and reaction in the weightlessness and vacuum of space. A small camera attached to the top of the airlock captured the smiling, laughing cosmonaut as he cavorted in mankind's first walk in space—the flight of the *Diamond* on March 18, 1965.

The flight plan called for ten minutes in space. For those ten minutes, across a distance of three thousand miles, the thirty-year-old lieutenant colonel of the Soviet Air Force, a skilled parachutist, fighter pilot, and world-class athlete, threw out his arms in rapturous joy as he floated and turned. Below him, passing 120 miles below, the earth rolled by at 17,400 miles an hour.

"I didn't experience fear," he explained later. "There was only a sense of the infinite expanse and depth of the universe."

After ten minutes he turned for a final look at the beautiful planet rolling beneath him and then slid into the airlock, feet first. He became jammed in the opening. The minutes sped by, and Belyaev cautioned him that he was running low on his oxygen supply. Leonov breathed deeply and slowly, concentrating. He found the problem. His suit had expanded and he was caught like a cork in a bottle. Slowly and carefully, he partially depressurized the suit and, using his great strength, pushed himself back into the airlock. Opening the airlock's inner hatch, he pushed himself back into the cabin. Belyaev dragged him to his seat, securing his harness. Both men checked the security of the inner hatch. "Let it go," Belyaev said. They worked the panel controls and the airlock disconnected and drifted away from the *Diamond*.

The first EVA (Extra-Vehicular Activity) was now history. Belyaev studied Leonov. "He looks like a man who has just been reborn, a man who has just come back from another world." In his excitement he banged a fist on Leonov's shoulder and shouted, "*Molodets!*"

His voice carried to the control center. Hearing the cry of "Great show!" from Belyaev, controllers cheered and applauded.

While the world heralded this new, spectacular Soviet space feat, the flight of the *Diamond* suddenly became a misadventure and a classic example of how many serious things could go wrong on one brief flight.

Trouble loomed the day after the space walk when it came time to fire the retro-rockets for return to earth. The automatic stabilization system failed, and Belyaev and Leonov went through the proper contingency maneuvers for Belyaev to take over manual control of the spacecraft. This took time, and they delayed the retro-fire one orbit. When the commander finally triggered the rockets, he did a magnificent job of guiding his ship through the harrowing reentry. But the extra orbit pushed their new landing site nearly a thousand miles east of where recovery forces waited, at the intended touchdown point.

The two cosmonauts caused the craft to overshoot the landing site by an additional two hundred meters when they had to leave their seats for visual confirmation of instrument and control settings while they aligned the craft for retro-fire. This would have meant nothing if they still had had automatic altitude control. But they didn't, and their movement slightly changed *Diamond*'s center of gravity. The craft crashed in the thick forests near Perm, in the Ural Mountains. *Diamond* was not equipped with Vostok-like ejection seats, and there were no personal parachutes on board. Only the spacecraft had parachutes. The cosmonauts were not able to escape the craft and remained inside as *Diamond* fired its landing rocket shortly before touchdown.

Official news stories stated that *Diamond* had landed in deep snow when, in fact, the spacecraft had crashed in the forest and had wedged itself tightly between two large fir trees.

Leonov and Belyaev remained inside the Voshkod, unable to open their hatch.

In the freezing night, the cabin fans were going at full blast, the electric system so twisted that the cosmonauts were unable to shut down the whirling blades. Ironically, while the cooling fans worked, the heating system had failed.

A recovery helicopter arrived at the scene two and a half hours after the landing, and official Moscow expressed great relief that the embarrassing affair would soon be over. But the deep snow and thick forest cover made it impossible for the chopper to land safely.

The helicopter crew dropped warm clothes to the shivering men, but the clothes fell into the higher branches, out of the reach of the cosmonauts, who would have been unable to retrieve them had they dropped right next to the hanging capsule, as the men were trapped inside the ship.

So the clothes remained in the upper reaches of the trees, *Diamond* remained jammed between the two firs, and the cosmonauts huddled through a frigid night. The next morning, a rescue crew entered the thick Perm forest on skis and wrestled the spacecraft free of the trees, releasing the freezing cosmonauts from their overnight prison. Hot food and warm clothes soon restored Leonov and Belyaev, who then skied out of the forest to a waiting helicopter.

With confirmation that the two cosmonauts were on their way to Moscow, Soviet officials began putting a positive spin on the flight, emphasizing the importance of the world's first space walk and downplaying the overshoot and the horrifying night in the forest. A government spokesman explained to the Russians, and the world, that walking in space must be mastered before great stations could be built to orbit the earth.

"But," he added, "our immediate goal, the target before us, is the moon."

A clear warning of the Soviet intention to get to the moon before Apollo.

That warning and the space walk galvanized Americans. Leonov's space walk had clearly demonstrated the Soviets' superiority in space. With commendable candor, NASA's Chris Kraft complimented the Russians and admitted to the American public that the space walk "was a tremendous surprise, and coming just five days before our first Gemini flight, it caught us completely off guard."

To make matters worse, the Russians trumpeted the new Voshkod as a spacecraft capable of carrying three men into space. They underscored that it would take at least another four years for the Americans to be able to launch three men within a single craft. One *Pravda* headline read, "SORRY, APOLLO!" and the accompanying article taunted NASA with the remark, "The gap is not closing, but increasing. . . . The so-called system of free enterprise is turning out to be powerless in competition with socialism in such a complex and modern area as space research."

The moment unquestionably belonged to the Russians, and in Washington there might have been more concern about the future had officials there known that cosmonaut Pavel Belyaev soon after his return began training for a circumlunar flight, a loop around the moon to be made in about two and a half years, in 1967. Nor were they aware that Alexei Leonov also had been assigned to the group training to go to the moon, would become commander of that program as well, and that Leonov and Oleg Makarov had been selected to make the first two-man circumlunar flight.

But the United States was not standing still. It no longer was stumbling in space, and its program was gathering steam and momentum. Schedules were set and were as much as possible to be met no matter what the effort demanded.

Gemini was next—the bridge between the fledgling Mercury and Project Apollo. Wernher von Braun and his team would sit out Gemini. They were busy perfecting the monster rockets for Apollo—Saturn I for earth orbit tests of the three-man spacecraft and Saturn V, the behemoth to carry astronauts to the moon. At North American Aviation in California, the moon craft was being developed, assembled.

America's course was set.

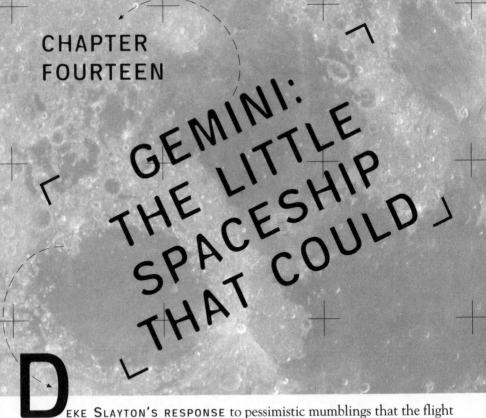

CHAPTER FOURTEEN

GEMINI: THE LITTLE SPACESHIP THAT COULD

DEKE SLAYTON'S RESPONSE to pessimistic mumblings that the flight of Voshkod II had erased much of Gemini's luster was not printable. One thought dominated his mind: *Let's get on with it! Don't bother me with bullshit!*

"When the going gets tough," explained a colleague about Deke, "that son of a block of granite digs in his heels, sets his jaw, squints at the future, lights the afterburner, and gets going." Slayton was the driving force that brought ever closer the scheduled launch date for the first manned Gemini mission, on the calendar for March 23, 1965—nearly two years after the last American had flown in space.

Problems arose. Problems had been expected. Deke's immediate and overriding issue was, of course, that he had to change astronauts in midstream. With Shepard taking a battering from the dizzy spells he was fighting off, there was no choice but to move the back-up team into the cockpit. Deke and Alan shared that decision, called in Gus Grissom and John Young and told them, in effect, "Suit up."

The decision was based on more than ranking in seniority because Grissom was judged to know more about the Gemini ship than any of the astronauts. In addition, Gus was as close to Deke as anyone in the group, including Alan, and from the program outset Deke had charged Gus to ride herd on the Gemini design team and assure that flight crew input and recommendations were transformed into actual hardware for the systems the astronauts, not the engineers, wanted.

"Mercury was essentially an automated system," Deke said in answer to queries about the new two-man spacecraft. "It was designed that way because it was the first of its kind. We flew two unmanned Geminis in which automated systems also attended to the flight requirements. Then we took the reins off Gus, and when he got through with changes and design improvements, there just wasn't any way Gemini could be flown without pilots at the controls. How did the astronauts feel about it? You could hear their applause a mile down the beach."

What emerged was a spacecraft that retained the same bell shape of the Mercury but which was three feet taller, wider at the blunt base, and at three and a half tons weighed more than twice as much. Two men would ride in a cabin that had about as much space as that in the front seat of a small sports car. Gemini carried an onboard propulsion system for extensive maneuvering in orbit, a guidance and navigation system, and a rendezvous radar. It also was designed to parachute to a water landing.

As the time neared for the first manned Gemini flight, which once had been his to command, Alan Shepard grew uneasy. He was chained to the ground, and increasingly he felt the weight of his folded wings. He began to take out his frustrations on the astronauts in training, coming down heavily on any errors or faults. When a man was assigned to a flight, he came under Alan's microscope. At weekly astronaut meetings, when everyone got together to thrash out problems, Alan berated his fellow pilots and would often lash them mercilessly.

"I felt I laid the pressure on them exactly the way I expected to be treated by anyone monitoring my progress before a flight," he said.

His reputation changed. No longer the easygoing, highly skilled pilot, he was the "icy commander." He had the incredible ability to switch his moods at will, be hell on wheels in the morning and the master of charm in the afternoon.

The astronauts began searching for a way to predict Shepard's mercurial mood changes. Newcomers were baffled as how to behave toward him. Finally his secretary, Gaye Alford, hit on a unique idea to prepare the pilots.

Each morning she would assess Alan's frame of mind the moment he entered the office. Then she selected one of two pictures to hang on the outside of the door. A visitor would see either a photo of a scowling chief astronaut or one of a pilot with a beaming smile. Those who walked past the scowl did so at their own risk.

"Al could be friendly, outgoing, warm, a good leader and companion," observed astronaut Mike Collins. "But he could also be arrogant and put down friends or foe alike, with a searing stare and caustic comment."

Buzz Aldrin, unquestionably one of the most brilliant astronauts in the program and a stellar mathematician who performed research on orbital rendezvous, had this opinion of Shepard: "Alan took his job very seriously. He was a stern taskmaster who did everything by the book. We even had to check in and out of his office when we traveled on business or to the Cape or to the contractors. Al wanted team players, not individuals, and his style was soon imprinted on the Astronaut Office."

Deke Slayton, who often said little but missed virtually nothing taking place about him, took note of the adverse effects of Alan's mood swings. After one weekly meeting he asked Alan to remain behind for a private conversation.

"Don't you think you're being a little too tough on the guys, Al?" he asked his friend.

Alan considered the question. "Maybe," he said. "You could be right, Deke. I guess I am a bit hard-nosed. But dammit, you know my way of doing things. If I'm harder on them than I should be, it's because I'm trying to run a tight ship. Maybe too tight."

"Anything else?" Deke prodded gently.

"Me?" Alan laughed without humor. "Man, I'm mad at the world, being grounded like this. Sending these guys out to fly while I'm wired to my desk, well . . . Yeah, I suppose I have been taking it out on them. It's just so damn frustrating."

Deke understood: "Look, Al, I know what's chewing at you. I went through the same thing when I was grounded. It took time, but gradually I knew I had to accept what I was doing. I'll be the first to admit it wasn't easy to swallow. It sort of gets stuck in your craw. But I worked at it, and I found out after a while that I'd actually mellowed. The bottom line, my friend, is that we're a unique family. We've got to get along."

From that moment on, Gaye Alford more often than not hung a picture of a smiling Shepard on her boss's office door.

Before Gemini 3 started its final countdown, Grissom created a flap within the NASA hierarchy. Gus had never been able to shake off the innuendos that the sinking of the *Liberty Bell Seven* Mercury capsule was due to his error rather than technical imperfections in the hatch design. So he christened Gemini 3 *Molly Brown* from the Broadway musical *The Unsinkable Molly Brown.* The press thought it was great, and there were broad smiles for Gus's ingenuity.

In NASA headquarters, Jim Webb and his top officials were aghast at the undignified name Gus had affixed to his spacecraft. Orders were issued that all official agency documents and references henceforth would identify the craft only as Gemini 3. The rest of the planet ignored Washington, and to the world *Molly Brown* it was, and *Molly Brown* it would stay.

Five days after Belyaev and Leonov found themselves jammed between trees in the Perm Forest, Grissom and Young roared off their Canaveral launch pad atop a Titan II rocket for a five-hour check-out of the Gemini craft. They tested the ship's systems, operated all its maneuvering thrusters, and brought the moon just a bit closer when, for the first time, they guided a manned space vehicle to higher and lower orbits and into a different orbital plane—maneuvers essential to going to the moon and ones that the Russians had not yet demonstrated. They pronounced Gemini a pilot's delight. "It handles easily," Grissom told flight control.

They came out of orbit with the machinery functioning as advertised and splashed down in the Atlantic. Gus then deviated from the flight plan. He refused to open the *Molly Brown's* hatches until Navy swimmers, dropped from helicopters, attached a flotation collar to his bobbing craft. It took time to do all this, and Gus paid a penalty for the delay. The *Molly Brown* may have been unsinkable, but it couldn't sail worth a darn. The capsule pitched and rolled in the waves, and Gus soon was seasick. "This thing is no boat," was how John Young put it to the recovery team.

What most people had never really appreciated was that John Young, a great pilot, also had an irrepressible streak of humor. Before the launch, some secret tests had been conducted by some of the astronauts on a corned beef sandwich, Gus's favorite, at Wolfie's delicatessen in Cocoa Beach. On several corned beef sandwiches, actually. After each was prepared, a cook would climb to the top of a tall ladder and drop the sandwich into "free fall." The one sandwich that held together better than the others was sealed into a package and delivered to Wally Schirra, who in turn gave the sandwich to John Young, who smuggled the "secret experiment" aboard *Molly Brown.*

During the mission Young presented Grissom with his tasty surprise. Gus laughed and took a few bites from the sandwich.

When the news later reached the medical teams of NASA, all hell broke loose. The flight surgeons complained to top NASA brass that those few bites had negated the mission's medical protocol. Engineers climbed on the bandwagon and said that even a few crumbs floating about within the weightless cabin could have fouled up electronic gear.

Incredibly, some members of Congress leaped into the fray, earning some newspaper space by shouting that NASA "had lost all control of its astronauts."

"All because of one lousy corned beef sandwich," was Alan Shepard's amazed reaction.

Deke couldn't believe what was happening, but he had to diffuse the situation. Wally and John hadn't broken the rules. Wally had told Slayton before liftoff of the sandwich gag, and Deke, who knew the pressure building on his pilots, judged it was a great way of relieving the tension the astronauts lived with day to day.

But now he had no choice but to respond to the frowning NASA officials and loose-lipped congressmen. That the two men had flown a great mission and proved the design of Gemini seemed to have been forgotten in the aftermath of the sandwich episode.

Deke signed and issued a memo that said in part: "The attempt to bootleg any item on board not approved by me will result in appropriate disciplinary action." NASA officials smiled approvingly and directed that in the future the astronauts would not assign names to their spacecraft and would use only numbers to identify each mission. Because the brass still had the power to knock a man out of a flight, the crews wisely kept their comments to themselves and, on the surface at least, complied with the new rules.

James McDivitt and Edward White II judged the entire affair as evidence there were some screws loose in Washington, but they kept their counsel to themselves. They already had named their Gemini 4 *American Eagle*, which seemed appropriate and patriotic. But orders were orders, and the second manned Gemini mission went into space with the simple, sterile name of Gemini 4.

The flight became one of the great steps in the history of man in space. The future flights to the moon required expertise in extravehicular flight—a man moving outside his spacecraft. NASA had not planned its first space walk until later in the year on Gemini 6 but, stung by the exhilarating performance of Leonov, NASA tightened its schedule, examined the risks and benefits, and Slayton informed space rookies McDivitt and White that the flight of the Gemini 4 would feature America's first space walk.

Everyone shifted into high gear. McDivitt and White were excellent choices as a crew. They were long-time close friends and former classmates at the University of Michigan, and they went into an immediate, crash training course. Ed White would be the man to "step outside."

Six weeks after Grissom had munched his corned beef sandwich, on June 3, 1965, Gemini 4 was in orbit, and on the fourth turn around the planet, White strapped an eight-pound pack of emergency oxygen to his chest. He attached a gold-tinted visor to his spacesuit helmet to protect him against the fierce sun glare. His suit had twenty-one layers of protective material that would insulate him from 250-degree temperatures when he was directly exposed to solar radiation and 150-degrees-below zero temperatures when he drifted into the spacecraft's shadow. He checked his twenty-five-foot tether, the ultimate of leashes, which would provide him with a steady flow of oxygen, a communications link to McDivitt, and assure he would not drift away.

High over Australia, the astronauts began depressurizing their cabin. Unlike Voshkod and its cumbersome airlock, Gemini had fighter plane-like cockpit hatches and, when the hatch on White's side was swung open, both men would be fully exposed to space vacuum.

Over the Pacific Ocean, White opened his hatch. The earth rolled by beneath them at three hundred miles a minute. Between Hawaii and Mexico, he gingerly eased into the hard vacuum of space.

It was an incredible, marvelous, exhilarating leap into the future. Ed gripped a hand-held gun, armed with pressurized oxygen, which he used to propel himself to the limits of the tether. With each spurt of pressure his body followed the ancient laws of Newton and moved in a direction exactly opposite.

That was the technical side. The human element was absolutely grand as the excited, buoyant Ed White somersaulted, floated lazily on his back, pirouetted, stood grinning like a kid on Gemini's titanium hull. He could "fly" a distance of twenty-five feet in any direction from the spacecraft, and he made the most of the reach of his golden tether. A touch of unreality entered the entrancing scenes when a thermal glove White had left on his seat drifted slowly up and away from Gemini into its own orbit, one of the strangest satellites, launched by itself.

The two astronauts were so intent on what was happening and White was in such euphoria that the twelve minutes scheduled for the space walk passed quickly. It was time for White to get back inside while they were still in bright sunlight. Gus Grissom was CapCom in Houston (this was the first operational flight for the new center), and he knew that the exhilaration White was showing could be dangerous. It was akin to "raptures of the deep" that scuba divers face, or that pilots find clouding their judgment at high altitudes.

"Gemini Four," Gus called in a stern voice, "*get back in*." McDivitt called to White, still frolicking outside. "They want you to get back in *now*."

Ed White didn't want to return to the cockpit. "This is fun!" he said exuberantly. "I don't want to come back in, but I'm coming."

As he moved toward the hatch, he found that maneuvering along a spacecraft is easier said than done. Without handholds or footholds, White found the return slow and difficult. He needed several extra minutes to work his way back into the cabin. McDivitt helped pull him down into his seat, he strapped in, and they closed the hatch and pressurized the ship. White had been outside twenty-one minutes.

"There was very little sensation of speed," he reported to Mission Control. "The view from up here is something spectacular. I could see much greater detail than I can from an aircraft flying at forty thousand feet. . . . I could see the outlines of cities, roads. I could see the wakes of ships at sea."

NASA was wildly enthusiastic about the mission. Medical teams confirmed the crew was healthy. Dr. Charles Berry, the astronauts' flight surgeon, stated: "It was far, far better than anything we could have expected. We've knocked down an awful lot of straw men. We had been told we would have an unconscious astronaut after four days in weightlessness. Well, they're not. We were told that the astronaut would experience vertigo, disorientation, when he stepped out of the spacecraft. We hit that one on the head."

The Project Gemini team would dispel many, many more myths and unknowns in the months to come as one by one the techniques needed for Apollo were mastered and perfected. There were many problems along the way and some near-tragic happenings. But Gemini would get the job done.

Gemini 5 stretched long-duration flight to eight days and Gemini 7 to fourteen. After spending two weeks in space with Frank Borman in a cramped spacecraft, Gemini 7 crew man Jim Lovell remarked, "It was like spending fourteen days in a men's room."

The tediousness of their mission was broken on Day Eleven when they received a welcome visitor from earth.

Steering by the constellation Orion and following the command signals of their onboard radar, Gemini 6 astronauts Wally Schirra and Tom Stafford tracked down and caught their Gemini 7 target in the first ever rendezvous of two manned spaceships.

"We've got company," Lovell reported as he watched Schirra maneuver the final few yards.

"There's a lot of traffic up here," Schirra responded.

"Call a policeman," suggested Borman.

For five hours the two Gemini spacecraft flew together in formation, doing fly-arounds and circling each other in lazy pirouettes. Schirra reported he closed to a distance of six to eight inches between the two craft, backed off, and came in again.

Schirra and Stafford returned to earth the next day. Borman and Lovell came home two days later after spending more time in space than all the Soviet cosmonauts who'd ever left the earth. When they walked across the flight deck of the recovery carrier, they were a bit gimpy-legged after fourteen days in zero gravity, but doctors pronounced them in amazingly good physical shape.

The rendezvous between Gemini 6 and 7 had not been in NASA's original plans. Schirra and Stafford were originally to have chased down and docked with an unmanned Agena satellite. But the Agena blew up on its way to orbit, and agency officials came up with the ingenious plan to launch Gemini 7 first and send Gemini 6 in pursuit.

Another milestone reached on the highway to the moon. But linking two ships in orbit still had to be proven. That job now fell to the Gemini 8 crew of Neil Armstrong and Dave Scott. Their Agena target satellite, boosted by an Atlas rocket and its own engine, shot into orbit on March 16, 1966, and they blasted off in pursuit ninety minutes later on what would become Gemini's most harrowing mission.

Using the rendezvous lessons of Gemini 6, Armstrong and Scott caught up with the waiting Agena in only five hours. For more than a half hour, commander Armstrong circled and inspected the twenty-six-foot-long satellite to confirm its stability, then with consummate care nudged Gemini 8's nose into a docking collar mounted on Agena. Clamps, electric motors, and connections clicked home, and the two craft were now as one.

"Flight, we are docked," Armstrong announced to Mission Control. "It's really a smoothie. No noticeable oscillations at all."

Next would come the powered-up maneuvers of the docked vehicles, using the Agena engine to fly to a higher orbit and to shift the orbital plane. It was not to be. Over China and temporarily out of radio contact with ground stations, Scott studied the movement of their Gemini/Agena combination. To him the stability they found early in docking was fraying noticeably. "Neil, we're in a bank," he said. The words were barely spoken when the heavy, long spacecraft "took off in roll and yaw," rolling around like a log in water, while the nose began to swing wildly.

In moments Armstrong and Scott had been thrust from a smooth flight into a struggle to survive what had swiftly and unexpectedly become the first real emergency in manned space flight.

They were terrifyingly alone, 185 miles out in space, out of touch with Mission Control and gyrating wildly with a powerful rocket loaded with four thousand pounds of deadly, volatile fuel. The Gemini/Agena had become literally a twisting, turning bomb waiting for the first chance to turn into a searing fireball.

Faster than the two men could believe, roll and yaw became even more extreme, an orbital equivalent of severe, uncontrolled turbulence in flight.

Armstrong, who'd flown the X-15 rocket plane, wrestled with the controls. Briefly he seemed to be regaining mastery of the ship. Then the Gemini/Agena went berserk. "The trouble uncovered itself again," Armstrong said much later, reviewing the events, "and the rates of tumbling increased to a point where we felt the integrity of the combination . . . was in jeopardy." The commander never let up fighting for control. He reduced the spin to a point that it was safe to break the Gemini free of the Agena. With a bang the ships disengaged, and the astronauts were astonished to find themselves spinning even faster. The seven-thousand-pound spacecraft was now whirling at better than a complete revolution every second.

They had thought the problem was with the Agena. But the fault lay with their own Gemini. One of the craft's sixteen maneuvering thrusters had stuck open, spewing fuel into space and imparting the wild spin. Unless they stopped it, regaining control would be impossible. The severe motions were on the edge of destroying the ship or whirling the men so swiftly they would soon be unconscious.

Far out in the Pacific, the *Coastal Sentry Queen*, a Gemini tracking ship, heard the electrifying message from Armstrong, fighting for both breath and spacecraft control. "We have a serious problem here. . . . We're tumbling end over end up here. We've disengaged from the Agena." The calm in his voice could hardly be believed against the battering they were taking.

Mission Control now had the bad news. How far was the Agena from the Gemini? Could the two smash into each other and doom men and ships? Were the two pilots on the edge of nausea or vertigo or both, or were they only seconds away from being hammered physically to the point where they could no longer function?

"It's rolling, and we can't turn anything off," Armstrong reported. Scott knew they were almost at the point that "we had to do something, or else it would be too late to do anything."

Armstrong threw away the book and decided to use the nose rocket thrusters normally used for reentry control. One by one he shut down the fifteen other maneuvering thrusters, immediately switched on the reentry rockets, and kept blasting away until slowly he regained control. He raced against time, hoping the reentry thrusters would function until the stuck thruster spewed all its fuel.

It took a half hour, but Armstrong brought the Gemini 8 under control. The old rules came back into effect, and the rules dictated that once reentry thrusters had been fired, for any reason, the astronauts were to return to earth as soon as possible. This was because continued use would deplete the fuel supply of the only system available for controlling the craft during the critical reentry.

But the Gemini 8 was now far from any main recovery area. Mission Control said to hell with standard procedure and ordered Armstrong and Scott to set up reentry for an emergency landing zone in the western Pacific. High over the African Congo, in darkness, retros fired to start a thirty-two-minute dash through the atmosphere.

During their darkened plunge, there were no tracking stations to aid them in position reports. Incredibly, the astronauts performed the reentry maneuver skillfully and on the mark. They splashed down 480 miles east of Okinawa after a mission lasting just under eleven hours. Soon an Air Force rescue plane roared overhead to disgorge para-rescue teams. Three hours more and the spacemen were safe aboard a Navy destroyer.

Examination of the spacecraft determined that the gremlin that had sent them spinning had been an unexplained flash of electricity that had jammed open the Number 8 thruster at full power.

Chris Kraft turned in a hair-raising report: "The spin rate was up as high as 550 degrees per second. That's about the rate at which you begin to lose consciousness or the capability to operate. Neil Armstrong realized they were in very serious trouble, and he took all the power off the Gemini to try to stop the spin, and then he figured the only way to recover was to activate the reentry attitude-control system. That was truly a fantastic recovery by a human being under such circumstances and really proved why we have test pilots in those ships. Had it not been for that good flying, we probably would have lost that crew."

Rendezvous and docking had now been demonstrated but not yet perfected. That was done on the final three Gemini missions—10, 11, and 12—when the crews chased down and linked up with Agena satellites, using their engines to maneuver through the skies. During the trackdowns they rehearsed all the critical moves Apollo moon walkers would have to make when they fired off the lunar surface to fly to a reunion with the mother ship in lunar orbit.

Space walking, surprisingly, became the hardest Gemini nut to crack. NASA should have been been alerted when Ed White had trouble reentering his capsule after his pioneering stroll. But because of the earlier ease he had had cavorting jubilantly from Gemini 4, mission planners were fully confident that space walking was a piece of cake.

A year had passed since White had taken his walk, and Gemini's Gene Cernan was tapped to be the second American to step into the void of space. So confident were the planners that they installed in the back of the Gemini 9's equipment module an AMU—Astronaut Maneuvering Unit—which Cernan would don as if it were some bejeweled Buck Rogers jet pack and use its thrusters for some fancy flying in space. The Air Force, which developed the jet pack, trusted its reliability and told NASA: "Just send him out with the backpack. He won't need that damn tether. All it will do is tangle him up."

NASA brushed off the Air Force, cited safety reasons, and said Cernan would test the gear using a tether 125 feet in length.

For his initial step outside the Gemini 9, once he and Tom Stafford were in space, Cernan remained attached to a twenty-five-foot lifeline. He looked forward to the freedom experienced by Ed White. He didn't find it. White had gone outside just to skip and flit about for a few minutes. Cernan was charged with specific duties, and he was to remain outside his ship for more than two hours. He discovered the two missions were vastly different, and he learned quickly he was as clumsy as a sloth climbing a greased pole.

To don the AMU, Cernan had to work his way to the equipment storage at the rear of his spacecraft. The AMU and the long tether awaited him there. He started back gingerly. He didn't fly and he didn't float, and he didn't have anything that faintly resembled a glorious experience. He did everything he could to clamp his gloved hands on something that would secure him to his ship. Without proper handholds, he kept slipping off the sleek hull of the vehicle and had to fight his way back. He could very much have used a jet gun like the one that propelled White about. A fifteen-foot traverse he had planned to last a few minutes took him nearly an hour of exhausting, frustrating, clumsy struggle.

But he made it. "It's a strange world out here! Whew!" he radioed Stafford. "Take a rest," Tom told him. Cernan didn't need that advice. He was grateful just to hang onto some equipment in the rear of the Gemini. After a brief rest he began to try to maneuver his suited body into the AMU, which resembled a hiker's backpack that would have fit comfortably on the frame of a grizzly bear.

Trouble remained his constant companion. The job demanded more than just slipping into simple straps. He needed to make electrical connections, and he found that every move was more time-consuming that he had counted on. When he seemed to have a job under control, he floated helplessly from the spacecraft. There was no way he could maintain a solid body position. There were a few footholds and handholds there, but they were woefully inadequate. He needed positions that would allow him to use leverage. Soon he was severely overworking his own chest pack, which circulated oxygen through his suit and also removed excessive moisture from his body. He perspired. Fog collected inside his helmet visor and froze, and he endured excessive heat, perspiration, and ice all at the same time. He was barely able to see through the visor, a potentially lethal situation for a man turning like a bloated rag doll in vacuum several feet from the security of his spaceship cabin.

Stafford called Mission Control. "We've got problems. Gene is fogging up real bad." Then he spoke directly to Cernan to check on his progress.

"I'm really fogged up, Tom," Gene told him. Stafford didn't like the sound of Cernan's voice, and he again called the team on the ground. "The pilot's visor is fogged over. Communications are very poor. He sounds like a large gargle. If the situation doesn't improve—" Stafford suddenly went quiet, then came back on line with his voice strained. "It's no go on the AMU! The pilot is fogged up completely!"

"We confirm, no go," Mission Control responded. Cernan's only interest now was to get the hell back inside the ship somehow. He began his return carefully, hand over hand, slipping, fighting every inch of the way, able to see only through small sections of his visor where his breath had melted away the frost. Ground-control doctors measured his heartbeat at 180 per minute. "He's in trouble," one said quietly, unnecessarily.

Tom Stafford suddenly felt an icy chill as he recalled a very private conversation he had had with Deke Slayton before launch. "Look, Tom, what Gene's going to do out there is pretty risky. If he's outside and he's in trouble and, if for some reason the spacecraft is in trouble, short of fuel, or something, and well, if there's no chance of getting him back in . . . " Deke's voice faltered, then picked up. "There's nothing written on this. No mission rule, but, well, I think you understand. If that kind of thing happens, *you've got to cut him loose.*"

Stafford could hardly believe what he was hearing. But Deke was dead serious. "If it's a hopeless situation, you've got to save at least one crew member and the spacecraft."

Tom never did learn what he would have done at such a moment. Cast off a fellow pilot to certain death? He couldn't—

But there was Cernan struggling back into the cabin, and he finally made it. The third man to walk in space, the second American, had been outside a record two hours and nine minutes, and all of it had been pure hell.

After the Gemini 9 returned, NASA leaders ordered future AMU missions scrubbed until space walking could be accomplished with a level of reliability and non-exhaustion. But problems persisted. Both Mike Collins on the Gemini 10 and Dick Gordon on the Gemini 11 experienced troubles maneuvering away from their craft. Collins, who used a jet gun to move over to a nearby Agena satellite, reported: "I found that the lack of a handhold is a big impediment. I could hang onto the Agena, but I could not get around to the other side where I wanted to go. That is indeed a problem." Gordon, like Cernan, became hot and sweaty and his visor fogged. "I'm pooped," he said simply after cutting his walk short.

All the many successes and extraordinary accomplishments of the Gemini still left NASA's leadership in a quandary. The question voiced in various expressions cut to the heart of the problem: "How can we send men to the moon, no matter how well they fly their ships, if they're pretty helpless when they get there? We've racked up rendezvous, docking, double-teaming the spacecraft, starting, stopping, and restarting engines; we've done all that. But these guys simply cannot work outside their ships without exhausting themselves and risking both their lives and their mission. We've got to come up with a solution, and quick!"

One manned Gemini mission remained on the flight schedule. Veteran Jim Lovell would command the Gemini 12, and his space-walking pilot would be Buzz Aldrin, who built on the experience of the others to address all problems with incredible depth and finesse. He took along with him on his mission special devices like a wrist tether and a tether constructed in the same fashion as one that window washers use to keep from falling off ledges. The ruby slippers of Dorothy of Oz couldn't compare with the "golden slippers" Aldrin wore in space—foot restraints, resembling wooden Dutch shoes, that he could bolt to a work station in the Gemini equipment bay. One of his neatest tricks was to bring along portable handholds he could slap onto either the Gemini or the Agena to keep his body under control. A variety of space tools went into his pressure suit to go along with him once he exited the cabin.

On November 11, 1966, the Gemini 12, the last of its breed, left earth and captured its Agena quarry. Then Buzz Aldrin, once and for all, banished the gremlins of space walking. He proved so much a master at it that he seemed more to be taking a leisurely stroll through space than attacking the problems that had frustrated, endangered, and maddened three previous astronauts and brought grave doubts to NASA leadership about the possible success of the manned lunar program.

Aldrin moved down the nose of the Gemini to the Agena like a weightless swimmer, working his way almost effortlessly along a six-foot rail he had locked into place once he was outside. Next came looping the end of a hundred-foot line from the Agena to the Gemini for a later experiment, the job that had left Dick Gordon in a sweatbox of exhaustion. Aldrin didn't show even a hint of heavy breathing, perspiration, or an increased heartbeat. When he spoke, his voice was crisp, sharp, clear. What he did seemed incredibly easy, but it was the direct result of his incisive study of the problems and the equipment he'd brought from earth. He also made sure to move in carefully timed periods, resting between major tasks, and keeping his physical exertion to a minimum. When he reached the work station in the rear of the Gemini, he mounted his feet and secured his body to the ship with the waist tether.

He hooked different equipment to the ship, dismounted other equipment, shifted them about, and reattached them. He used a unique "space wrench" to loosen and tighten bolts with effortless skill. He snipped wires, reconnected wires, and connected a series of tubes.

Mission Control hung on every word exchanged between the two astronauts high above earth. "Buzz, how do those slippers work?"

Aldrin's enthusiastic voice came back like music. "They're great. Great! I don't have any trouble positioning my body at all."

And so it went, a monumental achievement right at the end of the Gemini program. Project planners had reached all the way to the last inch with one crucial problem still unsolved, and the man named Aldrin had whipped it in spectacular fashion on the final flight.

Project Gemini was history.

It had made the next giant step forward with ten manned flights in twenty months. It had demonstrated every critical procedure the moonbound astronauts would need to make it all the way to their destination, a quarter million miles distant, and back.

And Gemini had come from behind in the increasingly fierce competition with the Soviets and had given America the lead. The best of the Russians to this point had been clumsy plodding in orbit compared with the free-wheeling antics of the Gemini crews and their spacecraft.

No Russian had yet mastered rendezvous. They had never docked one ship to another. They had conducted one space walk, but were far from even approaching the mastery and skills Aldrin had given the American program.

Apollo was next. This was *the* program. It would send man to another world.

It would enfold the future and wrap it like a flag to be placed in the hands of the nation and the world.

But America would pay a price.

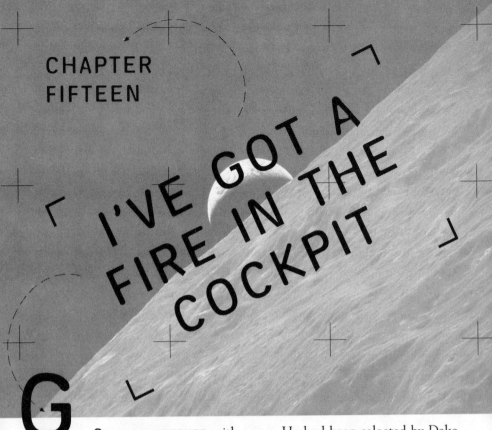

CHAPTER FIFTEEN

I'VE GOT A FIRE IN THE COCKPIT

GUS GRISSOM SEETHED with anger. He had been selected by Deke Slayton and Alan Shepard to command the first manned mission of the new three-man Apollo spacecraft, and nothing about the assignment was going well. The Apollo 1 launch had been scheduled for February 21, 1967, and now, in early December, he was still wrestling with an extraordinarily complex machine that hadn't yet reached final design acceptance and that revealed new problems at every turn. Grissom and his crew members, Ed White and Roger Chaffee, had been kicked from one scheduled launch date to another, and after months of delays their patience had been whittled down to raw temper.

There would be no flight, not the first chance of moving Apollo 1 an inch from the launch pad at Cape Canaveral until the endless glitches and failures and the malfunctioning systems of the spacecraft had been addressed and remedied, until Apollo 1 had been officially accepted as flight-worthy. Gus had begun to feel the damn ship would never leave the ground. Apollo 1 was an expensive, heavy, complicated spacecraft with thousands of systems large and small that needed to function and intermesh perfectly for the mission to succeed.

To assist the astronauts in their training for the mission, a nonflying mockup of the Apollo 1, a flight simulator, had been faithfully reproduced down to every knob, dial, handle, and control button. Grissom's wrath this day was focused on the flight simulator and on the man in charge of bringing the simulator up to safety and performance standards of the actual spacecraft. That man, Riley McCafferty, squirmed uncomfortably under the heat from Grissom.

The crew had just completed an inspection of the actual Apollo 1, or Spacecraft 012 as the capsule was known to its manufacturers, North American Aviation of Downey, California, where engineers were working around the clock, responding to NASA requests for design changes and other modifications. Within minutes of strapping himself into the simulator, Grissom was in a barely controlled rage. The simulator and Spacecraft 012 now differed in so many significant ways that Gus felt he was wasting his time. He summoned McCafferty, and he didn't mince words.

"Damn it, Riley," the Apollo 1 commander said scathingly. "This simulator is worthless! Why isn't it up to date? Too many things don't match the spacecraft. It's more trouble than it's worth."

"But—"

"Riley, it's a piece of crap!" Gus shouted as he and the other crew members climbed out of the simulator.

Riley explained that the Downey engineers had made hundreds of changes to 012, and it took time to incorporate them into the simulator.

Gus just stared at him.

The next day, his anger burning in him like hot coals, disgusted with the contractor team, Gus let them all know how he felt. The irate commander hung a lemon on the crew hatch of McCafferty's simulator.

It was impossible to keep the press from picking up on the constant delays and the tension between the Apollo crew and NASA staff. Word leaked out that Apollo was beset by countless problems. When a reporter questioned him about the problems, Gus offered a burst of harsh, unfunny laughter and tried to respond honestly without inflaming too much controversy that would endanger the Apollo program. "I've got misgivings. We've had problems before, but these have been coming in bushelfuls. Frankly, I think this mission has a pretty damn slim chance of flying its full fourteen days."

The problems with the simulator and preparing for a flight that had been repeatedly postponed weren't the only sources of pressure on Grissom. As mission commander for Apollo 1, he became the lightning rod for complaints and caustic remarks that the Apollo program was complacent and moving too slowly toward its stated goal of sending an American to the moon.

In late 1966, barely weeks after the spectacularly successful conclusion of the Gemini project, NASA's future adventures in space had never looked so promising. There were those in the agency who stated flatly that a man would be on the moon as early as 1968, a full year before the deadline set by the late President John F. Kennedy. In moving quickly to get Apollo off the ground, there were those who said NASA was responding to intense political pressure. Lyndon Johnson was a president besieged from all sides. Racial unrest scourged American cities from coast to coast. Massive protests against the unpopular war in Vietnam were fueled by rising death tolls, and Johnson was coming under additional fire for cooperating with a corrupt South Vietnamese government.

The year 1967 rolled in like a political garbage truck with its tires burning. This was the year Johnson would have to begin his fight for reelection if he was to remain in the White House. He needed a public relations miracle to recapture the hearts and minds of an electorate no longer in his grasp, and the prize Johnson wanted to offer America was Apollo. If he could get NASA to boost that program ahead of its own schedule and get Americans safely to the moon and back during the election-year struggle, his political fortunes would turn around. LBJ passed the word to his private staff that he would "very much appreciate" NASA getting off dead center and flying again. The message to NASA was clear: "Get off your asses, gentlemen."

With the technical staff working overtime, quickened by scathing insults from Gus Grissom and other leaders of the Apollo team, Spacecraft 012 received its final inspection and was moved to Launch Pad 34 atop its Saturn 1B rocket the first week of January. Additional tests, however, were required while the spacecraft was on the pad. One test considered essential for a thorough review of the ship's operating systems involved pressurizing the unmanned ship with 100 percent oxygen. When the test had been completed to everyone's satisfaction, a final test would be run with the three-man crew aboard the craft, suited up and with all electrical and communications live. Under the pressure of time, however, NASA decided to skip the unmanned test and go directly to the "full dress rehearsal" with 100 percent oxygen, the crew in position, and the spacecraft hatch sealed just as tightly as it would be for a launch. NASA posted Friday, January 27, for the all-up test.

On the morning of the test, Grissom, White, and Chaffee shared an early lunch in their crew quarters with Apollo spacecraft manager Joe Shea, Deke Slayton, and Wally Schirra, commander of the back-up crew for Apollo 1. "There had been a lot of communications problems with the spacecraft," Slayton recalled afterward, "and Gus wanted to hassle Joe a bit, chew on his ass about them."

Shea didn't need any hassling. He knew the Apollo craft had problems. He was a lightning rod, catching most of the heat, the man in NASA who had to ride North American Aviation to do better. Just six weeks earlier, in a frank and revealing news conference, he had admitted that, yes, there were problems with Apollo, that more than twenty thousand failures of one sort or another had plagued the program. Most were trivial, but there were enough important faults to cause deep concern. Said Shea, "We hope to God there is no safety involved in the things that slip through."

"Suddenly," Deke recalled, "Gus suggested to Shea, 'Goddamn it, Joe, if you think the sonofabitch is working, why don't you get your ass in the cabin with us and see what it sounds like?'

"Joe didn't think much of the idea. It didn't make sense to him, and he told Gus he'd monitor the test from the blockhouse. Then, after thinking about it for a while, I said, hell, maybe it would be a good idea if I got in there with the other three guys to see firsthand how the system really worked. In fact, that was my intent when we left for the launch pad.

"But when we got there and started loading the guys into the spacecraft, it became obvious this was a dumbshit thing to do because whatever communications hookup they gave me wouldn't be the same that Gus and the guys were using. So I opted at the last minute not to get in there. Instead I went to the blockhouse."

Just before Gus slipped into the Apollo, Wally Schirra held him back a moment. Wally hated the hatch of the Apollo. As far as he was concerned, it should have been built with a quick-opening explosive mechanism that operated swiftly like those in the Mercury. For Schirra, capsule 012 had an ominous feel. The hatch was double-hulled. It had to be opened manually, and to escape in an emergency it was necessary to open both hulls of the hatch and then release a third hatch that was part of a protective metal shield that insulated Apollo during liftoff. Engineers had designed it that way to avoid an accidental loss of the hatch en route to the moon or during the punishing reentry, when Apollo would come blazing back to earth at more than twenty-four thousand miles per hour. This design came about, ironically, after Grissom's Mercury capsule had sunk when the hatch inexplicably blew after he splashed down.

"Listen to me, Gus," Wally told his friend. "It'll take you a minimum of ninety seconds to get all those hatches open. If you have a problem, even a communications problem, get out of the cabin until the problem is cleared up. Got it?"

"Got it." Gus heard the warning, then filed it away somewhere in the back of his mind.

We're ready to get with the count." The words came through from the blockhouse and from crews working the huge steel gantry surrounding the 224-foot-tall combination of Saturn 1B rocket and Apollo 1. "Everybody get with it."

The teams initiated the countdown, rehearsing a dry run—an exercise called a "plugs-out" test. When the simulated countdown reached zero, Apollo 1 would move to its next critical phase. A forest of electrical, environmental, and ground check-out cables servicing the rocket-spacecraft stack were to separate to verify that both the booster and the spacecraft command and service modules could function on their own internal power.

That was the program, the dry run for everything except fueling and the actual launching. In full spacesuits the three men strapped themselves into the capsule, Grissom on the left, White center, and Chaffee on the right. Technicians closed and sealed the spacecraft hatch. The three were now plugged tightly within a ship from which they could not possibly remove themselves for a minimum of ninety seconds to two minutes.

The count rolled on. "Confirm hatch closed and sealed," came the call. Pure oxygen was then pumped into the cabin until the pressure gauges read 16.7 pounds per square inch, two pounds higher than the normal sea-level pressure of the nitrogen-oxygen atmosphere outside the ship.

NASA's test called for a capsule environment free of any contaminants. The extremely high oxygen content and cabin pressure within the ship would create this environment, and served no other purpose. Once the capsule was in space, cabin pressure would be reduced to a level of five pounds per square inch, less than one-third the thick oxygen atmosphere in the ship at the time of launch. Engineers had vetoed an earth-like nitrogen-oxygen mixture for Apollo because of the extra complexity of handling two gases, the extra plumbing that would be required, and the extra weight it would add.

The test was monitored from three separate outside locations. In the White Room, an enclosed platform on Level 8, the spacecraft level of the gantry, five technicians from North American Aviation were on immediate standby for any emergency should the astronauts need their assistance. They would never forget that day.

An elaborate blockhouse sixteen hundred feet from the rocket held a crew to run down the systems checks of the booster. Five miles away, in an operations building, electronic tendrils reached underground to the spaceship under the guidance of the ACE (Automatic Check-out Equipment) Control Room. Only one person at each of these two control sites was permitted to talk directly to the astronauts. Skip Chauvin, the electronics genius who served as spacecraft test conductor, had set himself up in the ACE facility. In the blockhouse it was Stuart Roosa, a recent addition to the astronaut corps who was cutting his teeth as a communicator with a flight crew. Sitting close to Roosa were Deke Slayton, Joe Shea, and Rocco Petrone, director of NASA launch operations. As a U.S. Army major, Petrone had worked with the von Braun rocket team at Huntsville. Enamored with rocketry, he had left the military and joined the space agency team preparing to send men to the moon.

Small problems appeared in the countdown. The astronauts called in, annoyed and obviously disturbed with a sour odor that filled their cabin. It seemed to come from the unit that controlled their cabin environment. Grissom didn't like it. Moments later, the sour smell was sucked away by interior controls. Then it returned. Angry muttering was heard from inside the Apollo. Again the cabin systems purged the foul odor.

It didn't take long for Grissom to wish he'd hung a lemon outside the spacecraft as well as the simulator. Despite the long and tedious months of problems, checks, repairs, and more repairs, and the assurances that the ship was now flight-ready, the communications system aboard Apollo started unraveling. The system had angered Grissom for months, and now its problems choked the lines between Apollo and its test control sites.

Static crackled painfully in pressure suit helmets. Something electrical was screwing up badly. The static went from annoying to unacceptable. Often the crew couldn't understand or even hear any calls from Chauvin or Roosa. But Gus's voice carried over the lines with no question of his mood.

"If I can't talk with you only five miles away," he snapped, "how can we talk to you from the moon?"

"Hold," came the call from Chauvin. Again and again the static, clearing up and then coming back like gravel tumbling down a metal chute, snarled voice calls. The countdown stretched on for hours.

Joe Shea began to pace back and forth in the blockhouse. He had a large team waiting for him in Houston, and he had to leave. He wrote down cryptic notes about whose ass he would kick over the communications lines foul-up and left the blockhouse to catch his flight to Texas.

Slowly, haltingly, stumbling from one glitch to another, the count went down to T-minus 10 minutes.

"Damn it, I can't even shut off my microphone now!" Gus complained. There was a promise of terrible trouble with that call. Communication over those lines was electric. The electrical system was a mess that kept getting worse.

Immediately Chauvin halted the count, standing by to define the problems and get the system back on line. It was already after six o'clock. The crew had been in the cabin more than five hours. Someone commented that John Glenn had flown three times around the world in less than five hours. Tempers were short.

An engineer contacted Chauvin. "Let's cancel out today. This could go on forever. We're better off if we shut down and do the full test again."

Chauvin shook his head. Time was more important. It was becoming damned critical with all the pressures mounting to get this ship on its way into orbit. Houston was antsy, and there was that unofficial pressure from the White House filtering through the NASA ranks. Fly the bird.

Chauvin reminded his team that communications, no matter how screwed-up, were not the primary reason for the test. He felt they were close to the "plugs-out" climax and a verification of the onboard systems and ordered the test to continue. They'd simply bypass the problem with Grissom's mike. The entire problem might amount to no more than a loose wire.

They scheduled the countdown to pick up again at 6:31 P.M.

No one saw it begin. No one knew then, or ever, just when it came to life. No one saw it, heard it, imagined it could happen, dreamed in their worst nightmare that it would happen. Or dared to contemplate what was virtually inevitable.

The catastrophe that was to engulf Apollo 1 at T-minus 10 minutes and holding had commenced hours earlier. A technician on the Saturn gantry had reviewed his check list of procedures and time lines for events to be activated. The hatch had been sealed, the astronauts secured in their couches, the spacecraft powered up. Internal cabin pressure began rising for the tight seal required for the "contamination-free" environment.

Valves opened, pure oxygen flowed into the cabin. The pressure went through changes. Ambient air of 21 percent oxygen, nearly 79 percent nitrogen, and a smattering of other normal atmospheric gases was flushed from the three-man cabin. Sensors confirmed the desired reading of 16.7 pounds per square inch of 100 percent oxygen. And the cabin, its equipment, wiring, plastic, Velcro, suits, instruments, anything and everything was soaked in pure oxygen.

If everything had functioned perfectly, the tragic events that overtook Spacecraft 012 might never have happened. But this was a ship beset by problems and one that couldn't even communicate properly with a blockhouse sixteen hundred feet distant.

This was the spacecraft that an Apollo quality-control inspector, Thomas Baron, had condemned as "sloppy and unsafe," the ship that spacecraft manager Joe Shea admitted had been plagued with more than twenty thousand failures in its construction and assembly. This was the same craft John Shinkle, Apollo program manager, castigated as missing at least "half the damn engineering work" that had been listed as completed, and that Rocco Petrone, director of launch operations, railed against as a totally unacceptable "bucket of bolts." This was the spacecraft that had been awash in a thick soup of 100 percent oxygen for more than five hours.

Pure oxygen under normal conditions is one of the most dangerous and corrosive gases known. In a short time it can corrode and transform iron and other metals into flaky garbage. Exposed to an ignition source, it is extremely flammable. It had been used in the Mercury and Gemini spacecraft without trouble, and NASA engineers had become complacent about the possibility of a fire.

For more than five hours the oxygen in the pressurized cabin of 012 had permeated the surface of everything in the cabin, everything from plastic to paper check lists, to nonmetallic insulation, to aluminum and fabric— everything. Pure oxygen ate into the material, squeezed under outer molecule layers, brought with its caress igniters waiting to be triggered.

Below the couch on which Gus Grissom lay ran bundles of wires. All kinds of wires performing all kinds of tasks. Some carried electrical current to different operating systems of 012. Others were hooked to the suits of the astronauts for medical monitoring and communications. The wires had not been brought together in sealed and protected tubing but had been laced together with plastic and other strapping. The wire was in lousy shape. It had been moved, shaken, pushed, shoved, squeezed, stepped on, and in some cases had lost its outer insulation to constant rubbing and friction. It was a mess.

Somewhere beneath the seat of the commander of Apollo 1 an open wire chafed. Insulation was torn. The wire, charged with electrical power, lay bare.

It sparked.

The spark exploded. In an instant faster than thought, the tiny flicker of electricity became a massive shock wave of flame, which fed on the oxygen-soaked environment of the pressurized capsule interior.

In the blockhouse, Deke Slayton and Rocco Petrone froze where they were, muscles stiffened, voices cut off in mid-sentence, their eyes staring in disbelief at the television monitors displaying the interior conditions of Spacecraft 012. There wasn't time to verbalize thought. Something horrifying, unbelievable, was rampaging in Apollo.

In that same moment, what had been the cabin of Apollo became an incredible whirlwind of fire raging and tearing at everything it could reach. As fast as the shock wave smashed back and forth against the three men caught helplessly inside, Slayton and Petrone knew their world would never again be the same.

The monitors and instruments were messengers from hell. As Deke flicked his eyes from one gauge to another, he saw a huge supply of oxygen flowing into the spacesuits.

The gauges showing electrical currents had gone mad. The flow surged wildly, needles flickering back and forth.

The cabin temperature gauges were all jammed hard against the upper limits of their pegs. A radar beacon died; in a split second it simply fell off-line.

Throats clinched with the awful shock. Deke, Rocco, the others with them, could only stare, wordless with disbelief. Deke's worst nightmares were alive and screaming.

On the medical monitors Ed White's pulse rate leaped crazily upward. The gauges showed sudden bursts of movement by the three men.

The first voice call was muffled and incredibly distant, as if it were coming from anywhere but a modern spaceship. But it rippled into headsets with all the impact of a physical blow.

"Fire!"

One word from Ed White.

Immediately afterward came the unmistakable deep voice of Gus Grissom in a cry to his closest friends:

"I've got a fire in the cockpit!"

Instantly afterward, Roger Chaffee's voice:

"Fire!"

Then a garbled transmission and then the final plea:

"Get us out!"

Another transmission, words no one would ever understand, a scream, and—

Silence.

In the blockhouse Deke leaped from his chair, his own voice an anguished cry. "What the hell's happening?"

Horrified eyes were glued to every monitor. One screen showed a view through the spacecraft window—Deke thought he saw a shadow moving inside. He couldn't be sure. Instantly he turned to another monitor. Muscles rigid, mouth open, he saw orange flames flickering about the area of the spacecraft hatch. He could hardly make out anything else.

The flames expanded swiftly, became a terrifying white glare. When the violent blinding flash subsided, Slayton could see more clearly. He didn't want to believe his eyes.

Hellish flames. Thick smoke boiling outward.

He grasped at hope. It might . . . maybe it was a fire in the White Room on Level 8 outside the spacecraft.

Then came the icy shock of reality. Those calls of fire, that final garbled scream—they had come from inside the Apollo.

Skip Chauvin and Stuart Roosa were trying frantically to talk with the crew. Again and again they called, desperate, their faces chalk-white.

No response.

At the moment Ed White had cried "Fire!" one man had seen more than a moving shadow. In a room near the base of the launch stand, technician Gary Probst, startled by the sudden cry, locked his gaze on a television monitor fed by a camera that zoomed in for a close-up look of the spacecraft porthole. Gary saw a partial view of the Apollo interior from over Ed White's shoulder. Flames flashed across the porthole. The next moment he saw White's gloved hands reach above his head toward the mechanism that secured the hatch. Grissom's arms came into view. The picture flickered with sudden intense motion.

Apollo 1 was a single, huge blowtorch. The polyurethane foam cushioning that spread over the spacecraft floor had absorbed so much oxygen it acted as an oxygen source, gushing upward in a wall of pure fire between the crew and the hatch that was their only escape. In those timeless seconds flames danced wildly and attacked hose lines feeding to helmets that in turn fed to noses and throats and lungs.

No human being could get that hatch open in less than ninety seconds under perfect operating conditions. These weren't perfect conditions.

Metal pipes and solder joints flowed like warm jelly, casting off even more flame as they dropped to the spacecraft flooring, bubbling and hissing like tiny lava pits, spattering molten drops of blazing metal about the cabin. The heat tore into oxygen sources, melting them open, and additional oxygen under tremendous pressure instantly transformed into screeching blowtorches of pure white fire.

The life-support system was caving in on itself, melting, bursting, collapsing. Cooling fluid spewed through the capsule. In the atmosphere of now-blazing oxygen and soaring temperature the cooling fluid poured forth like water from a hose, only it was flammable and it, too, burst into flames and sprayed its own fiery torrent across the cabin.

And yet, incredibly, the astronauts functioned as the well-drilled team they were. Immersed in flames curling along, about, under their bodies, splashing against their transparent visors, wrapping about their torsos, the astronauts attempted to perform their emergency duties.

Gus Grissom's most immediate task was to push hard against a lever that instantly would depressurize the cabin. That alone would have sent the heavy oxygen atmosphere, even engulfed in flame, gushing forth from Apollo like the afterburn of a powerful jet engine. Gus tried, but the lever itself was in flames.

Ed White twisted his body painfully, reaching over his left shoulder to pull with all the strength he had the release grip of the inner hatch. An instant later he could not see past the fire roaring about his helmet.

Chaffee performed in perhaps the most incredible manner of all. To maintain communications he had to remain in his couch, to lie there in the fire as it exploded steadily, relentlessly. He made the cry beseeching the gantry crew to get them out of the holocaust and, then, as he began another call, he reached his body across Ed White to help Gus with the hatch as the heat roared into his lungs. In his final moment a scream burst involuntarily from his seared windpipe and larynx. Before he could comprehend any further, the air in his and his crewmates' lungs was sucked out. The lives of Gus Grissom, Roger Chaffee, and Ed White had been snuffed out in eight and one-half seconds.

At that moment when the spark flared beneath Grissom's couch, there were twenty-seven men working at various levels of the gantry. This included the five in the White Room at the entry hatch to 012.

Pad leader Donald Babbitt was stunned to hear the anguished cries from the spacecraft. The words from Grissom, "I've got a fire in the cockpit!" galvanized Babbitt to instant action. His voice carried across the working level, "Get them out of there!" James Gleaves, closest to the capsule, spun about to help open the hatch. Babbitt already had rushed from his desk to slam his hand against the emergency call button to the blockhouse.

He had just enough time to hit the alarm and not a second more as a blinding sheet of flame burst from Apollo and an explosive shock wave slammed Babbitt to his knees. The five men on Level 8 staggered backward, battered by the concussion, running to escape both heat and roaring fire. They lurched and stumbled, dazed, along the swing arm. Secondary blasts pursued them as they reached the tower elevator. Here they stopped. Babbitt grasped a man wearing a headset and a mike and spun him around. Babbitt shouted: "We're on fire! I need firemen, ambulances, equipment, now!"

Struggling to breathe and choking on the fumes sweeping across the work level and the swing arm, yet without hesitation, the workmen with Babbitt grabbed every fire extinguisher they could see. If ever men ran willingly straight through the gates of hell, these were the men. Faces seared by the heat, hands singed by flames lashing out from Apollo, knowing explosives could engulf them in one last fiery crescendo, they dashed back to the spacecraft to spray the hatch and try to cool it off so they could open it with hands already shedding scorched skin.

Other men ran frantically to Level 8, wearing gas masks and carrying extinguishers. Another flaw in the system: the masks were designed to protect their wearers against fumes from a fuel spillage. They proved useless against the thick smoke. The workers stayed. Against the terrible pain, washed by waves of heat, toxic fumes, and choking smoke, they fought down the flames with their extinguishers.

The men fought as long as they could before they were blinded by fumes and smoke. Choking painfully for air to breathe, they fell back to allow others to get to the hatch. As quickly as those men were overcome, the original crews rushed back. Without a word of instruction they battled in relays to open the hatch and release the three astronauts inside.

Jim Pierce of North American Aviation immediately called the main office in Downey, California. His anguished descriptions were carried through loudspeakers in the company's conference room. Pierce's voice described what sounded like the end of the world.

"The whole thing could blow up any minute! I . . . there's fire spewing from the spacecraft . . . I can see molten metal falling away!"

Someone in the Downey office cried out, "Oh, Jesus!" His voice was barely heard over people bursting into tears.

On Level 8 the blaze was out. Babbitt's men, exhausted, some injured, finally managed to open the hatch to the spacecraft. They reeled backward as a blast of heat gushed out in equalizing pressure. Thick toxic smoke billowed forth. Nearly six minutes had passed since the first shouted cry of "Fire!" from inside 012.

As best they could, Babbitt's crew struggled to reach the astronauts. They might still be alive. There was only that hope. The rescuers tried desperately to see through the swirling ash and smoke. Blinded, they groped with their hands for survivors. Then they stopped. The bodies were unmoving.

In the blockhouse, Rocco Petrone stared, white-faced and trembling, at the televised scenes of the open hatch and the men fighting to reach the astronauts. He hit his microphone so they would hear him in the White Room. "Can you do something for the guys?" he asked.

A voice came back, choking. "N-no. . . it's too late." Rocco turned away from the television monitor, shattered. He looked at a man standing by him. "Turn off the television. Please. I don't want to look at it."

There had been a sliver, no more, of hope. Moments before Petrone was told, "It's too late," a voice on Level 8 was shouting over the radio loop,"Get a doctor out here, quick!"

Deke Slayton heard the call. You don't need a doctor for dead men. Deke grabbed two doctors standing nearby, Fred Kelly and Alan Harter, and all three ran from the blockhouse.

Deke lived a lifetime in that mad run to the launch pad. He and Gus had been close friends for years. They had hunted, fished, flown together every chance they had. All the way to the pad he was holding out that last fading glimmer of hope that maybe, just possibly, somehow miraculously, the guys could still be alive—and could have been protected by their suits. But he knew the odds were less than slim. They'd been in that fire much too long.

But he kept running, kept hoping, kept that silent screaming inside his head that *something* had kept them alive.

Suddenly he could think of only one thing, of what had happened years before when he and Gus had been in a water rescue exercise, when he had fallen off his raft and almost drowned because he had never really learned how to swim. It was Gus who swam to him. It was Gus who saved him.

Hang in there, buddy! he shouted in his head as his leg muscles drove him faster and faster to the pad. *Hang in there, Gus! Hang in there.*

Alan Shepard was in Dallas, about to make a dinner speech, when someone hurried to his side at the head table. In a hoarse whisper he told Alan that Gus, Ed, and Roger were dead, killed by a fire at the launch pad. The news hit Alan with the force of a sledgehammer.

Numbed with shock, he moved in a fog to the podium. He fought to speak, his voice a rasping, almost silent choking sound. "I . . . I have just been informed of the loss . . . the loss of my comrades . . ."

A long silence followed. Alan Shepard remembers little of what he said that night.

Deke and the doctors reached the pad, rode the elevator to Level 8, rushed along the swing arm to the White Room. The hatch had been open only moments.

What Alan Shepard could only imagine, Deke saw for himself. Saw the doctors lean into Apollo 1's open hatch, saw them pull out slowly. One turned to Deke, shaking his head. "They're gone."

Lola Marlow, secretary in the Astronaut Office at the Kennedy Space Center, was in tears, wracked with sobs, when Deke walked in minutes later.

"Every telephone in the office was ringing," Lola recalled. "The whole world was calling. Deke came in and he was shaking like a leaf. He had a cigar, but he couldn't light it. I tried to light it for him. He couldn't hold it he shook so badly. So did I . . . I never managed to light it."

Wearily, still in shock and pain, Deke phoned the Astronaut Office in Houston. Soon he was speaking with Wally Schirra, who had flown there from the Cape after the "plugs-out" test had begun. Wally had the news already, and he'd set in motion what he knew Deke was calling for him to do.

Wally had already assigned other astronauts to notify each family. They were to offer whatever help that might be needed. They reached Betty Grissom first and then Pat White.

When they reached Martha Chaffee, she was on the telephone, calling the Holiday Inn in Cocoa Beach to see if Roger was in his room.

NASA fought off the newsmen, many of whom were also close friends of the astronauts who had died. It was impossible to hide the story. The Canaveral press corps had too many sources directly on pad 34. That kind of news travels fast. NASA stayed tight-lipped until all the families were told. Then it released officially what the permanent press at the Cape already knew.

As midnight approached, after Deke Slayton and Chuck Friedlander had answered a thousand phone calls, had done all they could, the two men sat staring at the walls. Friedlander was the chief of the Astronaut Office at the Cape, and he moved to a locked cabinet. Alcohol wasn't permitted on the government installation, but Friedlander returned with a full fifth of scotch, and he and Deke sipped the whisky until sunrise.

For Deke it was the beginning of the time needed to cover the wounds inside him. Finally, the man who'd fought in combat and flown the new, tricky swept-wing jets in the high desert skies was able to find through reflection answers to questions that threatened never to let him be.

His first look inside Apollo 1?

"It was devastating," Deke related with that terrible sadness that never quite goes away from such an experience. "Everything inside was burned, black with ash. It was a death chamber. The crew had obviously been trying to get out. The three bodies were piled in front of the seal in the hatch. Ed White was on the bottom, and Gus and Roger were crumpled on top of him. The suits had protected them from the flames. None of them had any physical burns of any consequence. It was all that goddamn shit in the environmental control systems that got them, asphyxiated them. I had to turn my head away."

It was over.

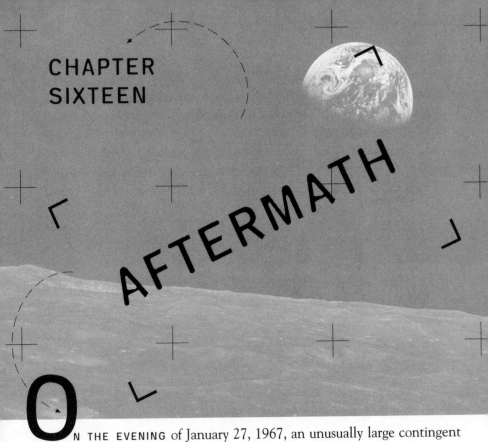

CHAPTER SIXTEEN

AFTERMATH

ON THE EVENING of January 27, 1967, an unusually large contingent of NASA officials and supervisors crowded with other government representatives into the International Club in Washington, D.C., to hear an address by Vice President Hubert Humphrey.

Earlier in the day, Anatoly Dobrynin, Soviet ambassador to the United States, had signed the Russian half of a new treaty with the U.S., an agreement that stipulated that henceforth space would be used only for peaceful purposes. Applause from both sides greeted the signing. Toasts were offered, and Lyndon Johnson seized the opportunity to wax enthusiastic about America's marvelous progress in its efforts to land men on the moon before the end of the decade.

Dobrynin smiled, fully aware of the new Russian program that would leapfrog the Americans in the race to the moon. As he listened to Johnson, he knew that several Russian spacecraft were already in their long pre-launch countdown and would slice several years from Russia's own manned lunar effort.

The group left the White House to reassemble for dinner at the International Club. Among the officials present were Dr. Wernher von Braun, who by now had been transferred from the Army missile team to NASA, Bob Gilruth, Jim Webb, and a large contingent of top aerospace industry officials. Despite the agreements signed that day, many of the Americans found it difficult to refrain from tweaking Russian sensitivities. Gemini had soared so high and far into the future that some Americans snidely commented that while the United States would be reaching out through the solar system, the Russians would be performing repetitive orbital yo-yos over the earth.

No comments could ruffle Dobrynin, who before leaving his homeland had been updated on the scheduled flights of the new Soyuz 1 and Soyuz 2 spacecraft, both technically superior to the older Vostok and Voshkod machines, and as capable as Gemini in its ability to rendezvous, raise and lower orbits, and change orbital planes. Furthermore their design would allow for more sophisticated space-walking activities. With a few changes Soyuz would be capable of zipping once around the back side of the moon and returning to earth with a single cosmonaut on board.

If the schedule held, in only three months the Russians would launch the two Soyuz machines into earth orbit. They would be directed by advanced systems, radar, and robotic controls, and travel to a rendezvous point while in each ship a cosmonaut would monitor his spacecraft's progress. Then, several hundred feet from each other, the pilots would take over manually and maneuver vessels to a docking. The two cosmonauts would then engage in a space-walking adventure. Each cosmonaut would leave his Soyuz and "hand walk" to the other's ship and occupy that craft. They would then separate, perform additional maneuvers, and then return to earth.

If all went as planned, Soyuz 1 would depart the world precisely at 3:35 A.M. Moscow time on April 23, and Soyuz 2 would follow exactly one day later. It would be a tremendous accomplishment for the motherland, Dobrynin mused, and might even quiet the prattling, boastful Americans.

Dobrynin turned to look at Vice President Humphrey, surrounded by the congressional space delegation. Most of them had never seen a rocket leave the earth.

Dobrynin's eyes suddenly narrowed as he saw the president of North American Aviation, Lee Atwood, in an excited conversation with another man. The ambassador watched with growing interest, first as Atwood left the main room of the club, then as he returned visibly upset.

As he passed Bob Gilruth, Atwood whirled suddenly to grab Gilruth's hand. A brief conversation ensued; Gilruth left the room swiftly. Dobrynin kept his eyes on Atwood, who was now leaning close to Jim Webb, his agitation greater than ever. Whatever he had to say to Webb turned the top NASA man visibly pale.

Dobrynin sensed something very important was happening, and the faces he saw told him that whatever had happened must be very bad indeed. Wernher von Braun left the room on the run. Other executives dashed to telephones. The buzz of excited, almost frantic conversation grew louder.

Dobrynin could hardly believe his eyes. Some of the men were crying!

In the midst of what was obviously a grim tragedy, club waiters started to serve dinner. Hardly anyone paid attention. The ambassador watched a man rush to the side of Webb; the NASA administrator strode rapidly away. He had already received something of terrible import from Lee Atwood. What could this new message be?

Webb took a telephone call from Julian Scheer, his deputy director for public affairs. "Jim, there's been an accident at the Cape. A fire—"

"I know," Webb broke in. "What's happened since?"

Scheer hit him between the eyes. There was no other way. "Sir, the Apollo 1 crew is dead."

Dazed, shocked, Webb forced himself to return to the reception room. He spoke privately, haltingly, with Vice President Humphrey. Then he motioned for attention. The room fell utterly silent. They listened, shattered, as Webb told them that the three astronauts had just died in a fire at the launch pad. He turned and left the International Club, rushing to his office at NASA headquarters, only several blocks away.

Webb knew all his skills would be required to manage this crisis—skills honed as a Marine pilot, as Harry Truman's budget director, and later undersecretary of state, and as director of such enterprises as McDonnell Aircraft, the Oak Ridge Institute of Nuclear Studies, and Sperry Gyroscope.

His key staff was soon at his side. Immediately they turned the tragedy over to Major General Samuel C. Phillips, a master manager, who had ramrodded the Air Force's Minuteman intercontinental ballistic missile program into fruition before being recruited by NASA to run the entire Apollo program.

Phillips called the Cape. "Impound all records, all tapes, every last piece of equipment associated with that fire, and I mean *everything*. Put the clamps down everywhere. I'm on my way."

In Houston, events were brutally personal, individual, starkly emotional. Deke, Alan, and other astronauts visited the women who had instantly become widows. They were gathered at the home of Patricia White.

"I was still badly shaken. Rattled and battered is a good way to say how I felt," said Deke Slayton. "Betty, Pat, and Martha were holding up better than I was. They broke the tension by making me a surprise presentation."

It was incredible. These three women had lost their husbands to that raging fire in the ship they had hoped to fly through space. Now they were concerned with Deke.

There was a long-standing tradition that each person selected as an astronaut was to receive a silver astronaut pin, its design a shooting star soaring upward, trailing a long, comet-like tail. After any newcomer reached vacuum above earth and became an astronaut in deed as well as name, he received a second, gold pin. The three pilots of Apollo 1 had decided on their own that because of Deke's heart problem the odds were he'd never fly in space. Yet without Deke they might never have had a chance to boost off their launch pads.

So Gus Grissom, Ed White, and Roger Chaffee had arranged to fly a gold astronaut pin aboard Apollo 1 and, after the flight, present it to Deke to wear. Gus also knew Deke would never wear a pin that exactly matched the pins worn by men who'd sailed around the earth, so the three astronauts inserted a small diamond in place of a star.

"Since what those guys planned could never happen now," Deke recalled, "the wives, for whatever reason, chose this, the saddest and grimmest occasion in their lives, to present that pin to me. I was absolutely overwhelmed. Flattened. It was a gesture I'll never forget."

In the days following the Apollo 1 fire, the fallen astronauts were being remembered in almost every home in the nation. The home of Frank Sinatra was no exception.

"Ten days before Gus, Ed, and Roger died," Alan Shepard recalled, "they were flying to the Downey plant in California in search of an updated Apollo simulator. They ran into minor problems with one of their T-38 jets and had to land at Nellis Air Force Base outside Las Vegas. While the jet problem was being fixed, they decided to take a Vegas break from their day-and-night training schedule.

"They took in a show in town, and Frank Sinatra was on stage. No sooner had Gus, Ed, and Roger appeared in the room than Frank had them brought up to a front table. They were wearing their astronaut flight jackets, and Old Blue Eyes took a shine to Gus, mission patches and all.

"Well, Gus, he just said, 'Here, take it,' and gave the jacket to Sinatra. Sinatra was so moved he cried before his audience. Ten days later, he cried a lot more."

Gus Grissom was buried at Arlington National Cemetery. Rifle volleys split the air, a bugler sounded the mournful, stirring notes of taps, and jet fighters thundered overhead in a final salute. Six men in uniform stood stern and rigid, at attention. They had been seven, the original team of American astronauts. After eight years of bold accomplishment, tragedy had removed one of their proud number. The president of the United States, grieving families, men and women high in government and in the space program stood with them.

Three hours later, there was again the salute, the bugler's plaintive notes, the roar of jets as Roger Chaffee went to rest at Grissom's side. On the same day, Ed White went to his final destination on a bluff overlooking the broad Hudson River at his beloved West Point.

It was over.

Now there was to be a new beginning.

Deke and Alan knew that with so many new astronauts in the business and newcomers about to join the ranks, neither of them could exhibit an excess of emotion at the loss of Gus and the men who had died with him.

"When you're in the flying business," Deke said to the other astronauts, "you run into this. I've been through my share. I've watched mid-air collisions, aircraft downed by flak and fires. An aircraft goes haywire, and you lose a fellow pilot, a friend. It's a tough break, but it's part of the business. But an accident on the pad like this, when suddenly everything goes to hell—it's so goddamned inexcusable! You're not supposed to get killed on the ground when you're a test pilot. Still, you shrug it off and continue flying. You just hope you're not the next to go.

"It's like being a bit paranoid, I guess. You keep looking over your shoulder to be sure they're not gaining on you. But never forget to look *ahead.*"

Vast changes swept through the ranks of the space administration. The director of NASA's Langley Research Center in Virginia was told by Jim Webb to immediately set up a board of review to "find out what the hell really happened." Floyd L. Thompson nodded, and brought in some of the toughest investigators and specialists in the business. Among them, representing the astronauts, was Frank Borman, who'd flown two weeks in Gemini 7. They put together a team from government and industry of fifteen hundred dedicated men and women, and they began the grisly task of examining the charred remains of Apollo 1. They studied everything from chafed wires to solidified blobs that had been molten metal. They traced through fifteen miles of wiring and began to sort out literally thousands of dials, switches, tubes, and connections. They built a duplicate of Apollo 1, swallowed hard, and set it ablaze. It shook up many of those involved so badly they went home to stare at walls. It was all too real, too painful. But they were learning. And they were changing the heart and the nature and substance of Apollo so that what had happened on Level 8 in January could never happen again.

From the outset of the aftermath, heads rolled at the top reaches of the space hierarchy. The search for failure, incompetence, disregard for safety, became a meticulous, driving force.

NASA knew it must succeed, because hanging over the agency's head was a genuine possibility that the Apollo program could be cancelled. The Apollo 1 fire had sparked a national debate over whether the country should be spending an estimated twenty-four billion dollars to send men to the moon when it had so many other problems to address: Vietnam, rising taxes, civil rights, and the environment. Public opinion polls found more and more Americans asking: Is this trip really necessary? The critics said the program cost too much, that the race to the moon was a political stunt. Many prominent scientists argued that less expensive unmanned probes could learn more about the moon than astronauts.

Apollo's proponents said the project's price tag was realistic. President Johnson said it would cost each American $120 over nine years. He said Americans spent much more a year on cigarettes and alcohol. The defenders said the program would produce untold scientific and technological benefits and would demonstrate the nation's ability to lead in an age of technology.

Wernher von Braun entered the debate, arguing Apollo was not just a project to land two men on the moon but to open the new frontier of space. "When Charles Lindbergh made his famous first flight to Paris," he said, "I do not think that anyone believed that his sole purpose was simply to get to Paris. His purpose was to demonstrate the feasibility of transoceanic air travel. He had the farsightedness to realize that the best way to demonstrate his point to the world was to select a target familiar to everyone. In the Apollo program, the moon is our Paris." The defenders won the debate, thanks mainly to the strong backing and clout of Lyndon Johnson.

There would be delay, but there was still a chance of landing Americans on the moon before the end of the decade.

On April 3, 1967, NASA's Grumman Gulfstream executive personnel transport wheeled away from the terminal at Washington National Airport and began taxiing for takeoff. Its destination was Houston, and in the cabin, among other passengers, were Bob Gilruth and George Low.

At this time Low was Gilruth's deputy at the Manned Spacecraft Center. Austrian born, forty-one years old, educated in private schools in Britain and Switzerland, he had come to the United States with his widowed mother in 1938. He studied to be a systems engineer and in 1949 signed on with the old National Advisory Committee on Aeronautics, NASA's forerunner. He worked for a decade on esoteric aeronautic problems and then shifted to the world of space and helped design the Mercury spacecraft. Modest and unassuming, he was a methodical man who sought harmony and order.

Jim Webb thought so much of Low's leadership ability, he tapped him to chair the agency committee that placed on President Kennedy's desk the report recommending a manned lunar landing. Low had met many times with Kennedy to answer questions about a moon program. His responses and candid explanations brought Kennedy to the national podium to commit America to the challenge of the century.

Now, six years later, Low raised his brows as the Gulfstream slowed, began an unexpected turn, and taxied back to the gate. The tower had relayed a cryptic message to the pilot to get back to the terminal, unload passengers, and ask them to wait in the pilots' lounge.

"Soon," explained Low, "there arrived administrator Jim Webb, his deputy Bob Seamans, George Mueller, the head of manned space flight, and Apollo program director Sam Phillips. Counting Bob Gilruth, everybody in the NASA hierarchy between me and the president was here.

"Jim Webb, using fewer words than usual, came right to the point: Apollo was faltering, the catastrophic fire on January 27 had been a major setback. . . . Time was running out."

Then Webb hit Low with an unexpected blow. "I want you to take over the job of rebuilding the Apollo spacecraft."

Low was stunned. Webb wanted him to replace Joe Shea as head of the Apollo spacecraft program. He couldn't think of a tougher job than raising the phoenix from its ashes. He ran through his mind the lay of the engineering land before him, the roads that forked left and right. Webb wanted him to perform a task that many people in his business now considered impossible.

George Low wanted America to reach the moon. He would take the job.

"No detail was too small to consider," he said. "We asked questions, received answers, asked more questions. We woke up in the middle of the night, remembering questions we should have asked, and jotted them down so we could ask them in the morning. If we made a mistake, it was not because of any lack of candor between NASA and contractor or between engineer and astronaut; it was only because we were not smart enough to ask the right questions. Every question was answered, every failure understood, every problem solved."

"What George Low did was instill a sense of dedication and purpose among those working under and with him," Alan Shepard recalled. "In the astronaut corps we marveled at the new Apollo spacecraft taking shape. We were gaining confidence all the while that, yes, they're creating something that will be safe for us to fly. After what happened to Gus, Ed, and Roger, that was saying a lot."

What really provided the momentum needed to get Apollo back on track was the stunning candor on the part of NASA in telling the nation—and the rest of the world listening—just what had gone wrong, and why it went wrong, and what it would take to make the fix. Seven days after George Low accepted the toughest job of his life, the Apollo Review Board issued its official report on the fiery death of Spacecraft 012. Critics had said NASA should not have investigated itself, and even they were amazed at the frankness of the report. The board chaired by Floyd Thompson did not mince words. It laid blame where it belonged.

The report was 3,300 pages long and weighed nineteen pounds. Because of extensive damage, the cause of the fire could not be pinpointed precisely. The most likely source was given as an electric arc in defective wiring; investigators found that insulation on wires under a small trap door in the environmental control system might have broken or frayed by rubbing against the metal door. Nine other possible sources were listed, all involving electrical failings in the same area, under Gus Grissom's couch.

The searing part of the report was a frank indictment of NASA and North American Aviation. This team was criticized for poor management, carelessness, negligence, failure to consider the safety of the astronauts adequately. "The board's investigation revealed many deficiencies in design and engineering, manufacturing and quality control . . . numerous examples in the wiring of poor installation, design, and workmanship," as well as poor welding and soldering of joints that carried flammable coolant through the spacecraft, the report stated.

NASA and North American Aviation officials agreed the report was fair, shouldered the blame, and answered to a new call to arms to bring a safe Apollo on the line. In addition to Low replacing Joe Shea, there were numerous shake-ups in NASA personnel. Top management of North American's Space Division was overhauled.

In the ensuing months, nearly a half billion dollars would go into the exhaustive redesign and rebuilding of Apollo, including a new hatch that a man could open in three seconds flat. The new spacecraft also would include extensive use of fire-resistant materials, a redesigned electrical system, better protection for plumbing lines, and use of a combination nitrogen-oxygen atmosphere system when the spaceship was on the ground.

What happened to the Apollo command module also could happen to the lunar module, so its builder, Grumman Aircraft, made extensive changes in that vehicle, too.

Commenting years later, Flight Director Chris Kraft said: "It was unforgivable that we allowed that accident to happen. The other side of it is had it not happened, we probably would not have got to the moon when we did. We made a lot of changes to the command and lunar modules as a result of that experience. I think we would have had all kinds of trouble getting to the moon with all the systems problems we had. That terrible experience also brought a new resolve and a renewed commitment to get the job done."

Deke Slayton spoke from a different viewpoint. He represented the pilots who would fly inside a spacecraft that left earth moon-bound on eight hundred feet of fire.

"I'm convinced," he said, "we would have ended up busting our ass in a number of ways before we got to the moon, and we may never have gotten there if it hadn't been for Apollo 1. We uncovered a whole barrel of snakes that would have given us a lot of headaches later on. We would have fixed them bit by bit and probably knocked off a few people in the process of working our way through it. The fire forced us to shut down the program and make a real end-to-end sweep."

NASA officials had estimated after the fire that it would be a year before a redesigned Apollo would fly in earth orbit. But this time there was no rush; the job, orchestrated by George Low, was done with painstaking care and attention to detail, and that year stretched to twenty-one months.

And while the recovery took shape, another incredible and terrible tragedy rushed to stage center. . . .

Eighty-six days after the Apollo 1 fire, on April 23, 1967, Cosmonaut Vladimir Komarov was launched from Baikonur within a splendid new Soviet spacecraft, Soyuz 1. It was intended for a long future as an earth-orbiting craft, and it was the spacecraft that would be modified to take cosmonauts around the moon. Komarov's mission called for him to spend a full day checking the Soyuz systems, and then watch another Soyuz lift off to join him in orbit and practice daring maneuvers culminating with a docking of the two ships (the word *soyuz* means union) and an exchange of vehicles by two space-walking cosmonauts, a capability the Soviets would need to reach the moon.

Komarov raced around Earth on a perfect first orbit, but the spacecraft almost immediately developed serious problems.

Unlike the American spacecraft, which ran on battery power, Soyuz was built so that, once it entered orbit, two solar panels on each side of the ship would spread wide like huge wings and draw a constant supply of electrical power from the sun.

The right panel extended.

The left panel remained jammed in position. Nothing Komarov could do would release the panel he needed to operate all the systems. By the second orbit, mission control was on full alert in "emergency red condition." Soyuz 1 was delivering barely half the electrical power Komarov required to control his ship.

The shortwave radio transmitter failed. His ultra-shortwave radio back-up remained operational, and he continued to communicate with controllers.

He received instructions to change the attitude of Soyuz to gain maximum power for the opened solar panel. This might give him enough electrical "juice" to free the jammed panel. The maneuver failed. In the next three hours Soyuz "began to shred itself" with one grave failure after another.

By the fifth orbit, mission control feared for Komarov's survival. He could not properly orient the new Soyuz. His power was failing. Communications began to break up. He shut down the automatic stabilization system and went to manual control with his rocket attitude thrusters.

The thrusters operated in balky spurts. Komarov felt the ship getting away from him.

Controllers instructed the cosmonaut to let Soyuz go into a drifting mode. He would soon be moving through a series of orbits during which he would not be able to maintain voice contact with his flight controllers. Between the seventh and the thirteenth orbits he would be away from tracking stations, completely out of touch with his control team.

"Try to get some sleep, Comrade Komarov," he was instructed. Then he moved out of radio range. He would be in the "radio dark" for the next nine hours.

The time passed slowly. At the close of the thirteenth orbit they heard Komarov's voice. His report sent chills through the control center. The electronic system for automatically stabilizing the spacecraft was gone for good. Manual control with sputtering thrusters was sporadic at best. Soyuz was rapidly becoming a careening, wobbling killer with its pilot trapped inside.

Controllers cancelled the launch of the second Soyuz. They told Komarov they had to gamble to get him back to earth. He must fire his retro-rockets for reentry on the seventeenth orbit and use all his strength and knowledge to try to manually hold Soyuz on a steady course through the harrowing plunge.

The flight control director said aloud what everyone knew. "He is out of control. The spacecraft is going into tumbles that the pilot has difficulty stopping. We must face the truth. He might not survive reentry."

The director picked up a telephone, issued orders. A powerful car pulled up before a Russian apartment complex. Two men rushed into the building, emerging moments later with a woman. The car roared off toward mission control.

The woman was Valentina Komarov, the cosmonaut's wife and the mother of their two children.

By the time she reached the control center, Soyuz 1 was tumbling through space. Komarov several times had become ill from the violent motions. He forced calm into his voice when flight control told him he could talk privately with his wife.

They brought Valentina to a separate console and moved aside to assure her privacy. In those precious moments Vladimir Komarov bid his wife good-bye.

Soyuz 1's retro-rockets slowed his speed, but reentry began with Komarov having little control. He fought the spacecraft with his experience, skill, and courage. He judged his position by gyroscopes in the cabin. Incredibly, he aligned the ship properly and held it firm until building atmospheric forces helped stabilize it.

First reports of the ship's landing indicated it had touched down about forty miles east of Orsk. It seemed Komarov had accomplished the impossible, fighting his failed spacecraft all the way through reentry.

But the Soviet officials had not witnessed what the farmers in the Orsk area observed as Soyuz fell to earth. Though Komarov survived reentry, he was fighting a ship spinning wildly out of control.

The main parachute had failed. His reserve chute fell away from Soyuz, immediately twisting into a large, orange-and-white rag, which trailed uselessly behind Soyuz.

At a speed of four hundred miles an hour, Soyuz smashed into the earth. Its landing system included powerful retro-rockets, which normally fired just above the ground to cushion the touchdown. But Soyuz slammed down hard and the rockets exploded, engulfing the capsule in flames. Farmers ran to the ship to throw dirt on the burning wreckage. An hour later they were able to dig through the smoldering ruins of the spacecraft to find the body of Vladimir Komarov.

The Soviet space program, like the American, had experienced a stunning reversal and entered a period of reexamination. Russia did not fly another Soyuz for eighteen months.

While the Apollo command and lunar modules were being redesigned and rebuilt, NASA's unmanned rockets were busy propelling a series of probes to the moon to compile information. The surface properties were largely unknown, and some scientists speculated a spaceship might be swallowed in a deep layer of dust, that electrostatic dust might leap up and destroy a craft's electronics, or that astronauts might fall into thinly covered crevasses. These questions had to be answered: How much weight would the lunar surface support? How steep were the slopes? Were there many rocks, and what size? Would the dust or dirt cling? Did the craters pose a threat? What was the precise size and shape of the moon? How strong was the gravity field into which the lunar module would drop?

The robots circled and landed on the moon, dug into its surface, photographed potential astronaut landing sites, and proved conclusively the moon was a safe place to visit.

There was more good news for NASA as 1967 drew to a close. Von Braun's massive Saturn V rocket, thirty-six stories tall with power in the first stage alone greater than that of five hundred jet fighter planes, had a spectacularly successful debut in November, hurling into orbit an unmanned Apollo craft, a dummy lunar module, and other attached equipment weighing a phenomenal 280,036 pounds, more than the combined weight of all the more than 350 satellites launched previously by the United States.

Many hurdles lay ahead, but the Apollo program had passed a major milestone, and NASA was feeling pretty good about itself.

Rocco Petrone, the launch operations director who presided over the launch, said, "I feel that the Saturn V, working the way it did, spacecraft and all, got us back in the right swing, where the American public and Congress could say, 'Yeah, those guys can do it.'" Von Braun was ecstatic. "I have always dreamed of a rocket which we could use to explore the solar system," he said. "Now we have that rocket."

Two months later, another major piece of Apollo hardware, the lunar module, the craft that would ferry two men to the moon's surface, was successfully tested in earth orbit.

NASA was definitely on the way back, and Jim Webb, at age sixty-two, felt it was time to move on. A presidential election was coming up. Lyndon Johnson, besieged by many problems, had said he would not seek reelection, and Webb said he was stepping down to smooth the transition to a new administration. Thomas O. Paine, Webb's recently named deputy, was appointed acting administrator and later was designated head of the agency by the new president, Richard M. Nixon.

"Webb was the glue that held it all together," Deke Slayton commented. "Without him we would have lost Project Apollo after the fire."

On October 11, 1968, a Saturn 1B rocket spewed bright orange flames as it lifted Apollo 7 into orbit. Halfway through the powered ascent on the first manned flight of Project Apollo, and the first manned liftoff for the Saturn 1B, Commander Wally Schirra radioed back, "She's riding like a dream."

The crew of Schirra, Walt Cunningham, and Don Eisele lofted into an elliptical orbit of 140 by 183 miles high. The first American flight with three astronauts was underway for a mission of eleven days during which, in the words of NASA's Sam Phillips, "the ghost of Apollo 1 was effectively exorcised as the new Block II spacecraft and its millions of parts performed superbly."

The astronauts tested the spacecraft's systems, conducted experiments, beamed the first extensive live television scenes from a manned orbiting vehicle to fascinated audiences around the world, and flew their ship longer than would be required for a trip to the moon and back.

They were impressed with the size of the ship relative to the cramped cabins of the Mercury and Gemini, which had confined the astronauts to their seats. The Apollo 7 crew could unstrap themselves and move around the cabin. If they wanted privacy, they could float into a closet-size area beneath the seats, which on later flights would serve as sleeping quarters.

The flight encountered only minor problems, and they were quickly resolved.

The biggest problem Mission Control had was with the crew. All three had nasty colds and were orbiting their world with stuffy noses. As the mission neared its end, the astronauts were in something less than the best of tempers and they became irritable.

The complaints started with the food and reached a peak on the ninth day when controllers decided to try some unplanned systems checks. The three astronauts reacted less than graciously, and read the riot act to the engineers for requesting these "Mickey Mouse tasks," which they classed as "ill-prepared and hastily conceived." Schirra shouted that the controller who had thought up one of the tests was an "idiot" and refused to accept any more changes.

Tempers were at their worst when it came time to start home. Schirra told his crew that they would make the reentry without their pressure helmets on. He was concerned that because of the colds, any sudden overpressure could damage their eardrums and cause other problems. Deke Slayton got on the loop to try to persuade them to wear the helmets, but that didn't cut any ice with Wally. He was in command of the spacecraft, and the pilot flying the machine always has the ultimate responsibility.

The behavior of the Apollo 7 crew, particularly of Schirra, made a lot of people on the ground angry. Back on earth, Wally received a tongue-lashing from Deke Slayton about the behavior of his crew. Cunningham, who had been making his first flight and felt he had to go along with his commander during the mission, summed up his feelings when he wrote later that "the entire Apollo 7 crew was tarred and feathered through the actions of Wally Schirra."

But Wally was too busy to be concerned about the complaints. Before the flight he had announced he would be retiring from the astronaut corps, and he didn't care what anyone thought. He and the others plunged into six days of intense debriefings to help prepare for the next flight—Apollo 8.

To Wally, only results counted. He was the only astronaut who had flown in all three of America's pioneering manned space programs—Mercury, Gemini, and Apollo—and even though their refusal to follow orders in orbit, together with Schirra's retirement, meant none of the Apollo 7 crew would fly in space again, Wally knew because of their performance in Apollo 7 the first lunar Christmas was just around the corner.

⌐APOLLO 8 AND A RUSSIAN PUSH TO THE MOON⌐

POLLO 8 COMMANDER Frank Borman was emerging from a deep sleep, and he resisted the wakefulness that tugged at him.

Where am I?

He lay absolutely still, suspended in what seemed like nothingness. Was he floating in water? Submerged, arms and legs suspended? But he was breathing normally, so this couldn't be water.

He refused to open his eyes. Sounds came to him, trickling, murmuring, whispering.

There, a hum, soft but persistent. Faster and faster he recognized specific sounds. He knew the wheezing, a bubbling mechanical brook from the clicking he heard.

Slowly he opened his eyes and focused his vision on the glowing circles of red, green, yellow, pale white before him. Numbers, letters, circles, squares, buttons, controls, dials.

I know where I am now. I'm two hundred thousand miles from home. . . .

He grasped the edge of a long fabric strip and pulled free the Velcro that had kept his body from floating away from his spacecraft couch. He glanced at the two other astronauts in the cabin with him, both still asleep. He smiled. He liked the thought of a few moments to himself. He leaned to one side and eased back the curtain covering a flat viewing window.

Awe and wonder swept through him. Apollo 8 was turning slowly so that the radiation heat from the nearest star, earth's sun, would be distributed evenly around the external surface of the three-man spaceship. A bright sphere eased into view, the steely glint of Jupiter resplendent in reflected solar glow.

Suddenly, the earth appeared before him. Not a vast horizon curving gently away from sight, but the whole globe, dominant with blue seas and white clouds, with bountiful rain forests and mountains rising above the surface. From here, as they eased toward the moon, the earth was perfectly round, machined by heavenly forces to a stunning sphere.

Apollo rolled, and the home of man slid eerily, silently, out of sight.

Tomorrow would be Christmas Eve. That Borman and his crew were here seemed impossible. They had left earth atop America's largest rocket. The mightiest energy machine ever built to lift straight up and away from the deep gravitational well of the planet. A monster of steel and ice and fire atop which no man had ever before flown—and they were risking everything to fly to the moon.

It was a gamble like few others known in history. The mighty rocket, the Saturn V, had flown only twice before. Unmanned. First success-fully, second with some failures.

The three astronauts within the cone-shaped, tiny world of Apollo 8 had the largest audience in television history. More than a half billion people watched television sets that carried sights they had never before seen and could hardly still believe.

Live views of their homes from a spacecraft more than halfway to the moon. Frank Borman acted as tour guide, describing earth's features. "What you're seeing is the Western Hemisphere," he said in a voice so matter-of-fact he might have been pointing out the Grand Canyon to the passengers of an airliner. "In the center, just lower to the center, is South America, all the way down to Cape Horn. I can see Baja California and the southwestern part of the United States."

Alongside him, crewmate Jim Lovell joined in. "For colors, the waters are all sorts of royal blue. Clouds are bright white. The land areas are generally brownish to light brown in texture." His audience two hundred thousand miles away appreciated the touch of the color artist—

Mercury astronauts Deke Slayton (left) and John Glenn sought shade under a makeshift tent they crafted from a Mercury capsule parachute during desert survival training in California's Mojave Desert in 1962.

Shortly after their selection as America's first astronauts, the Mercury Seven donned their military uniforms (left) and posed beside one of the Corsair 106-B jet aircraft they used for proficiency flight training. The astronauts are (left-right) Navy Lt. Malcolm Scott Carpenter, Air Force Capt. Leroy Gordon Cooper Jr., Marine Lt. Col. John Herschel Glenn Jr., Air Force Capt. Virgil Ivan (Gus) Grissom, Navy Lt. Cmdr. Walter Marty Schirra Jr., Navy Lt. Cmdr. Alan Bartlett Shepard Jr., and Air Force Capt. Donald Kent (Deke) Slayton.

The Cape Canaveral area in the early 1960s was a super-charged zone of high technology and high tension as scores of missiles and rockets were launched each month to uncertain fates. Partying and pranks were ways to let off steam after long hours at the launch pad. During one particularly rousing celebration at the Holiday Inn in Cocoa Beach (below), Deke Slayton was tossed fully clothed into the pool by his fellow Mercury astronauts in February 1961.

The Mercury Seven training at an Air Force survival school in the Nevada desert in 1960. From left: Gordo Cooper, Scott Carpenter, John Glenn, Alan Shepard, Gus Grissom, Wally Schirra and Deke Slayton. Dr. Wernher von Braun (left) developer of the Redstone and Saturn rockets and his wife Maria at their home in Huntsville, Ala., in the 1960s.

Deke Slayton (right) at the Capsule Communicator's position in the blockhouse at Cape Canaveral during the countdown for Alan Shepard's flight as the first American in space on May 2, 1961. Launch was scrubbed that day for weather but Shepard was launched successfully three days later on a 16-minute suborbital flight.

Skiing was a favorite sport of all the Mercury astronauts. In March 1960 (below), Alan Shepard, Scott Carpenter, Deke Slayton, Gus Grissom and Wally Schirra posed before hitting the slopes in Colorado.

Astronaut Alan Shepard (left) squeezed into the Mercury capsule he named Freedom Seven on May 5, 1961, and became America's first man in space as a Redstone rocket propelled him 115 miles high and 302 miles down the Atlantic tracking range. For five minutes the astronaut experienced the eerie sense of weightlessness, calling it "a wonderful feeling." The 16-minute flight ended with Freedom Seven parachuting into the ocean. A helicopter (above) raced to the scene, hoisted Shepard aboard and ferried him to a nearby aircraft carrier. "Boy, what a ride!" he told the crew.

President John F. Kennedy congratulated astronaut Alan Shepard and presented him with the NASA Distinguished Service Medal during a Rose Garden ceremony at the White House three days after Shepard became the first American in space. Louise Shepard, Alan's wife, stood at the left, with his mother. Other Mercury astronauts and

government officials observed. Seventeen days later, on May 25, 1961, in a special message to Congress, Kennedy challenged the Soviets in the space arena and put the United States on a course to the moon by declaring the nation should land a man on the moon and return him safely to earth before the end of the decade.

Project Gemini followed the pioneering Mercury program and during 10 flights of the two-man spacecraft in 1965 and 1966, astronauts perfected all the techniques needed to go to the moon - rendezvous and docking, spacewalking and long-duration flights of up to two weeks. Deke Slayton (above), chief of the Astronaut Office, puffs a victory cigar following the successful rendezvous in orbit of the Gemini 6 and 7 craft, the first such achievement in space. Edward H. White II (right) became the first American to walk in space, cavorting for 21 minutes outside while linked to the Gemini 4 craft with a 25-foot safety line. Gemini 8 approached an Agena satellite (below) and minutes later linked up with it in the first docking of two vehicles in orbit.

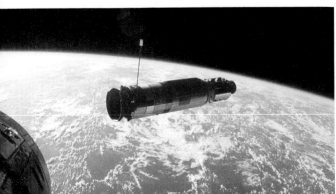

Modified military rockets boosted the early U.S. manned spacecraft. From left, a Redstone lofted an unmanned Mercury capsule on a suborbital test flight in 1961; an Atlas carried John Glenn on the first American orbital flight in 1962, and a Titan 2 hoisted the first manned Gemini flight in 1965.

Huge made-for-space-flight rockets developed by a team headed by Wernher von Braun were built for the Apollo program. From left, a mighty Saturn V barreled away from earth on its first test flight on November 9, 1967, gulping fuel at the rate of 15 tons a second; 20 months later another Saturn V propelled Apollo 11 astronauts on the first moon-landing mission, and a smaller Saturn 1B pushed Deke Slayton and two crewmates to a linkup with a Soviet spacecraft in 1975.

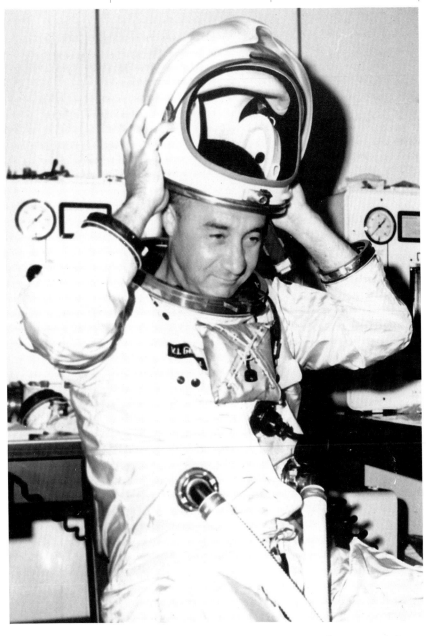

Tragedy struck the U.S. space program and slowed the march to the moon when fire flashed through the Apollo 1 capsule during a routine launch pad test on January 27, 1967, killing astronauts Gus Grissom, Edward White II and Roger Chaffee. Commander Grissom (above) donned a helmet to prepare for an earlier training session for the intended earth orbit mission.

The exterior of the Apollo 1 spacecraft was charred (above) when interior pressure burst the hull of the vehicle and spewed flames through the opening. A view of the interior (right) revealed the effects of the intense heat. Investigators said the fire was caused by defective electrical wiring.

The Apollo 11 crew (above), Commander Neil Armstrong, Command Module Pilot Mike Collins and Lunar Module Pilot Buzz Aldrin. Armstrong (left) rehearsed the descent to the moon in a Lunar Module training simulator. The Command and Service Modules for the mission (right) were moved from a workstand at Cape Canaveral's Vehicle Assembly Building to be joined with a rocket adapter section.

Massive Saturn V rocket (left) on Launch Pad 39A during a countdown test. Astronauts Neil Armstrong, Mike Collins and Buzz Aldrin boarded the spacecraft atop the 363-foot rocket for the rehearsal. On July 16, 1969, they rocketed away from earth and four days later Armstrong and Aldrin rode the Lunar Module Eagle to a touchdown on the moon's Sea of Tranquility while Collins stood watch in lunar orbit aboard the Command Module Columbia.

Aldrin (below) and Armstrong were on the moon more than 21 hours and spent two hours exploring outside their lander.

President Richard M. Nixon (left) greeted Apollo 11 astronauts Neil Armstrong, Mike Collins and Buzz Aldrin aboard the aircraft carrier USS Hornet in the Pacific Ocean following the spacemen's return from man's first moon-landing mission. To protect the earth from any possible contamination, the astronauts on recovery were placed in an elaborate quarantine trailer. So the president faced the three through a glass window in their isolation van., saying, "This is the greatest week in the history of the world since the creation. As a result of what you have done, the world has never been closer together."

At the Mission Control Center in Houston (above), NASA officials cheered, smoked victory cigars and waved American flags to celebrate Apollo 11's safe return. Flashed on a huge display screen were the 1961 words of President John F. Kennedy when he put America on course to the moon. His goal of a man on the moon before the decade was out had been achieved. Among those in Mission Control (in the forefront, left-right) were Chris Kraft, director of flight operations; George Low, manager of the Apollo spacecraft program, and Robert Gilruth, director of the Manned Spacecraft Center.

A major crisis developed in April 1970 when an oxygen tank exploded aboard Apollo 13. Astronauts James Lovell, Jack Swigert and Fred Haise were 200,000 miles from earth, rapidly losing vital oxygen and other supplies, when the blast hit. They sought shelter in the undamaged Lunar Module while Mission Control perfected the means to bring them home safely. Deke Slayton (below), demonstrated to controllers a makeshift device for removing stale air from the Lunar Module. Astronauts Haise, Swigert and Lovell (above) on the carrier Iwo Jima following recovery, while their spacecraft (right) was hoisted aboard.

Alan Shepard (above, center), was named to command Apollo 14, flying with Stuart Roosa (above, left) and Ed Mitchell. The Lunar Module Antares (right) carried Shepard and Mitchell to the moon's Fra Mauro highlands, where they made two outside excursions, in February 1971. Following their return to Houston, Mitchell, Roosa and Shepard (below) inspected some of the 92 pounds of moon rocks they collected on the surface.

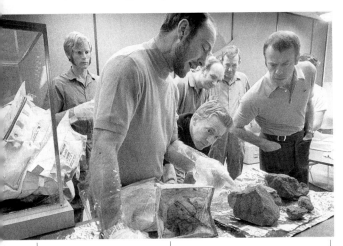

Alan Shepard made history of sorts when he became the first, and only, man to hit a golf ball on the moon. Using a makeshift club made of a six-iron head and the handle of a rock-collecting tool, he struck two balls. Hampered by his bulky space suit, Shepard almost missed the first and it plopped into a nearby crater. He hit the second squarely, and in the moon's weak gravity, it soared into the distance. "There it goes! Miles and miles and miles!" he shouted. During more than nine hours on the surface, Shepard and crew member Ed Mitchell set up a nuclear-powered science station, gathered soil and rock samples and erected an American flag. To lug their tools and collected rocks around, they used a hand-pulled cart they dubbed a rickshaw (shown here between the two men). The blastoff from the moon of the Lunar Module Antares was smooth and they rejoined Stuart Roosa in the command ship Kitty Hawk for the return to earth.

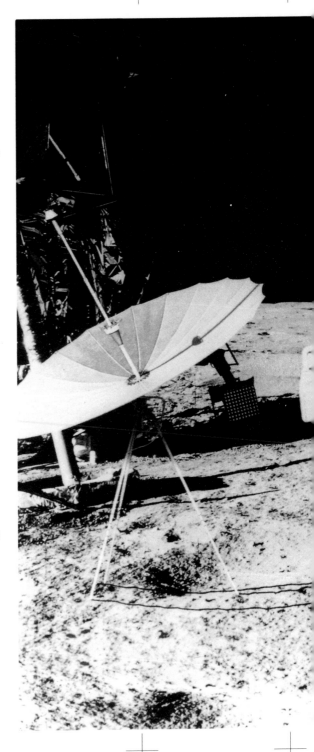

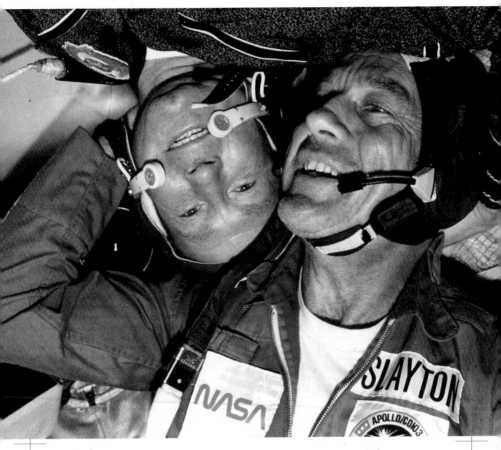

Sixteen years after he was named one of America's original Mercury Seven astronauts in 1959, Deke Slayton finally realized his dream of flying in space. He teamed with astronauts Tom Slayton and Vance Brand in July 1975 as they linked their Apollo spacecraft with a Soviet Soyuz carrying cosmonauts Alexei Leonov and Valeri Kubasov - a mission that would have been considered unthinkable just a few years earlier as the two superpower rivals maneuvered for superiority in space. A symbol of high-flying détente, Slayton and Leonov (above) cavorted in the weightlessness of the Soyuz cabin while orbiting the earth in the first international space flight.

A simplified artist's concept (below) showed the highlights of the Apollo Soyuz Test Project that linked American and Soviet spaceships. Soyuz was launched first from the Baikonur Cosmodrome in central Russia. Several hours later Apollo took off from Cape Canaveral and for two days Tom Stafford, Deke Slayton and Vance Brand flew complex maneuvers to track down the Soviet cosmonauts. Once joined, astronauts opened a connecting tunnel between the two ships and astronauts and cosmonauts floated freely from cabin to cabin.

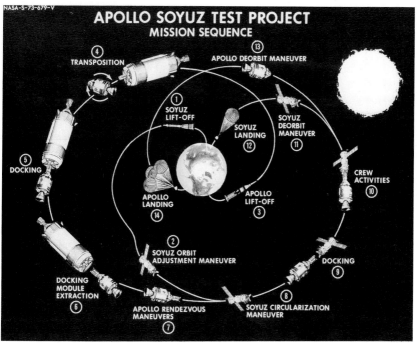

NASA-S-73-679-V

APOLLO SOYUZ TEST PROJECT
MISSION SEQUENCE

④ TRANSPOSITION

⑬ APOLLO DEORBIT MANEUVER

① SOYUZ LIFT-OFF

SOYUZ LANDING ⑫

⑪ SOYUZ DEORBIT MANEUVER

⑤ DOCKING

CREW ACTIVITIES ⑩

APOLLO LANDING ⑭

APOLLO LIFT-OFF ③

② SOYUZ ORBIT ADJUSTMENT MANEUVER

DOCKING ⑨

⑥ DOCKING MODULE EXTRACTION

⑦ APOLLO RENDEZVOUS MANEUVERS

⑧ SOYUZ CIRCULARIZATION MANEUVER

During training for the 1975 U.S.-Soviet space mission (right), astronauts and cosmonauts shared a meal in a mockup of the Soviet Soyuz cabin at the Johnson Space Center in Houston. Deke Slayton (at left) and Tom Stafford (third from left) sampled Soviet food with Alexei Leonov and Valeri Kubasov. Months later, in July, Stafford and Slayton (above) dined in the Soyuz cabin after joining their Apollo with the Russian craft in orbit. They hold containers of borsch (beet soup) which the cosmonauts had mischievously labeled "vodka." During the 47 hours the ships were together, astronauts and cosmonauts executed their own brand of shuttle diplomacy as they moved easily between the two craft, to conduct joint scientific and medical experiments, to give earthlings televised tours of the United States and Soviet Union and to exchange gifts, including flags and commemorative plaques. In a joint news conference, answering questions relayed from reporters at both control centers, Stafford said the union in space "opens a new era in the history of mankind."

television from deep space was still in black-and-white.

Then Lovell gave listeners something to think about. "What I keep imagining," he said in a tone reflecting his own deep thoughts on the matter, "is if I'm some lonely traveler from another planet, what I'd think of earth at this altitude . . . whether I'd think it would be inhabited or not."

The sudden somber mood broke when Mike Collins, CapCom in Houston's Mission Control, broke the spell. "You don't see anybody waving, do you?"

The third man in the Apollo, Bill Anders, was content at the moment to stare in wonder at the distant, fragile world with an onionskin-thin layer of atmosphere, which gave life to teeming billions on that resplendent, terribly isolated globe.

The transmission ended when Mission Control returned the astronauts to their duties aboard Apollo 8. Minutes later the spaceship slipped through an invisible veil between the planet and its satellite, the equigravisphere, the celestial twilight zone where the moon's gravity now dominated the continued flight of Apollo. Until this moment Apollo had been "coasting uphill" against earth's gravitational pull, almost as if the planet wished the craft to remain within its grasp. Now less than forty thousand miles from the moon, that sphere's gravitational attraction exceeded that of distant earth.

Once Apollo began to accelerate in response to the moon's pull, the mission took another giant step toward fulfillment. They would pick up speed, swoop closer to the battered lunar surface, and orbit the moon.

America watched and listened, rejoiced and celebrated a feat that promised to be one of mankind's truly prodigious steps into the future.

But not everyone shared the rejoicing. Many on earth were asking: "Why are they going to the moon during Christmas? Why couldn't they have flown the mission after the holidays?"

The astronauts of Apollo 8 knew the answer.

Zond.

Years before, the Russians had sent a series of robot ships to the moon to take photographs of the lunar surface and to perform an extensive variety of experiments that would prepare the way for cosmonauts.

America had its superbooster in the Saturn V. Russia had its superbooster in its massive N-1 rocket. But whereas Saturn V had progressed under Wernher von Braun, the Soviet superbooster program had stalled.

Test-flight delays had the N-1 chained to the ground, so the Russians regrouped and prepared several Soyuz spacecraft and a small fleet of tanker craft to be launched into earth orbit with their smaller R-7 rocket. Moscow urged its top rocket man, Sergei Korolev, to beat America to the moon. Korolev had launched the first satellite, and the Russians wanted the world to see the earth first through the eye of Russian television cameras and hear the excited voices of Russian cosmonauts singing from the tops of lunar mountains.

The plan was for Korolev and his team to send five unmanned tanker craft into earth orbit, each filled with rocket fuel. A group of cosmonauts would follow, gather the tankers like a flock of giant orbiting sheep, and herd them together into a single conglomerate—a lunar vessel of one manned spacecraft and the five fuel tanks. Simple and yet demanding, it would give the Russians a clear shot of beating the United States to a lunar landing.

As early as 1968.

That was their dream, their rallying cry. They brought their new Soyuz spacecraft to its launch pad to rehearse the rendezvous in orbit, docking, and space-walking maneuvers that would be required to get to the moon.

Then Vladimir Komarov died in Soyuz 1, delaying Soviet plans.

Korolov's team regrouped again. They began modifying the Soyuz spacecraft so it could carry one or two cosmonauts on a single pass around the moon's far side and a quick trip back to earth.

This would not be a landing, but it would be the first manned flight to the vicinity of the moon, and the Soviets would then crow that they had indeed beaten the Americans to that treasured goal.

They named the new, modified ship Zond, and in September 1968 they used a heavy-lift Proton rocket—smaller than the N-1 but larger than the R-7—to send an unmanned Zond around the moon and back.

Next they loaded a Zond with tortoises, flies, and worms, and in November sent it on a flight around the moon and brought the living creatures safely back to earth in what the CIA and other world intelligence agencies believed was the final dress rehearsal for the Soviets' world-beating, manned circumlunar flight.

But American intelligence was wrong. Russian officials wanted more tests, wanted more confidence in the reliability of the rocket and the spacecraft before committing a human being to such a risky undertaking. The cosmonauts fought to go.

Moscow refused to budge until more proving flights could be made. The moon must wait. There was still time to claim the prize. The Americans could stumble.

In mid-1968, intelligence agencies made NASA aware of Zond program, indicated that, if all went well with it, the Soviets could dispatch a single cosmonaut on a circumlunar journey in December or January. The fact the Russians couldn't land a man on the moon would not stop them from claiming they had reached that target first.

The U.S. officials hoping to fulfill the promise of John F. Kennedy saw failure in the race before them. Jim Webb, then the NASA administrator, told President Lyndon Johnson it was time for America to gamble, to consider putting astronauts on the next Saturn V rocket, the behemoth's first manned flight, and send them all the way to the moon aboard Apollo 8.

Some rocket experts at first scoffed at the plan, arguing that NASA was insane to commit human lives to such a flight so soon. On its last test flight, they pointed out, the Saturn V had suffered severe vibration problems, and three of its eleven engines had failed. The agency, they said, should stick to its orderly plan of using the Saturn V to send the Apollo command and lunar modules into earth orbit for tests, and then fly a lunar orbit mission, followed by a journey to land two men on the surface. There was still time to do all that and achieve the Kennedy goal before the end of 1969.

Webb told the outgoing president that it was the consensus of NASA engineers that they had corrected the Saturn V vibration problems, and that there was no need for an additional unmanned test. The Apollo 8 was to have been a test of the command and lunar modules in earth orbit, but the lunar module was behind schedule, would not be ready for testing in space for several months. A moon flight now could provide valuable knowledge about navigating to and around the moon, perhaps hasten a lunar landing.

Johnson saw the possibility of American astronauts reaching the moon—at least orbiting it—before his watch was over. He gave Webb the green light to continue planning for such a momentous mission and told him that he would support whatever final decision NASA made.

The idea for sending Apollo 8 to the moon had been hatched a few weeks earlier at a meeting in Houston. Deke Slayton had been among those there.

"George Low was the spacecraft program manager, and one afternoon he was talking with Bob Gilruth and asked him what could be done about getting around that particular problem with the lunar module," Slayton recalled. "They called in me and Chris Kraft to discuss it. Low asked if there were something we could do with the Apollo 8 command module. During a recent vacation, while sitting on the beach, George had thought about sending Apollo 8 on a circumlunar trip. He asked us if that would be possible.

"Remember now, we had not yet flown Apollo 7, had not flown a manned command module, and the last time we had flown the Saturn V it had almost come unglued because of the vibration problem. Shit, we didn't even have the software to fly Apollo in earth orbit, much less to the moon. It was still being developed.

"Chris was very surprised at this bold suggestion," Deke said. "He looked at Low and said, 'Good God, George, I don't see how we could do that.' I was very excited about the possibility, but, like Chris, I was very skeptical.

"George said he didn't expect an answer right away, that we should think about it awhile," Deke continued. "Chris called in his brains, and Alan and I talked the idea over with some of the astronauts. Al and some of the guys were really excited. After thinking about it for three days, we thought it was the greatest thing since sliced bread. Chris suggested that as long as we're going all the way to the moon, we might as well orbit it, not just pass by. We thought it was worth a shot."

There were naysayers, among them Jim Webb. But eventually all the key players became converts. But it all depended upon a successful flight by Wally Schirra's Apollo 7 crew in October.

During the discussions, Deke asked Jim McDivitt, who had been named the Apollo 8 commander, what he thought about the flight and whether he and his crew were willing to risk it.

"Jim and his crew had been training hard for that first test of the command and lunar modules in earth orbit," Deke said. "So he decided not to accept the proposed moon mission, preferring instead to keep training for the mission he was assigned. Jim reasoned that Apollo 7 would have to be perfect; otherwise, Apollo 8 would just be another earth orbit trip, without a lunar module.

"So I turned to Frank Borman, whom Alan and I had assigned to command Apollo 9—scheduled as a very high earth orbit test of the two modules. The sonofabitch almost turned handsprings when I suggested his crew swap with the Apollo 8 crew and that there was a possibility they would go all the way to the moon. His answer was an overwhelming yes."

Because of the uncertainty of the Apollo 7 outcome, Alan Shepard had Borman and crewmates Jim Lovell and Bill Anders working overtime in the Apollo simulator.

"They had to train for earth orbit, circumlunar and lunar orbit missions so they would be ready for whatever decision was made," Shepard said. "They were one busy crew."

Alan himself thought the idea of sending Apollo 8 to the moon was "a masterful stroke, a stroke of genius." He wished the mission were his. But because of his ear problem he could only dream.

Following the tremendous success of Apollo 7, Deke said: "We had a rapid review of all of Apollo 8's options. We debriefed Wally and his crew for six strenuous days. We reviewed the readiness of the Apollo 8 spacecraft and the Saturn V, of the deep space tracking network. The overwhelming consensus was we should go to the moon, not only go to the moon but orbit the moon."

On November 11, the new NASA administrator, Thomas Paine, approved the plan. He telephoned the decision to the White House, and Lyndon Johnson gave his blessing.

It was the single greatest gamble in space flight then, and since.

Time was important. Only a year remained before the end of the decade, and there was still the threat of a Soviet circumlunar flight. So, in spite of the approaching holidays, the first manned Saturn V, as large and as heavy as a U.S. Navy destroyer, heaved upward slowly, then faster and faster away from earth.

The mighty rocket burned perfectly through its three stages and sent Frank Borman, Jim Lovell, and Bill Anders winging away from their planet at 24,200 miles an hour on the morning of December 21, 1968. Jim Lovell sent a cryptic message back to Mission Control. "Tell Conrad he lost his record."

That was the height of 850 miles Pete Conrad and Dick Gordon had reached in their agile Gemini spacecraft. The new record would be on the order of 240,000 miles.

Zond was left standing on the launch pad in the Soviet Union. Bitterness replaced the usual holiday round of vodka and cognac toasts.

There was another message recorded this day when Lev Kamanin, top aide to Kremlin space officials and the son of the chief of cosmonaut training, wrote in his diary:

For us this [day] is darkened with the realization of lost opportunities and with sadness that today the men flying to the moon are named Borman, Lovell, and Anders, and not Bykovsky, Popovich, or Leonov.

No matter how well he had said it, Kamanin's sentiment was not the end of Russian efforts to reach the moon. Despite failures with their big rockets, the cosmonauts would continue to try.

But for the moment a marvelous product of science, technology, and engineering was on its way to the moon. The Apollo command module, crammed with instrumentation and equipment, moved swiftly toward lunar orbit.

It was Christmas Eve, and Apollo 8 would soon reach a point that a final decision would be made about whether to maneuver the craft into lunar orbit.

The three men in Apollo 8 did not know if they would be spending Christmas in lunar orbit, or retreating from that long-sought goal and returning to earth after one trip to the far side of the moon. In any event the flight would be a smashing success and would kick open the door to future manned exploration of our solar system, but everyone in NASA and the vast audience of the nation wanted Apollo 8 to go all the way.

Mission Control needed to decide if the big engine in the service module at the base of Apollo 8 was to be fired or to remain quiescent. If all systems aboard Apollo looked good to those in Houston, then Borman and his crew would receive the green light to fire the rocket that would slow their ship down and slide it into lunar orbit.

There were other considerations, each of which was absolutely critical not only to the mission plans but also to the lives of the three men. To slip the Apollo 8 into the desired orbit, its engine had to fire at full thrust for precisely 247 seconds. Following shutdown, the astronauts would use their attitude-control thrusters to point the nose of their ship in the direction of flight. If the engine burn faltered or failed early, the astronauts would soar past the moon on a flight path that would not return them to earth. If the engine burned too long, the spacecraft would not achieve its desired orbit and would speed downward to smash into the moon.

Despite all the circuitry and igniters and redundancy built into the systems, rocket engines sometimes do not fire. If the big engine failed to ignite altogether, Apollo 8 would be perfectly safe. It would swing around the far side of the moon, curving in its sharp orbit as if it were a celestial boomerang and, without expending an ounce of fuel, the astronauts would be on their way home. This was NASA's best of all insurance policies and a fact of centrifugal flight that clearly brought feelings of comfort to the astronauts and their families, and everyone else as well. This was the "Free Return Trajectory" inserted into the mission profile computer.

In Mission Control, every monitoring panel was "in the green." Apollo 8 was right on the money for its planned trajectory, and all spacecraft systems, monitored by radio signals constantly sending reports back to Houston, also glowed green.

Borman, Lovell, and Anders prepared for any contingency. Apollo 8 was about to disappear behind the mountain—fly above the side of the moon facing away from earth, and signals between earth and spacecraft would be blocked for more than twenty minutes.

CapCom Jerry Carr received the nod to send the message everyone had crossed fingers to hear. "Ten seconds to go," he signaled Apollo 8. His message, traveling at the speed of light, needed one and a third seconds to reach the spacecraft. "You are GO all the way."

Jim Lovell's voice, incredibly calm, came through headsets and speakers in Houston, "We'll see you on the other side." With those words Apollo 8 vanished.

Behind the mass of the moon it was as if Apollo 8 and its three men did not exist. There was no way to communicate with the spacecraft. No telemetry signals could be received. The mission had gone quiet.

At the precise moment dictated by its flight plan, the big engine fired in a soundless crash. For 247 seconds the engine blazed, a time period Lovell described as the "longest four minutes I've ever spent."

It was a splendid and epochal moment. Sixty-nine hours and fifteen minutes after throwing off its shackles from the launch pad, Apollo 8 locked into lunar orbit.

No one on earth knew that this had happened. This was a time of cliff-hanging suspense, a time to count the minutes and seconds that must pass before Apollo 8 emerged from the lunar back side to where it could send the desperately hoped-for signal of success. Jerry Carr kept up a persistent call of "Apollo 8 . . . Apollo 8 . . . Apollo 8 . . . "

After what seemed an eternity of intense clock-watching, headsets and speakers crackled. Smooth and calm as always came the voice of Jim Lovell:

"Go ahead, Houston."

Those three words—coming just at the instant they should have—sent Mission Control into a bedlam of cheering, whistling, shouting, and applause. Electronic signals flashed their message on the big viewing board. The Apollo 8 was in an orbit 60 by 168.5 miles above the moon. Later, on the third loop around the moon, the craft's main engine fired again and dropped the ship into the desired, nearly circular orbit of 60.7 by 59.7 miles.

But at the moment a thrilled global audience was waiting for the real story from space. It had nothing to do with numbers, velocity, or the technical details of the spacecraft, or its parameters of celestial balance around the moon. The interest of those on earth lay in one predominant direction:

What did it look like?

"Essentially gray, no color," reported lunar tour guide Jim Lovell, the first man ever to hold that job. He described the surface as "like plaster of Paris or a sort of grayish beach sand." In the first of two telecasts from lunar orbit, the astronauts relayed vivid pictures of a wild and wondrous landscape pitted with massive craters. "It looks like a vast, lonely, forbidding place, an expanse of nothing . . . clouds of pumice stone," Borman said. Lovell saw the distant earth as "a grand oasis in the big vastness of space." Anders added, "You can see the moon has been bombarded through the eons with numerous meteorites. Every square inch is pockmarked." Of the back side, Anders said most of it was too rugged for a manned landing. "It looks like a sand pile my kids have been playing in for a long time."

Lovell spoke eloquently. "The vast loneliness is awe-inspiring, and it makes you realize just what you have back there on earth."

Christmas Eve was like none other in the long history of celebrating that occasion. While millions of people brought families together in homes throughout the planet, the three men orbiting the moon continued taking sharp motion pictures and hundreds of clear color photographs that would later enable them to share with those on earth their fabulous adventure.

Then Bill Anders spoke, not just to CapCom, but to all the world listening to his words from so far away. "For all the people on earth," he said, his emotions unmasked, "the crew of Apollo 8 has a message we would like to send you." A brief pause, and then Anders stunned his audience as he began reading from the verses of the Book of Genesis:

"In the beginning, God created the heaven and the earth . . ." As Anders concluded the fourth verse, Lovell read the next four. Borman concluded by beginning his reading of the ninth verse, and then sent to the world a special Christmas message:

"And from the crew of Apollo 8, we close with good night, good luck, a Merry Christmas, and God bless all of you—all of you on the good earth."

Later, Borman would add a passage that would be repeated by the men who would venture to the moon, words spoken with stark emotion, sometimes with tears. As Apollo raced around the cratered world below, Borman watched the earth "rising" above the lunar horizon. "This is the most beautiful, heart-catching sight of my life."

Suddenly the concept of manned space flight clearly would never again be the same. The moon with its view of the distant, soft blue marble of life had become host to poets in spacesuits.

There was more work to do. The crew searched for and snapped hundreds of photographs of five landing sites under consideration for future Apollo missions, when men would descend to and walk and ride about that battered surface.

Then, all too soon, it had to end. Early in the morning of Christmas Day, Apollo 8 curled around the moon on its tenth and final orbit, again out of radio contact for the critical firing that would either start them on their journey home or leave them stranded in lunar orbit. At the scheduled moment Borman, Lovell, and Anders felt the flaring engine bell hurl forth a long stream of flame, emitting a wide plume of glowing plasma gas behind the engine.

On the 304th second the engine shut down, right on the mark.

The seconds dragged by maddeningly slowly for those who were members of a worldwide radio and television audience, as well as for those biting their nails in Mission Control. It was like a bad dream in which the dreamer is in quicksand, fighting for every step.

Finally, they heard Lovell's voice:

"Please be informed there is a Santa Claus. The burn was good."

It was better than that. After twenty glorious hours in lunar orbit, Apollo 8 was on the way home with unerring precision, "right down the corridor," the mathematical tunnel down which it had to continue to fly so it would reach a point four hundred thousand feet above the earth, at an exact angle, attitude, and speed to begin reentry with a rush greater than any man had ever known.

Fifty-eight hours after leaving the moon, earth used its gravity to drag Apollo 8 back into the atmosphere at twenty-five thousand miles an hour. The fastest spacecraft ever then began a brief life as a manmade meteor. Temperatures soared to what could be found on the surface of a star. Plunging downward with their backs to their line of flight, the astronauts knew their existence now depended upon how well their ship had been built.

No one saw Apollo 8 hurtling through earth's layers of protecting atmosphere. Only fire could be seen, intense, blinding white flame with an outer red sheath, a streamer of fire 125 miles long!

Apollo 8 traded off its tremendous speed for heat. The more fire flowing from the heat shield, the slower flew the spaceship. Then they were through the inferno of reentry. Two miles above the Pacific Ocean, just before dawn, in sight of—appropriately—Christmas Island, three large parachutes streamed away from the ship, opened partially for deceleration, then blossoming wide and full.

Splashdown, recovery, and return to thundering ovations. The world cheered, knew it never again would view the earth without an awareness of its beauty and fragility.

Three citizens of the planet had just completed what the New York Times described as a "fantastic odyssey."

The road to the moon had been opened.

CHAPTER EIGHTEEN

GETTING THERE, GETTING BACK

LOOK AT THIS SONOFABITCH!" The CIA intelligence officer scanned photographs, taken by the powerful cameras of an American reconnaissance satellite, of the Soviet Union's main launching center. "Look at this thing," the officer said to a companion as they studied the computer-enhanced picture of a huge launching complex. "It's bigger than our Saturn V, and the damned thing sure is a lot more powerful. You have those data reports ready?"

"Right here," came the reply. Everything began to dovetail into a clearer understanding of what was under accelerated drive deep inside the USSR. The photographs showed a rocket standing taller than a football field is long—as high as a thirty-seven-story building. Information gathered by intelligence agents added to the growing detail. The official title for the powerful monster was N-1, and it had one specific job. Get cosmonauts to the moon. Get them on its surface. Get them there and back before the American astronauts.

It was early 1969. Much of the Russian space program was shredding. Boosters rushed to flight readiness had proven to be unreliable and exploded on launch stands or in the air. The high-powered Zond program had been abandoned in the wake of Apollo 8. No need anymore for such a flight. Landing on the moon was the only mission that would count.

"You know what this means?" the CIA man commented to his companion. "They're going for the brass ring. If this mother works, they might still beat us to the surface of the moon."

The second man nodded, "Especially if we fail suddenly."

"Well, that's NASA's job. Not ours. Let's get these photos over to 'em." He added: "We'd better keep our recon birds going over that launch site. If they fly this booster successfully, they won't be sleeping much at NASA."

The now familiar countdown procedure came to life at Baikonur for the first N-1 unmanned test launch. Delays came and went, expected with so powerful and complex a booster. The moment of launch ground inexorably toward the long-awaited cry of *Ignition!*

At 12:18 P.M. Moscow time on February 21, 1969, engineers, technicians, controllers, and cosmonauts watched in awe as thirty powerful engines lit off. A Niagara of dazzling flame spewed down the curving flame trenches, sent fire whipping across steel and concrete.

The largest rocket ever built blasted free of its launch pad, heaving upward on ten million pounds of thrust. If this flight proved successful, the Russians were prepared to speed up their program to land at least one man, possibly two, on the lunar surface before the Americans could set sail.

N-1 hurled back a torrent of flame. Just as it cleared its support tower, engines number 12 and 14 "went dark"—their fuel shut off by an internal computer that sensed something wrong. Still N-1 accelerated, pushing into the area of maximum aerodynamic pressure. Right on schedule the twenty-eight remaining engines throttled back to reduce the shock waves of Max-Q. Then the booster was through the "shock barrier." At sixty-six seconds from liftoff the engines throttled back up to full power.

Instead of a smooth transition back to maximum energy, the engines kicked to full bore with tremendous vibration. The huge rocket shuddered and rattled so violently a liquid oxygen line came apart. Supercold, volatile liquid oxygen poured downward within the rocket body.

Fire leaped through the propulsion systems. An engine overheated. Computers were to shut down any engine with exceptionally high operating temperatures, but the computer safeties failed. Fire spread faster and faster.

Engines exploded. Turbo-pumps tore themselves to blazing wreckage in the death knell of the world's mightiest rocket.

High above the erupting flames and explosion, the escape tower attached to the payload snatched an unmanned model of a spacecraft intended to circle the moon and pulled it safely away from the fireball filling the sky.

A terrible eye of red flames appeared in the heavens, expanded in an instant to a flowering rose, and tore N-1 into tumbling, burning junk.

The wreckage fell thirty miles from the launch pad. In the stratosphere, where N-1 had met its enemy of aerodynamic pressure, flame still billowed and writhed, lofting upward in a mushroom cloud with a killer stalk and the final gasp of a dying colossus.

In launch control, there was no need to say what was obvious to all.

It would now take a miracle to keep Soviet Russia in the race to be the first to place men on the lunar surface.

Of all the people on the NASA team with a sense of unshakable confidence in the future, few could match the drive of Deke Slayton and Alan Shepard. From day one of the long journey from Project Mercury to the planned lunar landings, they had never faltered in their support of all members of the NASA teams. Through successes and disasters, through triumphs and tragedies, through their own extremely disappointing groundings, they had only one goal. And there was no turning back.

Together they selected the crews who would fly America's space missions. Early on, Deke had established a pattern: "I would assign the crew that backed up a prime mission crew to fly the third flight after that."

Thus, Deke and Alan called a meeting of the back-up crew that had supported the moon-orbiting flight of Apollo 8. Neil Armstrong, Buzz Aldrin, and Michael Collins gathered in a private room with them. Moments later, in a mixture of shock and exultation, they heard Deke offering them a chance to walk on the moon.

"I got right to the point," Deke said. "Because of Apollo 8's success, NASA management had set an ambitious schedule. There would be two more Apollo test flights, and then a landing would be attempted with Apollo 11—and that was a big if.

'You're *it*, guys,' I told them. 'You've got the Apollo 11 flight, and that means you get first crack at landing on the moon. That is, of course, if we pull off successful missions with Nine and Ten.'

"Neil, Buzz, and Mike grinned, thanked me for my confidence in them, and shook my hand. Two of them had the chance to be the first human beings in the history of mankind to walk on the surface of the moon. The third—in this case, Collins—would remain in lunar orbit, awaiting the return of the others. But they were realistic. All three knew two very big flights had to succeed if they were to get their crack at the brass ring."

If. The *if* covered all possibilities of delays, failures, changes in the heirarchy of NASA, and White House thinking. A long list of vulnerabilities jutting out at all angles. NASA had yet to fly that bug-eyed spidery creature it called the lunar module, which would take two men down to the moon's untouched gritty soil. It must be flown, and successfully, on Apollo 9. That meant firing the entire Apollo assembly of command, service, and lunar modules into earth orbit and simulating as many lunar flight procedures as could be squeezed into the mission. And if all that went well, there was still Apollo 10, which was going to repeat the moon orbit flight of Apollo 8, only with a much heavier payload demand on the Saturn V, which would send the entire Apollo package all the way to the moon for an even more demanding dress rehearsal.

If nothing went awry, if the two missions were successful, if the next great booster was ready, if the Apollo spaceships were ready to go, if no one broke a toe or came down with the flu at the last moment, if, if, if, then the first landing assignment would go to the crew of Neil Armstrong, who had pulled Gemini 8 out of its deadly spin; Buzz Aldrin, who had solved the problems of space-walking on the very last Gemini mission; and Mike Collins, the space walker of Gemini 10.

Mike didn't think they would avoid all the potential pitfalls. He figured that if the Las Vegas bettors were brought in, they'd say the odds of Apollo 11 getting the choice plum would be only one in ten, and the odds would improve to four in ten for Pete Conrad's Apollo 12 crew.

Deke Slayton was staying with a tried-and-proven system in selecting these three for Apollo 11. His strategy, which he employed with Alan Shepard, was to select three men who were personally compatible, who had skills that complemented one another's, and who would serve as a team—commander, command module pilot, and lunar module pilot—and assign them as a back-up crew. The strategy called for a back-up team to sit out the next two scheduled missions, drawing the prime assignment on the third.

"Well," Deke said grinning at his astronauts, "it's a bit more than sitting it out. You people are going to live in the simulators, and you're going to 'fly your mission' a couple of hundred times before you finally go out to the launch pad." He demanded everything of them except corporal punishment and a few hundred pushups. While two other crews flew, they'd practice in the simulators, follow every move the other teams made in their flights, listen to every detail of the debriefings, ask all the questions they could. They would become a great snowball rolling down a mountain, gathering data all the way for their own flight.

Standing on its four spindly landing legs, it looked like a creature from another world—hardly like a spaceship intended to land two men on the moon. This was the lunar module of the U.S. space program, LM for short. It was a sixteen-ton package of eighteen engines, eight radio systems, fuel tanks, life-support systems, and instruments, the product of six years of design and construction by NASA, Grumman Aircraft Engineering Corporation, and its subcontractors.

The LM was unique. It was the first vehicle designed to operate in airless space and as such was the first true spaceship. It was not intended to withstand the heat of reentry into earth's atmosphere and so lacked a heat shield and the sleek aerodynamic lines of the Apollo craft. Instead, with bristling antennae and four slender legs, it resembled an awkward giant bug.

It was the only major piece of Apollo hardware not tested by men in space. Assigned that critical task was the Apollo 9 crew: space veterans Jim McDivitt and Dave Scott and rookie Rusty Schweickart. Said McDivitt when he first set eyes on the craft: "Holy Moses, we're really going to fly that thing? It's a very flimsy craft—like a tissue-paper spacecraft. If we're not careful, we could easily put a foot through it."

Five days after Apollo 9 slipped into earth orbit, McDivitt and Schweickart opened hatches in the docking tunnel that linked the Apollo to the moon taxi, drifted in weightlessness into the LM, and sealed themselves off from Scott, who remained in the command ship. They ran down their check lists, tested their system, and "cast off" their ungainly vessel from Apollo.

In that careful retreat from the command ship, McDivitt and Schweickart became the first astronauts ever to fly aboard a spacecraft designed to operate only in the vacuum of space. Any lunar module fired away from earth was destined never to return. It could descend to a world without atmosphere like the moon, but racing into the atmosphere of the earth would transform the LM into a splash of flame. Anyone flying this machine simply had to make it back to the mother ship if they wanted to return safely to earth.

McDivitt and Schweickart were acutely aware of the unknowns they faced. Drifting free of Scott in the command module, they remained prudently close. If the lunar module failed in its propulsion or other control systems, they still had a way out of being helpless. Scott could close in and link up with the LM.

(Because two spacecraft were involved, to avoid radio call-sign confusion NASA allowed the astronauts to name their ships, lifting a ban imposed after Gus Grissom had tagged his Gemini capsule *Molly Brown*. Some officials were not too pleased with the names selected by the Apollo 9 team, considered them not worthy of this noble effort. Assessing the shapes of the two vehicles, the astronauts named the lunar module *Spider* and the cone-shaped command module *Gumdrop*.)

The first tentative stabs with attitude thrusters and different engines of *Spider* went according to schedule. It was time to bite the bullet, to fly away from *Gumdrop* with newfound confidence in the thin-hulled LM. Flame speared from the descent stage, *Spider* leaped away from *Gumdrop* as if flung from an invisible catapult, and in a move of great confidence the two men aboard the space bug sped 113 miles' distance from Scott.

Then came the moment to "break apart" the boxy main structure of the *Spider*. The bottom half, descent stage, whose engine in the near future would lower its crew to the lunar surface, was jettisoned, leaving a legless, seemingly helpless space creature with two men sailing through orbit. The ascent stage with the crew cabin carried an engine designed to lift this upper portion of the landing vehicle off the moon and carry it all the way to a rendezvous with a waiting command module. On an actual landing mission, the descent stage would serve as a launch pad and would remain on the moon.

McDivitt and Schweickart triggered this ascent engine and began executing a complex series of maneuvers nearly identical to those to be made by two astronauts leaving the moon to catch up with their command ship orbiting sixty miles above the surface. They flew with precision until *Spider* and *Gumdrop* were only a hundred feet apart.

Dave Scott moved his heavy command vessel in for docking. "You're the biggest, friendliest, funniest-looking spider I've ever seen," he told the two astronauts waiting for their ride home. After flying separately for six hours, the ships were together, and another milestone had been passed.

The ugly duckling of a lunar craft was looked upon with all the affection afforded the most beautiful of swans.

Two months later *Charlie Brown* and *Snoopy*, the spaceships of Apollo 10, eased into lunar orbit, this time on a mission not only to test the LM further, but also to perfect navigating around the moon and to confirm a future landing site. The Sea of Tranquility, so named by ancient astronomers who thought it was a smooth body of water, was their main objective. If one particularly level plain in that "sea" proved acceptable, then Apollo 11 would soon be aiming for that same area. But for now it was a matter of test flying and scouting.

Pilot of the big command module *Charlie Brown* was John Young. Commander of the mission was Tom Stafford, and when he and Gene Cernan drifted away in the lunar module *Snoopy* for its vital test, Young was all too aware that his own ship was the only ticket home his two friends had. Make a mistake and miss a rendezvous with these two pilots, and he'd be making a long trip back to earth alone.

This was the final dress rehearsal with everybody on stage center, a mission that had raised questions among impatient (and usually ill-informed) members of NASA, the public, and the press as to why another check-out was needed. If Apollo was going to lunar orbit and the two men would ride *Snoopy* down to some nine miles above the surface, why not "go all the way" and commit to a landing?

There were still too many loose ends, too many questions about precise navigation about the moon. This would bring them all together in a single tight package. Let there be no question that Stafford and Cernan were straining at the leash to take *Snoopy* down to knock loose a cloud of dust and land. But Apollo Program Director Sam Phillips nixed that from the beginning. To risk a landing after only one flight around the moon with a LM that had been run through a single flight test, Phillips said, would be "premature and foolhardy."

While yearning for an "all-the-way" flight, Deke Slayton and Alan Shepard agreed the decision was a prudent one. "We just couldn't risk it," Shepard said.

The press climbed all over Stafford trying to get some controversy going on the issue. Tom countered with an explanation that was smooth enough to be engraved in stone. "There are too many unknowns up there," he told the media hungering for a hot clash between the astronauts and their NASA bosses. "Our job is to eliminate as many of them as we can, and the only way we can do that is to take the thing down to nine miles or less and see how it behaves that close to the moon."

There was more than enough excitement in store for Tom and Gene without dusting Snoopy's landing legs. The countdown to fire the descent engine that would send the LM barreling toward the lunar surface began when Snoopy was over the moon's far side. Mission Control went through another bout of nail-biting as the astronauts punched through critical maneuvers out of sight and out of touch with mission monitors.

Then they heard the excited voice of Young from Charlie Brown; he'd appeared first around the limb of the moon, and as he reestablished radio contact with earth, he fired off the initial message of mission progress: "They are down there," he confirmed, "among the rocks, rambling through the boulders."

Moments later Snoopy appeared, and the exuberant voice of Stafford followed Young's tongue-in-cheek call of boulder-tripping. "There are enough boulders around here to fill up Galveston Bay. It's a fascinating sight. Okay, we're coming up over the landing site. There are plenty of holes there. The surface is actually very smooth, like a very wet clay—with the exception of the big craters."

Cernan's voice, too, rang with unrestrained excitement. "We're right there! We're right over it!" he cried as Snoopy whipped moonward to within the planned nine miles of the Sea of Tranquility. "I'm telling you, we are low, we are close, babe!"

Stafford's voice followed with its own infectious glee. "All you have to do is put your tail wheel down and we're there!" Snoopy swooped low over the moon, actually four miles south of the intended Apollo 11 landing site because of the navigational errors that planners had expected and intended to learn from. They raced on, and then it was time to prepare for the critical dismembering of Snoopy—separating the lunar craft so the legless upper portion would return them to Charlie Brown.

"Sonofabitch!" The curse from Cernan sent instant alarm through Mission Control. Before controllers could determine from their own instruments what could have caused the moment of obvious danger,

Cernan's own explanation followed.

Snoopy was wheeling around in wild gyrations, the snub nose both pitching up and down and yawing violently between left and right. This close to the moon the movements were terrifying and on the thin edge of lethal.

"We've got some wild gyrations," Cernan called out, his voice the measured tones of the veteran test pilot riding the razor's edge.

Fearful controllers in Houston studied their monitors and instruments. From what they saw, they knew Stafford was wrestling with the controls that would blow away the troubled landing stage from the upper ascent stage holding the two men.

Cernan's voice rang out over headsets and speakers. "Hit the AGS!" he yelled to Stafford to activate the abort guidance system. Stafford's hands flew across his controls; *Snoopy* calmed down. The past few minutes had been knife-edged flirting with disaster.

Had the two pilots not reacted as swiftly and skillfully as they did, in another two seconds *Snoopy* would have been locked into a dive toward the moon from which the astronauts never could have recovered.

"I don't know what the hell that was, babe," a relieved and puzzled Cernan radioed CapCom. "But that was *something*. We were wobbling all over the sky."

It was too close. Later they would find that an abort system switch had been, somehow and unnoticed, snapped into an incorrect position. That had sent *Snoopy* into a radar search for its mother ship, *Charlie Brown*, instead of stabilizing itself in preparation for separating its ascent stage from its descent stage.

It was a hell of a way to prove the ruggedness of the lunar module, which had performed far beyond what engineers believed would ever have been demanded of the skittish machine.

Finally separated, *Snoopy*'s upper-stage engine fired. With radar seeking the target and the astronauts flying their vehicle through intricate maneuvers, they rose in a closing maneuver to dock with *Charlie Brown*. Stafford continued his "it's a piece of cake" routine with the remark, that "*Snoopy* and *Charlie Brown* are hugging each other."

The three men, back together in the Apollo 10 command ship, made one more trip around the cratered landscape before beginning the homeward journey. Young took one last look at an alien world, and his words caught CapCom by surprise. "You've often heard the nursery rhyme about the man in the moon," he called. "We didn't see one here, but pretty soon there will be two men on the moon."

Charlie Brown fired up and went home.

Two months later, on July 20, 1969, eight years after President John F. Kennedy had promised to put a man on the moon, Neil Armstrong stepped from the lunar module and climbed down the ladder that would take him to the moon's surface. There was no hurry. Moving into the unknown demanded patience, caution, and readiness for the unexpected.

Armstrong stepped backward from the tiny opening of *Eagle*'s hatch. He headed where no human being had ever gone before, toward the surface of a world devoid of life, a surface punched with craters as small as one foot, as wide as thirty feet. Rocks and boulders littered this alien world with the rubble of cosmic bombardment.

Clumsy in his pressurized spacesuit, which restricted bodily movement, he felt strangely comfortable each time his boots touched a ladder rung, bringing him closer to alien soil.

More than a half billion people on the distant world of earth watched as television screens carried the ghostly sight of a spacesuited figure holding on with both gloves to the *Eagle*'s ladder.

They watched as he moved closer.

Another boot found a lower rung.

Closer.

He was on the bottom step.

Millions watched as Neil jumped the final three and a half feet to the moon's surface. Touchdown. They watched as his left foot pressed down hard on a fine-grained surface at 10:56 P.M.

He pushed his body slightly away from the ladder. Both boots stood planted solidly beneath him.

Immortal words spoken into his spacesuit radio and fired off to earth were heard throughout the world:

"That's one small step for man . . . one giant leap for mankind."

Fifteen minutes later, eager to follow, Buzz Aldrin climbed down the ladder. He stopped as he stood on the moon, "buoyant and full of goose pimples."

It had happened as promised.

Men were on the moon before the decade of the 1960s had ended.

Neil began to move along the ground. Every step was an experiment.

Every movement was an exploration. Every turn, walk, low-gravity jump was a first-time-ever adventure.

Despite the cumbersome spacesuits, both men found moving about in one-sixth gravity exhilarating and described the experience as floating. They would be on the moon for only a short visit, and they were in a hurry to try everything planned for them. Neil expressed it best when he said they felt like bug-eyed five-year-old boys in a candy store.

He and Buzz wanted to take advantage of lightweight gravity to make leaps inpossible on earth. (In their suits they would weigh 360 pounds on earth. On the moon, they weighed 60.) But while they weighed less, they still possessed body mass that restricted their ability to move as freely as they liked. When they leaped up, they found their efforts produced a loping movement. If they started to jog (running in these suits was impossible), the mass and velocity created kinetic energy. When they jogged at the fastest speed they could attain, their momentum made quick stops impossible.

The clock ran down swiftly, and there was much to do. Their check list of activities needed to be accomplished in the two hours they had before they were required to seal themselves back in the *Eagle*. NASA wanted no unexpected surprises on this first venture to the moon.

"The surface is fine and powdery," Neil reported to a fascinated world. "It adheres in fine layers, like powdered charcoal, to the soles and sides of my boots. I only go in a fraction of an inch, maybe an eighth of an inch, but I can see the footprints of my boots and the treads in the fine, sandy particles."

Alert for problems with balance, Neil extended a "collection sample rod" to gather several ounces of lunar surface material in a plastic bag, which that he stuffed into a suit pocket. In the event a sudden emergency forced an immediate departure from the moon, the soil sample would prove invaluable for research scientists back on earth. If they could remain outside for more than two hours, they'd have more than enough time to collect several buckets of lunar soil and rocks.

The clock ticked away steadily. "It's a very soft surface," Neil radioed back to Houston. "But here and there where I bored with the contingency sample collector, I ran into a very hard surface. It appears to be a very cohesive material of some sort."

Slowly Neil put aside the items on his check list, awed by the beauty of the alien landscape. He took a long, slow look at the moon's surface.

What he described astounded and fascinated listeners around the world.

"It has a very stark beauty all its own," he said slowly. "It's like much of the high desert areas of the United States. It's different, but it's pretty out here."

To Buzz, the moon's surface was: "Beautiful, beautiful! Magnificent desolation." He was struck with the shocking contrasts of color. There were many shades of gray, a pale tan, and areas of utter black where rocks cast their shadows along the airless surface.

Standing back from *Eagle*, they saw the silvery ascent stage and the gleaming crinkle-gold coating on the descent stage, the splayed spidery landing legs, and the wide semi-rounded footpads resting in gray dust.

The earth was a resplendent oasis of shifting colors, appearing far larger than the moon. And many times brighter as sunlight splashed off clouds and oceans.

The astronauts moved their television camera sixty feet from *Eagle* so that spellbound earth viewers could watch the dramatic moon expedition. They jammed a pole into the lunar ground, running unexpectedly into sub-surface soil so hard they could barely get the pole to remain erect with the American flag. A metal rod held the flag extended to its full width, since otherwise in the vacuum it would have draped about its pole. Armstrong and Aldrin then deployed a seismometer to radio to earth information on quakes and rumbles within the moon and to record the strikes of meteorites smashing into the ground. A device to measure the flow of radiation particles from the sun—the solar wind that blows through space—went into position. They also deployed a multi-mirror target for laser beams fired from earth by both American and Russian scientists.

There were protocols to meet for the historic occasion. On the lunar dust they placed mementoes for the five American and Soviet spacemen, Gus Grissom, Ed White, Roger Chaffee, Vladimin Komarov and Yuri Gagarin, who died in a plane crash in 1961. They unsheathed a metal disc on the descent stage with engraved messages to future moon visitors.

As Neil Armstrong read the plaque's words, his voice carried throughout the world. "Here men from the planet earth first set foot upon the moon, July 1969, A.D. We came in peace for all mankind."

The last phrase evoked both satisfaction and no small amount of cynicism on the home planet, where wars small and large raged across several continents. It might have been peaceful on the moon, but the message was almost a mockery of the reality on earth.

There was yet another small cargo—private and precious—carried by Neil Armstrong to the moon. It was not divulged at the time, but he carried the diamond-studded astronaut pin made especially for Deke Slayton by the three Apollo 1 astronauts and presented to him by their widows after that dreadful fire.

The two astronauts gathered fifty pounds of lunar soil samples, dust, and rocks, packed their precious find in sealed containers, and used a hand-powered pulley system to send the boxes up to the ascent stage. Working with untried equipment in vacuum, they struggled to get the boxes aboard their ship, kicking up a cloud of moon dust.

It was time to shut down the first moon walk. They worked their way to the ladder and squeezed into their "flight deck," and sealed and pressurized their cabin. They stowed gear not necessary for flight and went through their long check list, following every procedure worked out before leaving their launch pad. Buzz had walked and jogged about the moon for one hour and forty-four minutes, Neil for two hours and fourteen minutes.

They removed their boots, slipped out of the backpacks heavy with life-support equipment that had kept them alive on the moon, reopened the hatch, and dumped them along with crumpled food packages and filled urine bags onto the surface. Everything that kicked up dust was detected in Houston as slight moon tremors registered by the seismometer the astronauts had placed away from the ship.

CapCom reminded them that they needed to sleep for five hours before starting the countdown to liftoff. It was going to be impossible to sleep. They were cold in *Eagle*. Whatever had been set up to keep them warm on the airless world left much to be desired. Their available space was cramped and austere, and they were wound up tighter than alarm clocks with elation and excitement. For those five hours they slept fitfully.

They also were unaware of other dramatic events taking place before and during this period. On July 3, only thirteen days before Apollo 11 left for the moon, Moscow had ordered the launch of a second unmanned N-1 rocket, this time with a heavy Soyuz spacecraft in a dress rehearsal for a manned lunar touchdown. The Soviets could still be first on the moon—if Apollo 11 failed.

It was less than one hour before Russian midnight. A night launch is great for pyrotechnics, and this flight didn't disappoint those fascinated with skyrockets. Less than ten seconds after liftoff, just over five hundred feet high, a chunk of debris raced through a liquid oxygen fuel line. The system did not have filters to trap the debris, which lodged in the oxygen pump of the No. 8 engine. Control wires and electrical lines short-circuited, and the automatic KORD computer safety system shut down all thirty engines! At the same instant sensors flashed the abort signal to the escape tower, which ignited and safely raced away with a heavy Soyuz spacecraft.

The monster rocket, now "a mountain filled with fuel," smashed into its launch complex. The explosion was equal to that of a tactical atomic bomb, a blast so powerful and intensely bright it was detected by American reconnaissance satellites and set alarms ringing at the North American Air Defense Command.

The blast wave ripped outward from the devastated launch stand and smashed into another N-1 booster on the adjoining launch pad, caving in structural panels and seriously damaging the rocket.

The Russians would not attempt firing another N-1 for two years. There was no way to avoid the obvious—the Soviet Union finally had yielded first position in the moon race to the United States.

But there was still a chance to snatch some of the thunder from Apollo 11. Only three days before the Saturn V rose with its crew of three for the moon, the Russians gave it another try, but this time with a less powerful rocket bearing the unmanned Luna 15. If the mission worked out, then the six-ton payload would decelerate into lunar orbit, go through an exacting check-out, and then fire up its engines to soft-land on the moon. There a drill-like device would bore into the soil, collect samples, and load the lunar material into a sealed compartment. With its several ounces of cargo tucked away, an ascent engine would fire to lift the upper part of Luna 15 away from the moon and send it back to earth for reentry and landing. A successful Luna 15 mission would have given the USSR the opportunity to announce that it had collected the first soil from the moon and to argue that men weren't really necessary for lunar exploration. A bald-faced lie of which they were accutely aware in Moscow, but one that would suit their propaganda efforts perfectly.

Tracked closely by the Americans, Luna 15 slipped into orbit around the moon. As Neil and Buzz slept in fitful spurts in *Eagle*, the Russian probe fired up to begin descent to the surface. After fifty-two swings around the moon, Luna 15 inexplicably crashed into the Sea of Crises at six hundred miles an hour. Its wreckage-spewing impact was recorded in detail by the sensitive seismometer Neil and Buzz had placed on the surface.

Twenty-one hours after touchdown, Armstrong fired up the engine, and *Eagle* blasted free of its launch platform—the bottom half of the lunar module.

Insulation material torn free by the rocket blast scattered widely in a shower of debris. Neil Armstrong, watching the surface, saw the first American flag deployed on the moon yield to the whoosh of dust and debris and fall slowly over on its side.

That was all the time the astronauts had for sightseeing as they manned controls and computers and radar systems for the three and a half hour trip to rendezvous with Mike Collins and the command ship orbiting sixty miles overhead. They flew *Eagle* skillfully and precisely. Collins watched the LM drive "steady as a rock" and "right down the center line of final approach" toward linkup. Ten minutes later *Eagle* was firmly docked with *Columbia*. The two ships were once again one.

Buzz and Neil floated back into the command module, which soon echoed to the wild cheers of three astronauts whose flight would forever change man's view of his planet. At such a joyous moment, Collins related, these responsible, highly trained, extremely skilled men, who had just carried out the impossible, were "all smiles and giggles over our success."

Armstrong and Aldrin transferred their lunar booty into the command ship and then discarded their faithful *Eagle*, leaving it to orbit the moon for several weeks before lunar gravity tugged it down to a crash landing.

It would take sixty hours to make the return trip home. But by now Mission Control and the three men in *Columbia* knew the highway.

Just follow the trail locked into the computers by Apollos 8 and 10.

Buzz Aldrin wrapped up the mission with a deeply felt remark about the plaque left on the moon that stated so clearly that the crew of Apollo 11 had made the journey "for all mankind."

"I hope," Aldrin said, "some wayward stranger in the third millenium may read it and say, 'This is where it all began.' It can be the beginning of a new era when man begins to understand the universe and man begins to truly understand himself."

Meanwhile, the back rooms at Houston were the scene of intense planning and no small controversy.

One man was determined, come hell or high water, to get to that desolate little world of "stark beauty" 240,000 miles away.

No one gave him much of a chance of succeeding.

If it is possible to hate an inanimate object, then Alan Shepard hated his desk. Fighter pilots dread orders from higher command that chain a man to a desk, which has no chance of getting off the ground. Clip the wings of an eagle, and you wind up with a turkey.

And no matter what else he might be, Shepard in his heart and mind was and always would be an eagle born to fly. Even all the way to the moon and back.

Alan's grounding for almost five insufferable years remained strictly a medical problem. Ménière's syndrome had finally explained the sudden onset of dizziness, the giddy and helpless feeling of a room spinning before his eyes. Vertigo had been a constant threat. A pilot without balance doesn't fly. It's a one-way ticket to disaster.

More than a year before Apollo 11's history-making landing on the moon, Alan was going from bad to worse. As chief of the Astronaut Office he fought off melancholy as one by one he sent astronaut teams off to the launch pad to soar into space. By the summer of 1968 the hearing in his left ear had decreased and his balance had degenerated. While before that there had been some hope for recovery, medical teams now told him he'd never fly again.

Alan was a believer. In himself, and in doing the impossible. There was only one way to win back his wings, and that was to accomplish what the NASA doctors said could not be done: cure the Ménierè's syndrome that shackled him to his desk.

The long road back began with an exchange with fellow astronaut Tom Stafford. "Al, maybe I've got something for you."

Shepard listened. Tom Stafford wasn't one for idle conversation.

"The guys were talking about some doctor in Los Angeles who's pulling off cures for problems other M.D.s say can't be cured. They say he's a specialist in ear, nose, and throat. Nothing unusual about that, but the word is that he's developed a surgical procedure that can get rid of this Ménière's problem you've got."

"Surgical?" Alan asked carefully.

"Uh-huh. You're being treated by medication, right?"

Shepard nodded.

"Doing any good?"

"Sure. I'm just about deaf now in my left ear. You said he uses surgery?"

"That's what I'm told. More than that, whatever he does works. Why don't you go see the man? You've got nothing to lose."

"You're right," Alan replied. But he had a different outlook toward the possibility of release from debilitating affliction. He didn't judge the matter as nothing to lose. To Alan Shepard it was a case of his having everything to gain.

"Hell, yes," he said abruptly to Stafford. "I'll go anywhere, try anything."

He met with Dr. William House in Los Angeles. House examined Shepard, and they met for a straight talk.

"In my judgment, Mr. Shepard, you're what we call a classic Ménière's case."

"Thanks," Alan said dryly. He knew that already. Classic or common, it didn't matter. What counted was getting back into flight. "Can you help me?" he asked.

"I believe I can."

"Don't stop now, Doctor."

"Here's what I recommend. First, I'd go after the fluid that's building up such pressure in your inner ear. That's more than enough to ground any pilot. Turn the world upside down."

"Yes," Alan said noncommittally.

"I'd surgically cut a hole in the sac containing that fluid, after which I would insert a short tube between the fluid sac and a gland that leads into the spinal column. That will drain off some of the fluid, just for starters, so instead of the intense pressure within your ear, concentrated in a very small area, that pressure would dissipate over a much wider area. That, and your body doing its best to return to normal, just might do it."

Shepard didn't waste a moment. "Sounds like a hell of an idea to me. Let's go for it."

He returned to Houston to discuss the matter with Louise. His wife had watched him through five years of frustration and periods of bitterness. "But there's no guarantee in all this," he told her. "House will do everything he can, and I think he's great, but he can't promise he'll be successful. I'm burning up inside, Louise. I want so badly to fly again in space." He balled his hands into knotted fists of repressed emotions. "I'm willing to try anything."

"I know," she said quietly, fully aware that surgery involving the inner ear was delicate and dangerous at best, at worst bad enough to cripple for life the man she loved.

"Do it," she urged. "Go for it," she added in the same words he'd spoken to Dr. House.

He nodded. "All right." Relief flooded through him. "But we can't tell anyone," he cautioned his wife. "I want this done in absolute secrecy."

Alan Shepard swore Dr. House to absolute confidence. This doctor turned out to be as great a man as he was a surgeon.

"What name do we use for you?" was his response.

"Give me any name. Just so long as it's not mine."

House called his office nurse, one person who was in on the secret. She was of Greek extraction, and House kept that in mind. "Give me a name," he directed.

She pulled a name out of thin air. Medically, Alan Shepard wasn't in Dr. House's care. Victor Poulos was the name he used to check into the hospital. He remained there for two days for surgery and the follow-up examination.

Louise's eyes widened when she saw her man with a large bandage covering the left side of his head. Beneath that bandage a tube, thinner than pencil lead, had been inserted between his inner ear and his spinal column. He explained the operation to Louise.

"How soon will you know?" she asked.

"It could take months," he told here.

"It's already taken five years," Louise reminded him.

Back at work, Shepard might have well as tried to conceal a bright strobe light on the side of his face as to ignore the heavy bandage. No secrets there. So he asked, first, Deke Slayton and, then, the NASA flight surgeons to keep the true need for the bandage a confidential matter among themselves. They agreed, and then worked with Dr. House to monitor his progress as the months rolled by with agonizing slowness.

He improved steadily. He had waited five years for this surgery, and he waited another eight months while his hearing returned and he gradually regained his balance. At the end of that period he met privately with Dr. House.

"My friend, you're cured."

That wasn't quite enough. But when the NASA flight surgeons ran Alan through tests bordering on the edge of brutal, they grinned at

Shepard. "Dr. House is correct. You are cured. It's amazing. Full recovery."

Alan was elated, but still it was not enough. "Which means?" he asked.

They knew his hesitation. "It means," they told him, "you are fully qualified for space flight."

The spring in Alan's step wasn't just elastic. He almost went ballistic with untrammeled joy.

Suddenly the moon began coming closer. Alan went after the prize assignment like a hungry wolverine. He burst into Slayton's office.

"Deke, dammit!" he burst out, no pretense at protocol. "We've got to get me a flight to the moon!"

Deke was more than pleased for his old friend. They'd fought the same kind of battle for years, wore the same chains to a desk. He'd do anything to get Alan back "upstairs." It would mean one of the Mercury Seven would reach the moon, and as Deke sat back to consider the possibilities, his eyes widened. A flight for Alan could mean a slight crack in that closed door for Slayton. There just might be an unexpected silver lining in the cloud that had hung for so many years over both men. If a forty-five-year-old pilot like Shepard could rebound from a medical grounding, then there was still hope for himself. Like Alan's, until recently Deke's periodic medical exams had continued to paint a black wall for his future as an astronaut.

But what the hell. A whole passel of flight surgeons had time and again told Shepard, right here, that he'd *never* be at the controls again of anything that left the ground. And that included balloons!

"I'll get right on it," Deke said, breaking away from his thoughts.

Shepard was tearing apart the chains from his desk. "Hell, Deke, don't give me that getting on something. Let's do it."

Deke smiled. Alan was a thoroughbred trembling for the break at the gate. "Well, Pete Conrad's crew is training for Apollo 12," Deke said. "I'm just about to name Jim Lovell's crew to 13."

Alan knifed into the choices. "Okay, let's move Lovell back one mission and give me the shot at Thirteen."

Deke grunted. "Easier said than done, old buddy. I can't see Jim backing up for *anybody*. And under my normal rotation, he knows he's in line."

Alan grinned like a bulldog and sank his teeth in deeper. "You tell Jim yet he's got Thirteen?"

"Not yet."

"So make it official that I'm taking Thirteen," Alan pressed, "and at the same time you can assign Lovell to Fourteen!"

Deke said he would give it his best shot. "But until the smoke clears," he warned, "this is not official. It's too quick for that, Al. You start thinking about who you want as crewmates, make some preparations. I'm not promising you anything but to do everything I can."

Alan grinned and went to work. The whole affair was a sub-rosa crew assignment, but Shepard went at mission planning as if everything had been chiseled in granite. He selected two rookie astronauts. They were outstanding pilots and astronauts, sharp, capable, highly intelligent. They just hadn't cut the umbilical to free them for vertical launch beyond the atmosphere. Stuart Roosa would be his command module pilot, and Edgar Mitchell, whom most astronauts called "The Brain," would walk the moon with him.

His choice brought on a ground swell of grumbling, and even open criticism, from his peers. The innuendos were that Alan had selected two newcomers to the crew to lock in his own seniority among the group and guarantee command of the mission to the moon.

Shepard never let someone else's opinions derail his priorities. "Wrong!" he snapped. "I picked these guys for one reason only. They're tops. They can do the job. I want them with me. Do you guys read me?"

"Five by five, Al," they said carefully.

Before any crew selections were announced, the names-and-missions selections were kicked upstairs to NASA headquarters for final approval. Deke had never experienced a rejection of his crew selections, but there's always the first time.

The choice of Shepard and two space neophytes for Apollo 13 raised eyebrows to new heights at headquarters. Their caution seemed justified. NASA's top people questioned the wisdom of sending three men to the moon with a total space flight experience of but sixteen minutes— Shepard's popgun suborbital flight that had made him America's first in space.

They judged the selection, *at that time*, as an unjustified risk. Yet they also judged Shepard unexcelled in his ability, and they concurred in his selection of Roosa and Mitchell.

They stayed with caution without rejection. They told Deke to give the Shepard crew more time to train, to work the simulators, to be hair-trigger prepared.

The word went back to Houston. Jim Lovell would command Apollo 13.

Alan Shepard would command Apollo 14. He had his flight!

Faith, dedication, skills—unflagging for more than five years, they were making the comeback pay off in spades.

Fate also had stepped in, but no one yet could know the future. Being bumped back one flight to Apollo 14 was a godsend.

Apollo 12's Pete Conrad, Alan Bean, and Dick Gordon set sail on November 14, 1969, for the moon's Ocean of Storms. The Saturn V rose into what had been heavy rain but without lightning. But NASA quickly learned that a thirty-six-story rocket under full bore, climbing in rain, becomes a lightning generator from the escape rocket on top to the last jagged streamers of fire from its engines more than eight hundred feet long. Static electricity built up by its passage through the rain can suddenly discharge.

Soggy observers could not see the rising Saturn V through the rain and clouds. But suddenly the assembled thousands were surprised and alarmed when lightning flashed from the low clouds into the launch complex area. Someone in the VIP guest area cried out in shock, "My God, what's happening to the Saturn V?"

The three Navy commanders inside the command module they had named *Yankee Clipper* saw brilliant flashes and heard strange roaring noises that space veterans Conrad and Gordon had never before known. Lightning cracked against their spacecraft and tripped the main circuit breakers. Inside the ship, darkness fell, then came alive suddenly with flashing red warning lights.

Back-up batteries in Apollo 12 came on line to reinstate energy needs. Pete's first words seemed to drop the temperature in Mission Control by twenty degrees.

"I think we got hit by lightning," he said with astonishing calm. "We just lost the guidance platform, gang. I don't know what happened here." Even as he spoke, *Yankee Clipper* automatically brought itself back to light and power. The huge Saturn V, with a separate, independent guidance system not affected by the jolt, continued to accelerate the Apollo 12 assembly to earth orbit.

Conrad scanned his gauges. "We just had everything in the world drop out," he called CapCom. Dick Gordon quickly punched some circuit breakers to shift from the back-up to the primary electrical system. Everything came back on line.

Mission Control came back with relief. "We've had a couple of cardiac arrests down here."

The Saturn V's third stage pushed Apollo 12 into orbit about the earth and shut down on schedule. For the moment the astronauts were safe enough to take deep breaths and get their ship back in prime condition. Few among the ground teams believed the spaceship would even fire up for the moon. Every guidance, navigation, and computer control system had to be completely reset with updated programs and validated by computers aboard the ship and in Mission Control. Most engineers figured the odds as only one in a hundred for a moon landing.

The coordination between orbit and ground pulled off the near impossible. All systems checked out, the third stage flamed, and Apollo 12 was moon-bound.

Conrad planned to land within six hundred feet of an unmanned Surveyor robot ship, which had touched down to scout the Ocean of Storms landing site thirty-one months before. The descent from moon orbit in the lunar module Conrad and Bean had named *Intrepid* was a flight of "incredible accuracy and control." Pete told CapCom the Surveyor seemed to be waiting for them. He danced *Intrepid* about to select the best touchdown area and dropped gently to the surface.

"Outstanding!" Pete shouted into his mike. "I can't wait to get outside! Those rocks have been waiting four and a half billion years for us to come out and grab them. Holy cow, it's beautiful out there."

Pete was the shortest of the astronauts, just five feet six inches tall. He also was one of the funniest. He climbed backward from *Intrepid*'s hatch, worked his way down the ladder, pushed away for the drop to the ground, and sang out for all the world to hear: "That may have been one small step for Neil, but it was a long one for *me*."

The irrepressible Conrad and his teammate, Alan Bean, performed two four-hour excursions from their lunar lander, deploying scientific instruments and collecting seventy-five pounds of rocks and surface material. They jogged down the slope of a wide crater to Surveyor, then chopped and hacked fifteen pounds of material from the robot to take home for study. They also plowed their way through a surface markedly different from what Apollo 11 had faced.

"The dust!" Pete exclaimed. "Dust got into everything. You walked in a pair of little dust clouds kicked up around your feet." Dust clung to their boots and spacesuit legs so thick that during their rest and sleep periods they remained in their suits to keep vital parts from becoming clogged.

Back outside, they found unexpected sights exciting the scientists listening to every word. Conrad reported "a group of conical mounds, looking like . . . small volcanoes." They found green rocks and tan dust. The dust even covered the Surveyor, which proved *Intrepid's* landing engine had blasted away a dust storm for more than a thousand feet from the ship.

Soon, much too soon for Conrad and Bean, *Intrepid* left the Ocean of Storms and made a picture-perfect flight to rendezvous with Dick Gordon and *Yankee Clipper* for the return trip. They went home in a flight of continued dazzling precision.

John F. Kennedy's goal of landing men on the moon and returning them safely to earth before the decade was out had been achieved, twice.

The months rolled by with interminable slowness as Deke remained shackled to his desk while he sent his friends on history-making adventures to the moon. No one could miss the frustrated mood that sometimes hung over him like a dark cloud. Nor could they fault him.

He had felt a sense of victory in getting his close friend, Alan Shepard, command of a lunar mission. But that was an aside to the stonewalled urge to get "out there" himself.

His skipping heart hung about his neck like a great dead albatross. He was almost convinced he'd have the most unenviable record from among the original Mercury astronauts. He'd be the only one never to fly in space.

Deke was the most experienced test pilot in the group. Rated to fly anything with wings. Which meant he had to be ready for the totally unexpected.

Early in 1970 it happened. The dark and dismal future cracked open. Barely enough to send a glimmer of hope in Deke's direction.

He was at the Cape to prepare the Apollo 13 crew for its mission. The miseries of clipped wings worsened when Deke came down with a cold. He reported to the flight surgeons for help.

"Vitamins," they told him, fully aware that neither they nor anyone else really knew what caused the common cold. "Take vitamins, a hell of a lot of vitamins."

Deke grumbled and began cramming large doses of Vitamins B, C, and E into his body. He knew very well the old adage that if you treat a cold it goes away in seven days, and if you ignore a cold it goes away in a week. But his was a whopper, and he kept ingesting the vitamins day after day.

Ever since he'd been bounced off the second Mercury orbital flight, he had had his heartbeat irregularity on an average of once every month.

Then he realized that had changed since he had begun taking vitamins. He rushed into the office of Dr. Charles Berry, the astronauts' chief flight surgeon.

"Chuck, listen to me," he said excitedly. "I haven't had one of those heart episodes for a hell of a long time. Something's happening!"

Berry studied his long-time friend. "You doing anything different?"

"Hell, yes. I'm a vitamin junkie."

Chuck Berry nodded slowly. "Deke, keep taking them. Keep dancing with the one who brought you. If you continue to go along without any heart episodes, well . . ."

"Well, what?" Deke demanded.

Berry looked up as if he could see beyond the sky, then back to Deke. "Maybe, my friend, just maybe . . . "

Maybe . . .

CHAPTER NINETEEN

LIFEBOAT!

IT WAS TIME TO STIR the frigid broth deep inside Apollo 13. Four large circular tanks contained super-cold liquid oxygen and liquid hydrogen, the "soup of life" for the ship and its crew of three. Apollo 13 was a vessel of long-range exploration, and tiny fans stirred the tanks of liquid oxygen and hydrogen that kept its three astronauts supplied with breathing air, drinking water, and electricity for their ship. Astronaut Jack Swigert stretched slowly. It had been a good day, and he had completed most of his scheduled assignments. Swigert reached out to his control panel and flipped a switch, activating the tiny fans within the tank.

Two wires supplying electricity to one of the fans touched. A spark flashed, and the Teflon insulation soon glowed with licks of flame. Fire quickly sped toward the tank's oxygen supply. The intense heat caused a build-up of internal pressure, which quickly expoded the tank's structure. Its dome blew outward with the effect of a shotgun blast, destroying vital lines and systems. Apollo 13 was a ship torpedoed from within. Valves twisted shut, blocking the critical flow of vital liquids as the blast shredded everything in its path. The side of Apollo 13's service module blew out, and the spacecraft began to die.

Until this moment, fifty-five hours and fifty-five minutes after Apollo 13 had been launched from Cape Canaveral, Commander Jim Lovell had judged their Apollo flight as "the smoothest flight of the program." It had been so uneventful that only a few hours before, CapCom Joe Kerwin had radioed Lovell, complaining, "We're bored to tears down here."

As the side of the service module containing the astronauts' critical fuels and rocket power exploded, the crew felt a sudden *bang!* Two hundred and five thousand miles from earth all hell broke loose. Linked together like a small train, the entire Apollo 13 assembly—service module, command module, lunar landing craft —rocked violently.

Inside the command ship *Odyssey*, alarm systems rang shrilly amid an eruption of flashing warning lights. Jack Swigert contacted Mission Control, his voice filled with alarm. "Houston, we've got a problem here."

Flight controllers snapped to full alert. One called out, "What the hell's the matter with the data?"

"We've got more than a problem," another controller announced as he stared in disbelief at his monitor.

Shift manager Sy Liebergot notified the full team. "Listen, you guys. We've lost fuel cell 1 and 2 pressure. We've lost oxygen tank 2 pressure." He took a deep breath. "*And* temperature."

Temperature was vitally important. The liquid oxygen had to remain at a super-cold 297 degrees below zero, and the liquid hydrogen even colder, 423 degrees below, if the fuel cells these gases fed were to continue supplying power to the ship.

Those were the things flight controllers knew, but no one knew yet what had happened. Were their instrument readings correct? It didn't seem possible with a ship that had been functioning perfectly. Fred Haise, the third member of Apollo 13, said to his crewmates, "Maybe we got hit by a meteorite."

The entire spacecraft continued to vibrate badly. All signs indicated that Apollo 13 was breaking apart. The clamor of alarms and flashing lights continued while the crew, and Mission Control, under the direction of veteran Flight Director Gene Kranz, clung to the belief that electrical system glitches were the cause of the crisis indicated by their instruments. They couldn't accept that Apollo 13 had flown into mortal peril.

The astronauts reset their switches, expecting to bring everything back on line.

It didn't work. Swigert sent the grim message of "No joy."

The Apollo 13 assembly oscillated wildly, rolling and pitching up and down, swinging from side to side like a sailing vessel being tossed by hurricane winds.

Lovell, puzzled, looked through a porthole. Thirteen minutes had passed since the jolting shock of the explosion, and he stared out at what he knew was "potential catastrophe."

Mission Control froze with his words. "We are venting something out into the . . . into space," he radioed.

The Apollo 13 commander saw his crew's oxygen spewing away from the only tank that still held the life-sustaining gas, and he suddenly felt helplessly alone.

The cloud erupted from the spacecraft with such force and volume that amateur astronomers on earth, 205,000 miles away, saw the expanding sphere turn into a twenty-mile-wide cocoon of gas and debris.

Lovell felt a knot tightening in his stomach. From this moment he abandoned all thought of ever landing on the moon in the lunar module they had named *Aquarius*.

All that mattered now was staying alive. They were still plunging toward the moon, their oxygen almost gone, their electrical power dying. They knew the powerful engine would no longer fire, starved as it was of the electrical energy needed for ignition and burn. Without it they couldn't get into orbit about the moon. More importantly, without it they wouldn't be able to get home.

Without some kind of miracle they would be marooned in an orbit that would take them hundreds of thousands of miles away from the earth.

Apollo 13 was in its death throes.

If Lovell and his crew were to survive, Mission Control had just hours to perform calculations and make engineering recommendations that would normally have required weeks.

Gene Kranz immediately set to calming his shocked team. "Okay, now let's everybody keep cool," he announced. "We've got the LM still attached. The LM spacecraft is still good, so if we need it to get back home . . ." His voice trailed away for an instant before his mind snapped back to full alert. "Let's solve the problem, but let's not make it worse by guessing."

His words pulled the team together. Barely in time. Minutes later the second power system aboard Apollo began to fail. The second oxygen tank, damaged in the explosion, began losing pressure.

It was bad, and it was going to get worse. The ship would become a lifeless derelict and its crew doomed unless they could retain the electrical power that was draining away.

Leibergot and Kranz instructed Lovell and his crew to start powering down their ship immediately. Reduce to the absolute bare minimum whatever they needed to keep alive and retain communications with Houston.

There wasn't a moment to lose.

Kranz's team maintained a steady exchange with the Thirteen. Another group in Houston ran to telephones to call leaders and specialists. Chris Kraft, now the deputy director of the Manned Spacecraft Center in Houston, was in the shower when the phone rang. He threw clothes onto his still-wet body and broke all speed laws driving to the control center. He went to Kranz's console on a dead run.

Kranz didn't mince words. "We're in deep shit," he told his mentor.

Alan Shepard and Deke Slayton were in their respective offices, buried in paperwork, each with an ear tuned to the squawk boxes carrying the running conversations between CapCom and the Thirteen. The call from Swigert brought them upright. By the time Swigert finished that one brief "Hey, Houston, we've got a problem here," both men tore out of their offices and headed for Mission Control.

Deke hit the control room slightly before Alan. "The damn ship sounds like it's coming apart," Deke told him.

"Let's not waste any time," Alan nodded. "Better round up the guys for the simulators."

"Good, let's go."

During every mission a group of astronauts was always kept on standby in the event something went wrong. Their task was to learn the problems and immediately start working out solutions in the simulators that duplicated the craft in space.

Alan and Deke realized quickly that this was going to be a bitch to solve. Shepard put astronauts in both the command and lunar module simulators. The teams settled in to study the emergency, and Alan couldn't help thinking that this could have been his flight. He'd fought to get

Apollo 13. It could just as easily have been him, Ed Mitchell, and Stu Roosa up there, hanging to survival by a thread.

Shepard was also a realist and knew that if they failed to bring the crew of the Thirteen home alive, the chances of any future moon missions would be seriously jeopardized.

Gene Kranz, Chris Kraft, Slayton, and Shepard considered the options. They could go for a direct abort, turning Apollo 13 around so that the main engine could be fired to reverse the direction of the spaceship's path toward the moon. Shepard had been living with the Apollo systems now for months. "It's too risky," he said, countering the idea. "From what we've seen, electrical power is almost gone. The chances are the engine won't even ignite and, even if it did, we can't be sure of a long burn."

Kranz agreed. Their best course was to keep the Thirteen on its way, let the ship loop around the moon, and start the swing back to earth. On the first three lunar flights, Apollos 8, 10, and 11, the spacecraft had been programmed so that the final engine burn launched the ship into a "free return trajectory." Once the craft looped around the moon, it would be on the correct course for its return trip to earth. No additional engine firings would be required. If the astronauts on those missions had run into propulsion problems, all they would have had to do was to sit tight and wait out the ride, knowing their course for home was already set. While the "free return trajectory" minimized the chances that the Apollo craft would go off the course required for successful reentry, it also limited NASA's options in selecting launch dates and landing sites on the moon. With Apollo 12 NASA had tried something different; the spacecraft burned its final stage for a "hybrid free return"—the ship would still loop around the moon before swinging back toward earth, but a subsequent engine burn would be needed to put the capsule on the course necessary for a successful reentry.

Flight controllers had known the "hybrid free return" increased the odds for problems, but NASA had judged the added flexibility in selecting launch dates and in picking landing sites worth the risk.

It had worked fine on Apollo 12. But the same hybrid moon loop could be a disaster for Thirteen. On their current path the three astronauts would be pulled by lunar gravity into a safe whip around the moon and sent back toward earth. Except that they'd miss the planet by four thousand miles.

Deke scratched his head. "Looks like we better have a good burn with the LM, then."

They all shared the same question. Could the undamaged lunar module engine, with only half the power of the main service module power plant, have enough energy to "push" the combined weight of the command, service, and lunar modules onto the course needed for an accurate return?

The guidance officer calculated the exact velocity needed to correct the free-return path. He determined the burn time and angle data, which were sent to the astronaut crews in the simulators. They set up the simulated burn, fired the engine, and checked the resulting speed and direction. Then they did it again. Finally one astronaut held up a thumb. "They can do it," he said simply.

While these calculations were being made and the simulator tested, the question of the survival of the Apollo 13 crew became critical. Oxygen and power continued to bleed away from the ship, which was dying with every passing minute.

Exactly fifty-one minutes after the explosion had rocked Apollo 13, Gene Kranz turned to those gathered in mission control. "Two hours from now, unless we come up with something that's never been done before, those guys are going to be in a derelict ship. Except for three short-life batteries and their reserve oxygen supply. And we can't use them. They must save them for reentry."

"If they get that far, dammit," Deke cursed.

CapCom Jack Lousma started to relay the message. But Swigert got on the radio first, asking about the oxygen status in *Odyssey*.

"Oxygen is slowly going down to zero. We're starting to think about the LM as a lifeboat," Lousma informed him.

"That's something we're thinking about, too," came the reply from Swigert.

Flight Director Glynn Lunney and his team relieved Kranz and moved into the Mission Control trenches. The decision to transform *Aquarius*, the lander, into the first space lifeboat would be his to make. There wasn't much to talk about. *Aquarius* was the only chance those three men had.

The astronauts in the simulators worked furiously to come up with answers to new questions. Lovell and his crew had to shut down *Odyssey*. Not kill everything in the command module, but put it into a form of hibernation so that it could later be restored to life for reentry. At the same time they had to power up *Aquarius* without using electrical power from either the command or service modules. It had never been done before.

"It looks like we've got about eighteen minutes," a controller said. "The last fuel cell is going fast."

Lovell and Haise pulled themselves through the connecting tunnel into *Aquarius*. The landing module was designed to transport two men to the lunar surface and then to launch them to rendezvous with the command module. Under ordinary circumstances it would be called on to perform for approximately forty hours. The module now needed to serve as the astronauts' home. And somehow its systems must be extended to support three men for nearly four days of an earthbound journey.

Swigert remained in the dying command vessel while the others powered up the LM. One by one he shut down *Odyssey*'s systems. He continued working by flashlight when darkness enveloped the cabin. Before he joined the others in the LM, he transferred the precise alignment of the ship's guidance platform to a similar guidance system within *Aquarius*.

As Swigert worked in the ship that Lovell now judged to be "forlorn and pitiful," Lovell and Haise took apart Haise's moon-walk pressure suit to rig up a ten-foot hose from the LM oxygen system. Running the hose into the command module enabled Swigert to remain there long enough to completely power down the ship they would later need to return to earth.

As cramped as the command module had seemed, it was spacious compared with the lunar module. If they were to be sustained by the LM, the astronauts would need to return to the cold, damp, hibernating *Odyssey* for food and bathroom facilities. It would be an extremely uncomfortable three days.

Watching Lunney's team calculating flight trajectories, Deke told the others that things weren't as bad as they seemed, a conclusion that seemed mad under the circumstances. Had the explosion occurred while Lovell and Haise were on the moon, or if the blast had come after the astronauts left the moon, linked up with *Odyssey* and were on the way home, nothing could have saved the men, Deke pointed out. *Aquarius* would be gone, and their only hope for survival would have been the crippled and dying command module.

"If this had to happen," Deke observed, "it couldn't have come at a better time in this flight. Because it happened when it did, we just may get these guys home."

Kranz's team had worked a maneuver that would provide the best chance for Apollo 13 crew to make it back alive. "We'll go for a brief burn a few hours from now before they reach the moon. That will give them the free-return trajectory. Then, we'll do a second burn later to drop them into the slot for reentry. That brings them home in four days. So our job after the first burn is to figure out how we keep them alive before they run out of everything."

Five hours and thirty-five minutes after the crippling blast, the astronauts fired off the *Aquarius* descent engine for a thirty-one-second burn. Timing and engine thrust, along with alignment, were perfect. A critical milestone was behind them. "Okay, Houston. Burn's complete," Lovell confirmed. "Now we have to talk about powering down."

They had more than enough oxygen for the long trip home, but electrical power still hovered at critical levels. Without electricity to operate the fans that kept cabin air moving, the cabin would soon become fouled with the astronauts' own exhalations. You don't last long breathing carbon dioxide and water vapor.

Under normal circumstances, lithium hydroxide canisters scrubbed carbon dioxide from the cabin. But the useful working time of the canisters aboard *Aquarius* was computed on the basis of what was required for sustaining two men for some forty hours. But now there would be three men, and the hydrogen peroxide would be consumed one-third more quickly. The canisters needed to sustain breathable air for three men for four days—or Lovell, Haise, and Swigert would choke on what Jim Lovell called the "exhaust of our own lungs."

There were more than enough lithium hydroxide purifiers in the command module, but they were square-shaped and would not fit the round openings on the LM's system. Mission Control turned to Deke Slayton, the original "fix-it" man from combat and test pilot days.

Thirty-six hours after the astronauts retreated to their lifeboat, a lunar cabin light flashed a warning that carbon dioxide was nearing dangerous levels. Loss of consciousness and death would follow quickly. Slayton led a group of engineers who came up with what they called the "Wisconsin dairy-farm fix." Using pressure-suit hoses, batteries, tape, plastic, and cardboard from a flight manual as a jerry-rigged scrubber, they managed to hook an *Odyssey* purifier to the LM system.

CapCom led Swigert through a step-by-step assembly of the improvised device, which sucked carbon dioxide from the cabin, ran it through the makeshift scrubber, and returned fresh oxygen from the waste gases back to the cabin for the astronauts to breathe. Lovell, working with Swigert, said it was like building a model airplane.

At forty-two years old, Lovell had three previous space flights under his belt. He had orbited the moon in Apollo 8 and had logged 572 hours in space-flight time before Apollo 13. Fred Haise, thirty-six, a rookie with exceptional qualifications, had played a key role in developing the lunar module. A leading expert on its systems, he had experience with the LM that would ultimately pay life-saving dividends. His experience in space was matched by Jack Swigert, thirty-eight years old, who had not originally been scheduled for this flight. The command module pilot who had trained with Lovell and Haise for two years for this flight, Ken Mattingly, had been exposed to German measles shortly before scheduled liftoff. He had no immunity to the disease, and the NASA front office yanked him from the flight. Lovell accepted his back-up, Swigert, who with only two days of training proved extraordinarily skilled in his duties.

Apollo 13 swept around the moon to disappear behind its battered and cratered surface. Lovell, concentrating on the maneuvers that awaited the crew on the return trajectory was startled to see Haise and Swigert snapping photographs of the lunar surface as fast as they could adjust the apertures and speeds on their cameras.

Lovell shook his head in disbelief. "If we don't make this next maneuver correctly, you might never get those photographs developed." They responded by pointing out that Lovell had been here before, they hadn't, and they figured this was the only chance they'd ever get to shoot such pictures.

Apollo 13 looped around the moon. When the astronauts emerged from the other side, they resumed communications with CapCom and were told to prepare for a long burn of *Aquarius*'s engine. That moment would come two hours and six thousand miles away from the moon, and they needed every minute to guarantee the accuracy of their guidance system.

On any other flight, proper guidance and alignment would have been confirmed by using a space sextant to sight a suitable navigation star and feed data into the computer, which would verify that all was set to fire up the engine. But Apollo 13 was on its way home in the midst of a huge cloud of explosion debris that orbited the damaged spacecraft, reflecting sunlight and creating a field of "false" stars, making it almost impossible for the astronauts to align the sextant. They were desperate to find an astronavigational fix that would allow them to determine the precise corridor for reentry.

If a star wasn't available, Lovell figured, why not sight on the sun? It worked.

Right on schedule Lovell fired off the engine. For five minutes he ran the Apollo 13 on manual control. The automatic firing system couldn't be trusted at this point because of diminished power. Not to worry. Lovell blasted the service module, the command module, and the lunar module assembly perfectly on the desired course.

The crippled spacecraft chugged on, and another milestone was behind them. The odds for survival and successful reentry looked better and better.

Except they were running out of water. Apollo 13's systems generated tremendous heat and demanded steady cooling. That cooling came from the water supply. Haise quickly calculated that they'd run out of coolant five hours before beginning reentry. Apollo's systems would overheat and fail just at the time they needed them most.

Haise knew his ship intimately. He told Lovell they would run out of coolant water 151 hours into their flight. The crew recognized that with computers and guidance platforms overheating they would not survive.

Recalling earlier Apollo missions, Haise suddenly remembered that Apollo 11 had not sent its LM ascent stage crashing into the lunar landscape but had left it in orbit to send back telemetered data to Houston. Eleven's ascent-stage guidance system had survived nearly eight hours without water coolant before it began to fail. This fact gave the Thirteen a solution to the dilemma of how to provide coolant water for those last five hours before reentry. Knowing the systems would work up to seven or eight hours without coolant meant that for the five-hour period they were without coolant their systems wouldn't fry. They could make it all the way with a precariously thin but acceptable margin of survival.

To be safe, Lovell decided to put the crew on strict, personal water rationing of six ounces a day. Under ordinary space conditions an astronaut loses his thirst for water and dehydration can become a serious problem, but Lovell decided that for the trek home the risk could be set aside.

They were still far from being out of the woods. As soon as the critical engine burns and maneuvers ended and they had worked on the schedule for regenerating their cabin breathing air, the astronauts addressed another critical problem. Exposed directly to solar radiation, Apollo 13's surface temperature soared to 250 degrees Fahrenheit, while on the opposite or dark side of the ship, temperature plunged to more than 200 degrees below zero. *Odyssey* and *Aquarius* were blazing hot on one side, frozen on the other, and the astronauts lacked the electrical power to warm the interior of their lifeboat. They welcomed sunlight as it streamed into the ships, but it was more bright than warm. Without fans constantly moving the air around the two vessels, moisture increased. They were enveloped by walls that perspired steadily, and that thickened with accumulating moisture. Windows looked as if a rainstorm had lashed the ship from inside. Temperatures dropped steadily, and the astronauts shivered in a craft heated to only six degrees above freezing. Jack Swigert suffered especially from the wet cold. The moisture soaked his feet, and he lacked the protection provided Lovell and Haise by their lunar walking boots.

The astronauts were unable to don their spacesuits because of the limited space in the capsule, which was not much larger than a telephone booth. Their Teflon coveralls turned slimy and freezing to the touch. They longed for some thermal underwear.

Cold and wet and drifting for hours in the battered and crippled spacecraft, the astronauts experienced unexpected loneliness. In Mission Control, Deke Slayton, Alan Shepard, and the rest of the team felt growing concern for the emotional states of the thirteen's crew.

The men were sleeping only in short snatches. Their weariness became evident in their voices and what they said. "Those guys are worn out, they're hungry eating only cold food, they're sucking water rations, they're cold and wet, and they're dehydrated, and they don't know for certain whether they'll even get back alive," Deke said, summing up the problem. "On top of that they're not sleeping. We've got to get them to rest. If they don't sleep, they'll move into the slot for reentry when they're not at their best to handle their controls and systems. That's two days ahead of us."

Jim Lovell couldn't have agreed more. He described the situation in space as "three men cold as frogs in a frozen pond."

It was time to break the rules which stipulated that, with the exception of special cases, only CapCom could communicate between Houston and a space-borne crew.

"This is sure as hell one of those times," Deke told the controllers around him. "My judgment is we're gonna get 'em home safe, but only if they're alert, if they can handle the final critical hours. They've gotta sleep."

Deke moved into the CapCom's chair. "Hey, guys, this is Deke," he began. "Just wanted to let you know we're gonna get you back. Everything's looking good. We think you guys are in good shape all the way around. Why don't you quit worrying and get some sleep?"

This was their boss talking. The man they trusted implicitly. Deke's personal call eased tension, took away the sense of being utterly remote, removed some of the uncertainty.

"We think that's a pretty good idea," Lovell replied, and those in Mission Control smiled. They knew Deke's direct involvement would do the trick. Soon all three astronauts were fast asleep, performing a function as vital as breathing and eating.

When they awoke, it was to face a new and potentially lethal problem.

Odyssey and *Aquarius* were towing the heavy mass of the now-useless service module. The old pros judged it safest to keep the service section attached to protect the heat shield of the command ship from the cold, which could make it brittle and subject to cracking under severe reentry heat.

A problem arose because the wreckage of the service module, a shattered mass of wires, plumbing, and torn tankage, was still venting small amounts of gas. The escaping gas gently imparted a thrust that gradually pushed Apollo 13 out of the perfect corridor into which it had fired earlier. The astronauts were no longer headed for the intended splashdown target in the Pacific Ocean. Alarmed trackers warned that without another engine burn on an exact alignment, Apollo would streak along the upper heights of earth's atmosphere, gain a lifting force from the maneuver, and literally skip off and away from the planet. The danger of being left in perpetual orbit was real.

But could the overtaxed LM engine fire again with perfection? They had no choice but to try. Mission Control worked out the numbers, and the ship's commander fired the engine for only several seconds. That was enough. Lovell was pleased with what he had done. It was perfect. Apollo 13 skidded back into the center of the homeward highway.

The three astronauts were unaware they had become the center of attention for almost the entire planet. More than a billion people listened avidly to every scrap of news about the extraordinary effort to save three men in a disabled spaceship far from home. People filled religious centers as the world gathered to pray for Lovell, Haise, and Swigert. NASA announced the splashdown target as near American Samoa. There the aircraft carrier USS *Iwo Jima* waited to recover the crew and their spacecraft. But in the event the crippled *Odyssey* landed off target, British warships, a French aircraft carrier, and Soviet whalers drove voluntarily at flank speed to broaden the potential recovery area.

Apollo 13 accelerated steadily under the gravitational tug of the earth. Jack Swigert roused himself from a night of "rotten sleep" and floated forward from *Aquarius* to start the "reincarnation" of the *Odyssey* command module. He drifted into what had been a familiar spaceship cabin, now transformed into a cold and clammy tin can. Everything was covered or soaked in moisture. His immediate fear was that water had seeped into electrical harnesses and circuits. Alert to the possibility of sudden arcing and fire, he moved switches to return life to his ship.

As Deke Slayton had said long before this moment, what was learned from the fire on pad 34 was the best insurance for future crews and flights. Despite the moisture in *Odyssey*, the vessel came to life immediately and without incident.

Finally, with ship stirring like an awakening dormant robot, he switched on the three batteries essential for the reentry. Two were fully charged; one was weak. Swigert went back to the LM, returning quickly with a long cable, and recharged the weak battery from *Aquarius*'s power supply.

Fred Haise got on the radio with Houston. "What are you guys reading for cabin temperature in the CM?"

"We're reading forty-five to forty-six degrees," CapCom replied.

"Now you see why we call it a refrigerator."

"Uh-huh. Sounds like it's kind of a cold winter day up there. Is it snowing in the command module yet?"

"No," Haise grinned. "Not yet."

"You'll have some time on the beach in Samoa to thaw out after this cold experience."

"Sounds great."

It was a tension-relieving exchange as the astronauts slipped into the final hours. Early on Friday morning, April 17, just slightly more than five hours before predicted splashdown, Lovell fired Aquarius's small steering jets, hoping to improve his landing-target accuracy.

An hour later, Swigert threw a switch that ignited explosive charges. The battered service module separated. Lovell quickly fired the LM's jets for maximum separation from the service section to prevent a possible collision as both craft rushed earthward. Lovell had time to look at the service module and snap several photographs as it drifted away. The force of the explosion days before shocked the astronauts. "There's one whole side of the spacecraft missing," Lovell reported. "The whole panel is blown out almost from the base of the engine . . . it's really a mess."

Three hours later, just one hour from punching into atmosphere, fifteen thousand miles above earth and well into an accelerating dive toward the planet, Lovell and Haise moved into the restored Odyssey. They closed the double hatches of the connecting tunnel, triple-checked the seals, and pressurized the tunnel. They fired charges to separate their lifeboat. The pressurized tunnel added impetus to the separation. Just as advertised, they notified Houston, Aquarius popped away like a champagne cork.

"LM is jettisoned," Swigert confirmed.

"Farewell, Aquarius, and we thank you," CapCom called back in a salute to the faithful craft with the spidery legs.

"She was a good ship," Lovell said with deep emotion.

Four hundred thousand feet high, the command module Odyssey plowed into the atmosphere at better than 24,500 miles an hour. The three astronauts felt the first pressure of deceleration, the opening touches of gravity force.

Suddenly, it rained inside the command module.

The astonished astronauts looked about them. As gravity forces built up, water droplets covering the interior of Odyssey broke free in a sudden shower, collecting along the bottom of the spaceship.

Then Apollo 13 was deep in the fiery heat of reentry. For three minutes the ship was encased in heat of more than five thousand degrees. The ionized sheath forming about the plunging cone shape cut off all communications.

The world waited fretfully.

Clocks seemed to slow to a maddening crawl.

Mission Control was silent.

Squawk boxes crackled. A tracking aircraft over the Pacific Ocean was on-line. It had picked up a radio signal from *Odyssey*. No one cheered, not yet. Everyone had the same question: *What about the parachutes?*

Apollo 13 descended through a cloud deck two thousand feet above the Pacific, suspended beneath its three, huge, orange-and-white chutes. Mission Control, watching on television, went mad with relief, applause, cheering. Those in the trenches burst into tears of joy, hugged one another, pounded backs and shoulders.

Incredibly, Thirteen splashed down for the most accurate landing of all the Apollo missions to date. The *Iwo Jima* was just three miles away. The astronauts deployed flotation bags and assured the recovery ship and the world they were safe and sound.

As they were lifted by helicopter from rafts onto the *Iwo Jima* deck, sailors cheered, and the carrier's band swept into a rousing rendition of the most fitting of all musical tributes—"The Age of Aquarius."

Medical teams hurried the astronauts into sick bay for immediate inspection. There was less of the three men now than at the time they had begun their cliffhanging ride through the void. Among them the crew had lost nearly thirty-two pounds in the six days of their voyage, and Fred Haise was suffering from a mild urinary tract infection. Otherwise, considering all they'd been through, they were in good physical condition.

For a while it seemed the cheering and applause, the clamor of church bells, honking of horns, blasts of ships' whistles had become a single lifting strain echoing across the planet.

President Richard Nixon made a singularly touching and pertinent comment, which seemed to capture it all in one sentence. "You reached the hearts of millions of people by what you did," he told the crew of Apollo 13.

Jim Lovell had his own conclusion. "Our mission was a failure," he judged, "but I like to think it was a successful failure."

A touch of humor is always a fitting end to a cliffhanger. It was best expressed by the Grumman Aircraft Company, which had produced the lunar module.

Grumman sent a bill for more than four hundred thousand dollars to North American Rockwell, a "towing fee" for dragging the command and service modules through space for a distance of more than three hundred thousand miles.

A single message came from the public at large: Well done!

Alan Shepard knew that the nearly-fatal flight of Apollo 13 would delay his own upcoming mission to the moon. There could be no other way. Every facet of the design, construction, and functioning of the service module would need to undergo inspection, review, and design improvement.

But Apollo 14 would fly, and he would command the all-out effort to perform the mission as a full-blown and meaningful scientific study and exploration of the moon.

As darkness fell over his neighborhood after his world settled back to normal crises and problems, and as the three men of Apollo 13 reunited with their families, Alan left his home for a long walk alone in the warm spring night. He moved beneath trees and through winding paths, which eased his troubled thoughts.

Years before, the first Russian manned space flights had mocked America's stumbling efforts to ascend to earth orbit. Alan was to have led the way as the first man in space, but the stumbling blocks then had been weak minds and political jockeying, not the reliability of the rockets to be flown, and he had had to settle for being the first American in space.

He had been called upon then to save the manned spacecraft program from the myopic vultures in Washington, who were so eager to quit in the face of what they judged to be Russian superiority that could never be matched, let alone exceeded.

Though the three astronauts aboard the Apollo 13 had been safely returned to earth, Washington opponents of NASA regarded the mission as a clear failure and an unjustified and inexcusable waste of nearly four hundred million dollars of taxpayers' money. They were ready to shut down the expensive missions to the moon.

Alan Shepard had more than a space flight to make. He now carried the full burden of resurrecting the Apollo program, of reviving America's confidence and support for future trips to the moon. If his mission succeeded, he would share the accolades with everyone involved in the program.

If Apollo 14 failed, he alone would bear the burden.

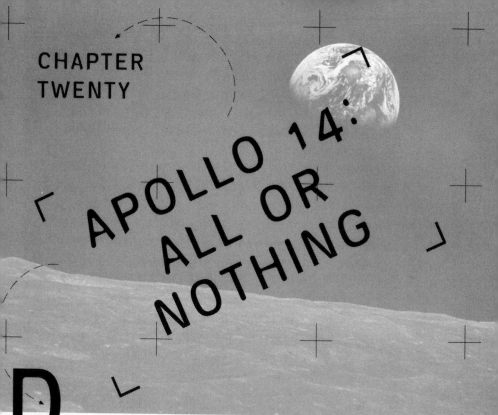

CHAPTER TWENTY

APOLLO 14: ALL OR NOTHING

DEKE SLAYTON took the long elevator ride to launch pad 39's "penthouse" nearly four hundred feet above ground level. He rose vertically in a steel cage through lights of varying colors and intensity, flashing past steel beams, thick cables, and work platforms.

The elevator swayed and rang with metallic sounds and then stopped. Deke stepped through the gate onto a railed catwalk spanning the precipitous drop toward a massive steel platform far below. He moved along the walk, halting as a strong ocean breeze forced him to grip the guardrail for support. He felt the catwalk sway beneath him. The effect was strange and a bit discomforting—this mountain of steel so enormous and massive, quivering as if it were balanced on springs. The catwalk stretched before him like a long arm. In a few hours it would swing to one side, pulling away from a monster rocket pregnant with explosive energy.

But that moment wouldn't come until the crew was on board. Not until the final countdown, which would begin forty-three minutes before the five most powerful rocket engines ever built would thunder to life and the Apollo 14 mission would be launched into space. Deke Slayton had knots in his stomach. Soon his good friend and two other astronauts he knew well would traverse the same catwalk to enter the command module named *Kitty Hawk,* the ship that would carry them to the moon. Blazing fire would send Alan Shepard and his crew away from earth at nearly fifty times the speed of a pistol bullet. It was almost beyond comprehension that men could hurdle away from their world at so blinding a speed.

Deke lifted his eyes. The shoreline of coastal towns along the Atlantic glowed and sparkled, long iridescent pearls of color that dotted the horizon. He shifted his gaze to the ground beneath him. Incredible, he thought. If you stood a football field on its end, it would still be sixty-three feet shorter than the Saturn V rocket, now quiescent, soon to be brought to life.

Deke was "walking the tower," an instinct born of a lifetime of pre-flight checks of the machines he flew. It wasn't quite the same as kicking the tires on a car, but the sense of inspection was the same. For a thousand reasons he wished he could be strapped atop the Saturn V. Not that he didn't want Alan to have his shot, to walk the moon. Something horrific could always go wrong and . . .

Memories of Gus Grissom flooded his mind. It had taken Slayton a long time to accept losing that man. Gus had been closer than a brother. Now, Shepard would ride an even bigger dragon. They all knew the odds. Millions of parts must work perfectly, in concert, or . . . He remembered Apollo 13 and its close call, and he cringed.

He shook his head. He didn't know if he had it in him to bear the loss of Alan Shepard. He wasn't at all sure that he could get over the loss of another so close. Alan filled that special space reserved for best friend, and here Deke was, "walking the tower," inspecting the rocket at every level as if it were some evil giant that in a searing burst of flame could kill Alan and—

A plane swept overhead, the throb of powerful engines matching the pulsating pain of his thoughts. He flashed back to when he had flown B-25 bombers in World War II. A hell of a ship. Two engines hammering out 3,400 horsepower. A brute for its day. Numbers flashed in his head. You could fill the sky with more than five hundred of those bombers, throttles mashed to the gate, and all of them combined didn't have the power to heave this rocket from earth.

At this moment, as Alan Shepard slept to rest for his lunar mission, fears gnawed at Deke. He had to know that everything at the pad was ready. He began his slow walk down the stairs, stopping at each level, checking and rechecking the thirty-six-story rocket, making sure everything was in place. Making sure . . . Good Lord, just making sure . . .

For Alan Shepard the early afternoon of January 21, 1971, provided an eerie moment of déjà vu. He had just left the van that transported him along with Ed Mitchell and Stuart Roosa to the launch pad. Walking toward the elevator for the long ride skyward, Alan stopped. He leaned his head back and looked up at the giant rocket towering over him. This was the same scene he had etched in his mind ten years earlier as he began the walk to the Redstone booster for America's first manned launch. Then, in pre-dawn darkness split by brilliant searchlights, he had studied the rocket that would take him into space.

"How much bigger this monster is . . ." he thought, not even certain he had spoken the words aloud. But the sight was staggering. The Saturn V loomed nearly five times taller than the Redstone, and Alan reminded himself it also was a hundred times more powerful. Even the escape rocket atop his Apollo spacecraft had twice the power of the Redstone booster.

There was almost an eerie silence about the launch pad this afternoon. The silence was the absence of voices. During those times he'd been here for training, the launch stand and the service towers had swarmed with activity, workers in every direction laboring to prepare the giant for its moment with destiny. Now the human beehive was almost deserted, and Saturn V was filled with more than a half million gallons of volatile fuel. Most members of the launch pad crew were gone; only those needed to complete the pre-launch orchestration remained.

"It's like a ghost town." Alan wasn't sure if Mitchell or Roosa had made the comment, but it fit. Making the scene even more dramatic were white clouds streaming just above them. The clouds came from enormous pressure within the massive fuel lines, oxygen streaming from pressure-release vents, whipping in the breeze.

By Alan's side stood Deke Slayton, who had been strangely quiet in the van. Now he stopped, looking long and hard at the three crewmen. During the ride to the pad he'd made a few comments about rain clouds in the distance. "There's a chance they might move into the launch area before you're set to go," Deke had told them, and again went silent. His words told them much more than he'd said. No one could forget the lightning that had struck Apollo 12.

And no one said good-bye; no one really knew what to say. "Watch your ass and have a good trip," finally came the farewell.

From the base of the launch pad, Deke looked up and focused on the narrow steel walkway leading from the elevator to the White Room, where technicians would make final checks of the astronauts' equipment and assist them in entering the spacecraft.

At last, the three spacesuited figures appeared, tiny dolls from this distance nearly four hundred feet below. Deke saw Shepard stop to stare down at him. He had the feeling of many years compressed to this single moment. Up there was the only one of the original seven Mercury astronauts who was about to realize the dream of one day going all the way out. To the moon. Alan Shepard, in a very real sense, was going for all seven of them—going not only for him and Gus Grissom, but for Wally Schirra, Gordo Cooper, John Glenn, and Scott Carpenter. Deke gave him a thumbs-up.

Three hundred technicians and controllers gathered at the Launch Control Center. Backing up that team were three separate firing rooms, each with five hundred people on constant alert for any contingency. Hundreds more waited through the countdown at Mission Control in Houston. Thousands of other men and women were on duty at tracking stations around the world, in tracking and recovery ships at sea, in tracking aircraft already airborne.

Alan Shepard knew the memories of the troubled Apollo 13 flight were still fresh in the minds of every one of those people. The tension riding the countdown was almost a physical presence in the spacecraft. Yet there was a new level of confidence that they had the tiger by the tail. As far as the three men in *Kitty Hawk* were concerned, the problems that had bedeviled Thirteen were fixed. Their moonship now carried three oxygen tanks instead of the two for each Apollo, and the extra tank was kept isolated from the others. There was another extremely welcome addition—a spare battery of four hundred amperes of power, which would supply their craft with enough electrical energy to handle all their needs from any point in the mission.

At T-minus 110 minutes, the count went down to business with the abort advisory system checks. Shepard's thoughts as he and his crew went through the check lists mixed the immediate work at hand with constant self-reminders that a lot more than going to the moon was riding on this flight.

Project Apollo was under direct fire from a growing number of critics, who were sniping at the whole manned lunar program. NASA had sold Apollo as a race to the moon with the Soviets for national prestige. With the race over, the public, concerned about other problems besetting the nation, had generally lost interest, and politicians were taking notice. Shepard knew that one more failure, even if the crew survived, would doom the entire project. Since Thirteen's cliffhanger ride, the political naysayers had already cancelled three scheduled flights to the moon. Apollos 18, 19, and 20, planned for extensive exploration, had been scrubbed.

Fresh from that kill, the doom merchants were already aiming at more Apollo scalps to hang from their office walls. To the dismay of the astronauts, Congress had slashed NASA's budget to its lowest level in ten years. What really shook the astronaut team was that even one of their own, Bob Gilruth, had suggested that it would be to everyone's interest to kill all lunar missions after Apollo 14.

Gilruth offered the theory that by landing on the moon the country had achieved the political goal laid down by President Kennedy. So why risk any more lives? Forget the moon, he urged top government officials, and get cracking with a manned space station in earth orbit. That way, if any serious problems occurred, returning astronauts to earth would be simpler and safer.

So Apollo 14, Shepard knew, had to be better than merely a successful mission. It had to be a superb effort that would keep Apollo afloat.

T-minus 43 minutes. No problems, no unexpected stumbles. The three men were in great spirits, feeling confident about their mission's success.

The count sailed smoothly down through arming the escape system. Range safety went to "green all the way." Launch Control tested the systems for power transfer to the booster; everything checked out perfectly. At T-minus 20 minutes the lunar lander they had named *Antares* came alive with its own internal power.

Fifteen minutes to go and *Kitty Hawk* was now on its own power systems.

Ten minutes. Perfect!

Eight minutes.

Hold!

Apollo 14 was still right on the money, but the storm front to the west bore down with heavy rain toward the launch pad. From within *Kitty Hawk*, with its protective shield, the storm wasn't visible.

CapCom gave them the word. "Looks like we'll hold for a while for the storm to pass."

Shepard groaned. "Christ, not again," he muttered to himself. Unpleasant memories flooded his mind. His first attempt to fly in Mercury-Redstone had gone down the tubes with heavy rain, and now he sat on the pad for his second launch, delayed because of another cloudburst.

Every previous Apollo manned flight had lifted off exactly on schedule. Shepard chafed at this unexpected delay. He'd waited too many years for his crack at the moon to be held up now because of rain.

In Launch Control, listening to Shepard's voice, Deke Slayton judged correctly the surge of frustration in his friend. Immediately Deke went on the communications loop to speak directly with Alan. "Hey, the way it looks, this storm is going to go right over us and out to sea," he said. "At least, it's more comfortable for you than in your old tin can."

"Yeah, right," the impatient Shepard growled. "Tell you what, Deke, let's get on with it."

"Hang tight, buddy."

The storm lashed the launch pad, swept out to sea. The count resumed and eased right through all final status checks. The launch team armed the destruct system, and the access arm leading to *Kitty Hawk* swung back out of the way.

T-minus 3 minutes 10 seconds. Apollo 14 was now on its automatic sequencer, the long-awaited "Initiate firing command," which slipped the rest of the count into computers.

T-minus 50 seconds. Saturn V went to full internal power. The dragon was stirring.

Butterflies swirled in Shepard's stomach, the same twinge of nervousness he had known before liftoff of his first flight almost a decade before. Few people would have believed that anything could gnaw at this supremely self-confident man, but to Shepard this was the kind of normal, sensible sensation almost everyone knows before a critical juncture in life. No matter how extensive the training and preparation, there's always that nagging thought that you've forgotten something or made a mistake. Alan had no concern about the equipment around him; there, his confidence was unflagging. His concern now, as it had been back then, was about one man: himself, that he might screw up.

He pushed the doubt away. He felt the first distant whispers of Saturn V flexing its sinews, the rush of thousands of gallons of propellants hurtling downward through their lines, turbo-pumps spinning.

He had the wild thought that the giant rocket was ten inches shorter than before fueling. How could they get to the moon with a booster that had shrunk? The fuel, of course. Millions of pounds of cold fuel contracting the rocket, bending metal.

T-minus 8.9 seconds.

The countdown call echoed across the spaceport . . .

"Ignition sequence start!"

She clasped her hands to her chest as the first savage pinpoint of fire snapped into view. Louise Shepard felt her heart leap into her throat. Three miles from the launch pad, she heard no sound, only saw growing flame; the roar would not reach her for fifteen seconds. She had been the first wife to endure that private, special agony of watching her husband rocket away from earth. Back then she had watched her television screen as the toy-like, slender Redstone sent Alan skyward.

Nothing could keep her away from being present now. By her side were family and friends, clasping hands, holding one another for strength.

Everything except her husband faded from her thoughts. She suddenly recalled a heart-rending moment the night before, the last time she had seen Alan close to her. How cruel such moments could be! They had been forced to remain on opposite sides of a glass partition, behind which the moon-bound astronauts had to remain in quarantine against any possible germs or virus. So close, so close . . .

They both pressed their lips against the glass, then stood back slowly. "I won't be making my usual phone call tomorrow night," he told her. Then, with a bittersweet smile she knew so well, he added, "I'll be leaving town."

Far below Shepard, Mitchell, and Roosa, a torrent appeared instantly, exploding beneath the five mighty engines of the first stage. Twenty-eight thousand gallons of water per minute smashed into curving flame buckets to absorb the downward blaze, bursting to each side of the high complex, cascading steam hundreds of feet out and upward.

The Saturn V roared and screamed, anchored to its launch pad by huge, hold-down arms chaining it to earth until computers judged the giant was howling with full energy.

Two seconds remaining . . .

Flame increased in fury.

An ice shower fell steadily. Chunks and sheets and flakes of ice poured from the rocket. An ice storm in Florida, coatings of ice that formed about the supercold oxidizers and propellants of the huge fuel tanks, tumbling, falling—the incongruous sign of the space age that the monster was ready to leave.

"All engines running . . .

"*Zero!*"

Alan Shepard was prepared, body tensed, for the hammering jolt he knew must come as the hold-down arms sprang back to free the Saturn V. Mercury and Gemini boosters had shaken the teeth of their crews with battering vibration. Men who had been to the moon told Alan, "Relax. Just lie back and enjoy the ride. It's a dream."

It couldn't be that easy. "When a rocket this big fires up, it just cannot be smooth," thought Alan. "I was wrong. Was I ever wrong!"

Shepard and his crewmates could hear the rumble of the five first-stage engines. But that was it: a rumble thirty-two stories below *Kitty Hawk*. The sound reached them like distant, muted thunder.

In its blaze of flame and torrents of cracking ice, Apollo 14 began its mission.

The astronauts felt a gentle sense of motion, mildly jerky at first, but to their surprise "a very gentle rise." It wasn't that way outside.

The Saturn V roared, bellowed, and shrieked, hurling out ear-stabbing sonic waves and a crackling thunder which sent birds flying and wildlife fleeing, and which slammed into people miles away, fluttering their clothing and causing them to step back uncertainly. The sound was so great its shock waves tumbled and mixed together, swirling deep thunder with an acetylene-torch cry that continued impossibly long, echoing from clouds, rebounding across the ground, seeming to split the very sky asunder.

The earth shook, a feeling akin to the jellied trembling of an earthquake. The longer the first stage burned, the farther its stentorian bellow hurtled outward. Scientists judge the savage roar of the Saturn V akin to the explosion of the volcano Krakatoa, which tore apart the islands of the Sundra Strait in 1883.

At three minutes past four o'clock on the afternoon of liftoff, Saturn V tore free of its hold-down arms and sounded Gabriel's trumpet of the space age. Eleven hundred miles distant, the overpressure in the atmosphere of Apollo 14's liftoff shook atmospheric instruments at the Lamont-Doherty Geological Observatory in Palisades Park, New Jersey.

It seemed the rocket mountain was held back it rose so slowly, more than seven and a half million pounds of blazing thrust straining to push more than six million pounds of rocket upward. Ten seconds passed after liftoff before the first-stage engines cleared the launch tower.

A ponderous, slogging, slow-motion beginning, but not for long.

Thirty seconds passed. The g-load was only one-half again what the astronauts felt before the engines ignited. But going to the moon is a matter of constant acceleration.

Fifteen tons of fuel converted to fiery thrust every second.

Now, with telling effect, g-forces increased as the giant continued to accelerate. Saturn V slammed into the area of maximum aerodynamic pressure—that jagged reef in the heavens where sonic waves hammered and pounded against the great body, trying to rip inside.

The Saturn and its load pushed through. The world outside saw a fire river eight hundred feet long. Shock waves formed like ghostly dervishes dancing along the circular flanks of the rocket. A mist appeared and expanded upward above the howling engines: ionized gas, shock waves, plasma in maddening motion.

Inside *Kitty Hawk*, the speed of sound now far behind, it was eerily quiet. Had the crew not heard the humming of electronic equipment in the command module, they might have been in a simulator on the ground.

They weren't.

"Stand by for the train wreck," the astronauts called.

Two and a half minutes from first motion, g-forces made them weigh four times more than they did at liftoff. The five great engines of the first stage had compressed the entire rocket like an accordion until first-stage shutdown. Without constant acceleration and with the sudden cutoff of stage one, the three men jerked forward in their seats. The accordion stretched out and then compressed again; the fuels sloshed, and the astronauts felt a series of bumps just like a train wreck.

They heard metallic bangs and assorted noises as explosives separated the now empty stage.

They were nearly forty miles high and nearly sixty miles downrange from their launch pad, climbing faster than six thousand miles an hour. Tongues of flame lashed briefly outside *Kitty Hawk* as solid rockets in the first stage ignited to push the stage back and to the side of Apollo. They didn't need a "highway-in-the-sky collision" at this point.

The crew heard sudden bangs from below as ullage rockets fired with brief but powerful bursts, kicking the second stage into motion to settle the propellants in the tanks. Five engines ignited right on schedule, and the three men were squeezed back in their couches from suddenly increasing acceleration.

Earth's atmosphere lay below. This is difficult to believe, Shepard judged their ride. Their continued ascent was "very smooth and strangely quiet."

Abruptly a new sound reached them. As if someone were pounding with a rubber hammer against the spacecraft. The sound came from a fiery rocket blast as the escape tower ignited automatically, jerking free the tower and the protective shield around *Kitty Hawk*, uncovering their windows for the first time. The ignition rocket and smaller rockets threw the no-longer-needed escape tower away to the side of their flight path.

They strained for a view through the windows. No luck. They were pointing upward, and they could see only the blackness of space.

Eleven minutes.

One hundred and fifteen miles high.

Faster and faster. Stage two emptied its tanks and went silent. Again the men snapped forward in their harnesses; again they were pushed back as the third stage lit off.

Two minutes later, still heavily loaded with fuel, the third stage shut down. Apollo 14 began its orbital race around the planet at 17,400 miles an hour.

"We settled into that wonderful, wonderful feeling of weightlessness," Shepard remembers.

He exchanged broad smiles with Mitchell and Roosa, almost boyish grins, and knew he'd make the most of the upcoming week. He released his harness, floating freely as if he were a feather with invisible wings.

"That alone," Alan said, "was almost worth the entire trip."

Apollo 14 sped around the earth for two and a half hours, nearly two full circuits, while the astronauts in space and Mission Control in Houston took the pulse and status of all elements of the three-stage spacecraft. Fourteen was proving to be a textbook mission, and CapCom sent up the message that came through like music aboard *Kitty Hawk*:

"Fourteen, you're GO for translunar injection."

Translunar injection. Their tickets to the moon. Ed Mitchell held up a gloved fist in celebration. They ran through a final check list, and once again lit the fire. Saturn's third stage reignited, hurling back a magnificent plume of red and pink and violet flame, and they were on their way.

At nearly seven miles a second—24,500 miles an hour—the rocket stage shut down. For the fourth time the three men "tumbled forward" against their harnesses, then once again they were weightless.

"We're on our way, guys!" Shepard exclaimed.

Stu Roosa raised a clenched fist. "Right on!" he cried.

There was no sensation of movement, only the delicious freedom of floating without weight. Immediately they rid themselves of the bulky, movement-confining pressure suits and donned comfortable flight coveralls. They were grinning like kids at a country swimming hole. They felt terrific.

An hour later, like a house of cards on a table shaken wildly, their world seemingly came tumbling down about their ears. Fourteen would suddenly be in deep trouble and, just as quickly, the scheduled moon landing would seem millions of miles away.

Stu Roosa now began the procedure of separation, turnaround, and docking *Kitty Hawk* nose-to-nose with the lunar module *Antares*. The maneuver had been accomplished on the previous Apollo flights with ease. *Antares* was parked in a "garage" attached to the top of the now-empty third stage. Stu would separate from the stage, fly a short distance away, and maneuver back for a hookup with the moon ship. Extending forward from the top of *Kitty Hawk* was an arrow-like probe which Roosa would connect with a cone-shaped drogue receptacle on *Antares*'s nose. The hook system assured a firm connection between the two ships and a tunnel through which the astronauts could float from one craft to the other.

Three metal probe latches, each the size of a cigarette, would lock inside the drogue and inch both spacecraft close enough for twelve larger and stronger latches on matching docking collars to clamp together tightly.

"Let's do it," said Shepard quietly.

"Houston, we're in a position to proceed with the docking," Roosa informed CapCom.

"We copy. You have a GO, *Kitty Hawk*."

Command Pilot Roosa fired his small control thrusters with the touch of a surgeon. *Kitty Hawk* and the heavy service module nudged gently toward the docking target.

The arrow slipped into the waiting cone. "Hot damn, you hit it dead center, Stu," Mitchell confirmed. The astronauts listened for the click of latches catching, watching their control panels for the green light that would confirm capture. No click. No green light.

They were astonished to see the docking probe rebound from *Antares*. "What the hell is going on?" Shepard asked with impatience. Stu had made a perfect dock, but it had refused to "take."

"Houston, we've failed to secure a dock," Stu called earth.

CapCom hesitated before answering. In Mission Control tension joined the duty team. Then a call, "Roger, *Kitty Hawk*. You've got a GO for another attempt."

Apollo 14 sailed alone and distant through the endless ocean toward the moon. Exuberance turned to concern. Stu turned to Shepard. "Al, what do you think?"

Shepard kept it cool. "The position was perfect. So were you. Let's go again."

Stu nodded, notified CapCom. "Houston, we're going in one more time."

Shepard and Mitchell could only watch as Roosa maneuvered the heavy command and service modules with his exquisite touch on the controls. In Houston, the huge control-room drone of noise and voices fell to the hiss and crackle of the open radio frequency to *Kitty Hawk*. Controllers either sat on the edge of their chairs or were standing, listening, waiting.

A whispered voice carried across the room. "If they can't dock, there goes the farm."

Far out in space, Roosa tweaked his thrusters. Perfect alignment. Probe and cone met, snubbed tightly.

And separated.

Roosa's voice carried through Mission Control, flat-toned, frustrated. "Houston, we do not have a dock. We're going to pull back and give this some thought."

"Roger, *Kitty Hawk*. We'll be doing the same down here."

The standby teams in Houston, kept on alert for emergencies with a flight in progress, came running. A sudden, loud cry burst through the groups. "Where in the hell are the probe and drogue?"

"We've always had a docking probe and drogue available in the control center," Chris Kraft said with swiftly fading patience, "as well as experts on the system." There were frantic calls for assistance, and the absent docking system had to be hurriedly located to help engineers understand what might be going on thousands of miles out in space.

In the spacecraft cabin, concern became a fog of gloom. Failure to dock successfully would cancel the scheduled moon landing by Shepard and Mitchell. Alan was swept from frustration to anger, the latter emotion all the worse because the crew had to wait for Houston's recommendations. No one needed voice aloud the penalty for a docking failure. If the lunar landing went into the bucket, Apollo 14 would fly an alternate mission, orbiting the moon for two days to map and study the surface with new cameras and instruments.

The whole idea galled Alan. Worse, he thought, their failure to land could once again unleash the critics and doom what was left of Project Apollo.

"No way," he told himself. "I've got to take this trip all the way down to landing. We've got to fix this thing."

Apollo 14's mission now clung precariously to mechanical latches that needed no hydraulic pressure, no electrical energy, no pneumatic drive. Metal pieces were supposed to come together with utter simplicity, go click, and lock. It was maddening to be blocked by one of the simplest mechanisms of the entire Apollo program.

In Houston, the probe and drogue teams met with top officials of Mission Control. Flight controller John Llewellyn, Bob Gilruth, Chris Kraft, Deke Slayton, and others huddled together, shoving the probe again and again into the drogue receptacle.

It worked every time on earth, but had failed every time in space.

The components were exactly the same. One man ran his fingers over the smooth metal in Mission Control. "If a piece of debris, or even dirt, lodged in this mechanism, the one that's on Fourteen, then it's possible it prevents the latches from depressing. That prevents, as well, their snapping into place."

"Hell, if that's the case, then Stu should keep at it. Coming back again could dislodge whatever is blocking the hard dock."

"Tell him to stay with it."

Three times in the next hour Stu Roosa brought *Kitty Hawk* in with his precise maneuvering. Three times the arrow slid into perfect position.

Three times the message came down from space: "No joy."

A new problem loomed. Roosa had been maneuvering *Kitty Hawk* so many times that he was eating into the limited fuel reserves of the spacecraft. Mission Control passed the word. Stu could make one, perhaps two more attempts, and then the docking maneuver must be abandoned.

And then they'd fly a limited, secondhand, blah mission.

Shepard got on the line to Houston. To hell with this again-and-again business, he told CapCom. It's not working. He recommended the crew get back into pressure suits, depressurize the cabin, and he'd exit *Kitty Hawk* far enough to reach the top of *Antares* and pull the two ships to a hard dock.

Mission Control teams discussed the matter briefly. Too dangerous, they said. They told Alan to remain in *Kitty Hawk*. Deke Slayton knew the frustration, the disappointment his friend must be enduring, but he concurred with the decision.

Several men rushed back to Mission Control from working in the simulators. "We think we may have a way," they said. CapCom passed the word to Roosa.

Change the procedure. Come in faster and harder for the docking, ram the docking probe as deeply as possible into the cone. If the closure rate was great enough, and absolutely accurate, the latches could telescope from the impact. That would drive *Kitty Hawk* hard up against *Antares* and hold it there long enough to engage the twelve latches of the docking rings directly. If the smaller latches, which normally made the first connection, were faulty, then they could be bypassed and a hard dock achieved.

If.

It had never been done except several minutes before in a simulator. And the simulator was a hell of a distance away from *Kitty Hawk* and *Antares*.

Roosa wasted no time preparing for what he and his crewmates realized was a do-or-die attempt.

"Houston," Shepard called, "we've got the LM on the television for you."

"Roger, *Kitty Hawk*, we see it. "There are scratches on the drogue . . . probe is entering on target."

Roosa went on line. "Houston, requesting a fuel status."

None of the astronauts liked the response. "*Kitty Hawk*, we'll reevaluate following your attempt to dock."

"Roger, Houston."

Suspense built as the gleaming Apollo rode on invisible rails toward the moon lander.

"Houston, we're going in," Roosa said quietly.

"Good luck, *Kitty Hawk*."

"Luck, hell!" thought an angry Shepard. It was time to stop screwing around with gentle maneuvers and worrying about fuel usage.

Time to ram the throttle forward, push the power right to the fire wall. "Stu!"

Roosa looked at his commander.

"Stu, just forget about trying to conserve fuel. This time, juice it!"

Kitty Hawk charged ahead.

CHAPTER TWENTY ONE

NO TURNING BACK

THE APOLLO 14 COMMAND SHIP, *Kitty Hawk*, appeared as if it were a thick-bodied creature about to devour the lunar module *Antares*. Its crew members sat three across, readying the spacecraft for its final shot at docking. Stu Roosa took his cue from Alan Shepard. He dismissed his concerns over fuel quantity in the tanks and brought his thrusters up for a full-power charge.

Three men braced themselves. Roosa brought *Kitty Hawk* alive with a powerful blast from the thrusters. Spears of transparent flame streaked back. Vibrating and shaking, *Kitty Hawk* burst forward. Shepard's command to "Juice it!" still rang in Roosa's ears as the probe at the tip of *Kitty Hawk* slammed into the lunar module cone.

Both ships rocked from the impact. No one dared to breathe, afraid the two spacecraft might once again rebound away from each other.

C'mon, grab, grab . . . grab you . . .

If Shepard's force of will could have kept the two ships joined as one, it—

No rebound!

They listened with fierce intensity, knowing not only the future of their mission but also the whole Apollo program was hanging on just one sound.

Click! The capture light came on!

"Got it!" three men yelled as one, pent-up breath bursting from their lungs. As quickly as exultation swept through them, they were silent, their eyes riveted to Shepard's data panel.

Alan turned his face slowly. The furrows that creased his forehead were swept away like smoke before the wind. His teeth flashed with that Tom Sawyer grin.

"We have a hard dock," he said quietly.

Roosa jammed his transmit button. He wanted to shout, but soft words had the same effect. "Houston, we've got a hard dock."

"Roger that, *Kitty Hawk*. We confirm." They could hardly hear the words of CapCom. In the background Mission Control was a bedlam of whistles and cheering. Shepard took a deep breath. They hadn't won this race yet. And they'd almost lost everything because someone had not done his job. A piece of dirt or debris had stuck in their docking mechanism, thwarting their plans and training.

He shook off the euphoria of the successful docking. "Houston, we're ready for business up here."

But Houston had applied the brakes. Controllers couldn't keep Apollo 14 from its energy-impelled drift moonward, but they refused to clear Shepard and his crew to continue with the planned lunar touchdown.

They were justifiably edgy about making premature conclusions. Flight operations boss Sig Sjoberg told his team leaders he was going to be damned sure "that this thing is indeed satisfactory for docking— again—before we can commit to the moon landing."

From his position, Sig was right. But the commander in *Kitty Hawk*, way out in the middle of nothing and on a planned collision course with the moon, was ready to chew his nails. He knew what Sig was doing, looking at all the possibilities. Suppose Alan and Ed made a great moon-landing mission, boosted back up, and then failed to dock with the command module? Their only chance for survival, let alone making the mission pay off, would be a never-tried formation flight between *Kitty Hawk* and *Antares*—slinging rope tethers between the two ships, and performing an impromptu space walk, also moving boxes of moon rocks from one ship to the other.

Well, hell! Shepard would have paced back and forth if cramped quarters in zero-g would have allowed it. Any test pilot worth an ounce of salt knows you can't predict all the problems and possibilities. That's why funereal smoke had drifted upward from test fields from the beginning of aviation. You knew the chances you were taking before you even slipped on your flight boots. It was no different now.

Besides, he and Mitchell had tested that jury-rigged ship-to-ship transfer in the astronaut water training tank. They had had fits, frustrations, and snarls, but they'd done it. And they could do it again. They didn't need another guaranteed hard dock. They could do the mission without—

CapCom reminded Alan and his crew they'd been up nineteen straight hours. "Get some rest," they ordered.

That's the great thing about flying on a centrifugal slingshot. They went to sleep, and *Kitty Hawk* kept chugging to the moon.

Shepard slept fitfully, alternating between brief periods of deep sleep and half-awake restlessness. Aware he faced unexpected problems, he was working out the solutions and alternatives before they happened. That way he could react to virtually any scenario instead of wasting precious time trying to figure out who or what was trying to take down his craft.

Several times he awoke and was drawn to the viewing ports of *Kitty Hawk*, where he could convince himself again that he was actually living what for so long had been a dream. It was folly beyond comprehension even to consider backing out of this mission before they'd gone all the way. To have vanquished odds that would have felled a hundred men and then quit because of some mechanical hang-up went against every fiber of his being.

It wasn't just his experience as a fighter pilot and test pilot that accounted for these feelings. Shepard believed absolutely that in almost every respect a man's adult character is formed in his early years. He considered himself to have been blessed in many ways, not the least of which was having as a father a man with an innate ability to understand machinery. Making things work had been as commonplace to the young Alan Shepard as opening a book on flight, and he'd done that many a time. Even as far back as age four, when Charles Lindbergh had made his solo nonstop flight from New York to Paris. That's when the love for aviation first had begun to grow inside him.

And what had been given as this gift by his father began to reap rewards while still in his teen years. Alan Shepard became, in the classic sense, an airport bum. There is no deprecation in this term. The airport bums were those kids who hung out at airports, doing odd jobs, seeking eagerly to sweep out hangars, to wash airplanes, to run the fuel hoses and change oil and clean windshields, all to hear the hallowed words, "Hey, kid, want a ride?"

That's where it began, and from sweeping and cleaning, the local pilots learned that young Shepard had mechanical magic in his hands. He could repair broken fuel and oil lines, torque replacement spark plugs just so. He could be trusted to taxi planes from one part of a field to another. Through this process began the skills of knowing how to listen to an engine, how to feel what a winged machine is telling its pilot.

"The Fix-it Kid." That was young Alan Shepard.

"A born pilot." That, too, was the praise from the experienced flyers.

The teenager walking, bumming a ride, or pedaling between home and airport was beginning to realize his dreams. He would be among the best.

Yet his wildest dreams could not match the reality of this moment. The kid who had thought breathlessly of flying at a hundred miles an hour had made the long upward march from rag-wings to metal, from propellers to jets, and exchanged his greasy jeans and coveralls for g-suits and flight gear and the helmet over his head and went supersonic.

And in spite of the years of fighting his inner-ear problem, it had happened. He was here, moon-bound, and he wasn't going to let some mechanical burp or a piece of scratched metal screw up *his* mission. Not if he had to step outside and change the oil in this thing, if he had to fix a "flat" on Apollo 14, he was taking his lunar ship to a touchdown in the Fra Mauro highlands on the shore of the moon's Ocean of Storms.

"To hell with worrying about docking after the moon walk," Shepard told himself. If Roosa were unable to complete a successful linkup on the return from the lunar surface, Alan had made up his mind to break all mission rules and fix the broken machine himself. He would have Stu hold the ships tightly with thrusters and in his pressure suit he'd climb into the tunnel between the two craft and manually pull them together.

He couldn't tell controllers what he had planned. They would have ordered him not to proceed with so foolhardy a move. But he was determined to do whatever was necessary to go on with the mission.

Houston greeted them at the end of their ten-hour break. "How'd you guys sleep?" asked CapCom.

"My mattress was hard," complained Roosa. A neat trick in weightlessness, but Stu carried it off with a straight face.

Alan was in no mood for casual banter. He glared at the cabin speakers until he heard Mission Control say, "You are GO."

The moon landing was back on the schedule. Tests on the ground had proven even a small chunk of debris could have fouled Roosa's docking maneuvers, that the astronauts could repeat their earlier "charge it" maneuver if it became necessary. Shepard lightened up when he heard the news. "Hot damn!" he exclaimed.

They sailed without a quiver through the halfway mark toward the moon. Gravitational pull from earth had slowed their speed to little more than a tenth of what it had been but, they were still moving at a heady 3,200 miles an hour. To the impatient Shepard they were "crawling" through space.

As the long hours passed they marveled at the glow of a moon growing ever larger in their view. During their scheduled rest periods Alan Shepard took every opportunity to study both the diminishing earth and the expanding details of their destination. "*Kitty Hawk*, how big a moon are you seeing?" asked CapCom.

"Sort of half," Alan replied. "It appears about the size of an orange held at arm's length. The moon is starting to take on a little bit of brown and grayish colors about this point, as opposed to being as bright as it appears from earth. You can start to see a little bit of texture."

When they awoke from their second sleep period, Alan and Ed floated into *Antares* for a meticulous check-out of their lander. Two hours later, they notified Houston their bird was "immaculate" and ready to go.

Terrestrial gravity diminished, and the moon's grip assumed dominance, a steady acceleration toward the small world now less than forty thousand miles before them. Global size required new thinking. The diameter of the moon just about equaled the distance from Los Angeles to New York.

"The moon is out my rendezvous window right now," Mitchell updated Houston. "We're running downhill very rapidly toward it."

The next day, they swept around the lunar far side into the thirty-three-minute radio blackout with earth. Thirteen minutes later, Roosa fired their big engine to reduce their speed by two thousand miles an hour.

"We've got capture orbit," Roosa confirmed as they emerged from the far side.

"This is really a wild place," Shepard sang out.

"Fantastic!" exclaimed Roosa. "You're not going to believe this, but it looks just like the map."

Mitchell gazed down with awe. "That's the most stark and desolate-looking piece of country I've ever seen," he added.

"Let's get to work, troops," Alan said, breaking up the sightseeing tour.

Roosa dropped them into an elliptical orbit with its low point ten miles above the surface. This would enable Shepard and Mitchell to save fuel for the critical phase just before touchdown. It was their best shot of making a bull's-eye landing.

On the twelfth orbit, Alan and Ed, moonwalk suits pressurized, unplugged from *Kitty Hawk*. Stu Roosa watched every move as the two spacecraft separated. "Okay," he called to *Antares*, "you're moving out. You seem real steady. I'm going to back away from you."

For the next four hours Shepard and Mitchell studied the lunar landscape on passes over their highland landing area at Fra Mauro and ran through their spacecraft systems and computer programs before being cleared to fire their descent engine and head for the surface.

As they swept above Fra Mauro, excitement grew in the lander. "There it is, big as life," Mitchell radioed. "The sun angle looks real good for the next time around."

Shepard was ready to drop the hammer with rocket fire. But not yet. He and Ed would take advantage of every allotted minute of check-out time, coordinating with Houston, to confirm *Antares* was in perfect shape for descent.

As they continued their sweep over their landing site, Alan recognized nearby craters. "I have Cone Crater, Triplet, and Doublet," he told Mitchell. Both men watched more details flash past. "Star and Sunrise. Right down there . . ."

"On the nose," Mitchell confirmed.

"Got 'em!" Shepard said, excitement rising in his voice. "Yep, sure do. Hoo-ha! I think we'll know them next time."

Mitchell called out their check list items, watching every move Shepard made, backing him up on every detail, missing nothing. Then it was time to begin the complete dress rehearsal of their computer-controlled descent program.

"Final pre-landing check," he announced to Alan. "Time to punch in."

The rehearsal was a full practice run for the lander's computers, for all its systems that would be used to fly the descent flight path to Fra Mauro.

"Got it," Shepard replied. He activated the simulated lunar descent profile sequence just as he'd done so many times with Ed in the ground simulators. Only this, hopefully, would be the final, final run-through.

Numbers flashed by as the critical sequence moved through the computers. If there was going to be a problem in the system, now was the time to discover the glitch and get it fixed.

"Practice descent has started," Shepard announced. "The computer is beginning the practice descent."

"On the mark," Mitchell confirmed.

They would do everything except fire the descent engine for this rehearsal.

But no sooner had the simulation begun than something did not click. Their monitors should have indicated that the computers had simulated engine ignition and that they were beginning their descent as if they were really on their way.

"Oops!" Mitchell exclaimed. "Hey, we're not showing a descent sequence."

Disbelief was clear in Shepard's voice as he called Houston. "Hey, our abort program has kicked in!"

Everyone in Mission Control leaned toward monitoring consoles. They all shared the same thought. If this had been the moment of truth, if they actually had been trying to fire the engine and that abort program kicked in, Shepard and Mitchell would never reach the moon. The abort program called for a rapid-fire sequence of events. The ascent-stage engine of the lunar lander would hurl out fire, the two stages would separate, and the computers would set Alan and Ed on a rendezvous course with *Kitty Hawk*, the mother ship.

"We copy, *Antares*," Houston responded. They had the currency of time in the bank for the fix. That's why they have the rehearsal. "Try your descent program again."

"Roger, Houston, we, ah—" Shepard's voice cut off as their consoles indicated that this time their engine would have fired on schedule. Mitchell glued his eyes to the panel. His instruments showed the simulation was underway. "We show simulated engine start. Everything came on line."

"Descent program commencing," Shepard confirmed to Houston. "It's starting down."

One man in Mission Control held up crossed fingers. "Now if it just runs like this when it's time to really light the fire."

"You bet."

The simulation suddenly was hung up again. Shepard's words burst through 240,000 miles of space. "Houston!" They could tell the vexation in his voice. "Our abort program has kicked in again."

Mitchell turned to Shepard. "Al, are we snakebit?"

Alan studied his instrument read-outs. They had a great ship, but somewhere in the innards of their computers a spurious signal was loose, like a virus, leaping from its assigned circuitry and kicking in the abort signal.

Suddenly that long pre-descent check-out period had become an invaluable blessing.

CapCom was as baffled as the two men swinging over rugged mountain lands. "This is Houston. You sure someone up there doesn't have a thumb on the abort button?"

Both astronauts checked. The "panic button" to kick in an emergency abort was securely encased in a plastic shield that prevented accidental tripping. Yet that mysterious signal kept hamstringing their computer.

"Nobody's on the abort button," Shepard radioed.

Mitchell was scanning every switch, gauge, and control, looking for the spurious-signal needle in their electronic haystack.

Everything checked out perfectly. Except that their mission was coming unglued before their eyes.

Mitchell's voice remained the cool Mr. Unflappable that Shepard knew so well. "According to everything on the panels, Al, we're smack on. Everything checks out normally."

"Houston, come in," Alan called. "What's wrong with this ship?"

"Stand by, *Antares*."

The two astronauts exchanged knowing looks. They'd have to wait for an answer. The established routine to handle a problem like the one they faced was to gather the best brains in Mission Control and hammer out a solution.

They were right. Top officials and engineers went into a huddle. Eyes kept looking to the timers on the walls. There was a finite period in which to produce a solution. Little more than three hours.

They went with the theory that if *Antares*'s computers were picking up a short circuit in the abort switch, it would cause the very problem being faced by the two astronauts. They soon isolated it to one set of contacts of the switch on the lunar module's instrument panel. Recycling the switch, or tapping on the instrument panel, removed the signal from the computers.

The experts realized they could reprogram *Antares*'s computers so they would ignore the abort command. But this killed their automatic abort capability. That was taking one hell of a chance. It meant putting the two pilots strictly on their own.

"They're the best," Deke Slayton said with steely authority. "Get with it!"

A telephone call roused Donald Eyles, M.I.T.'s computer whiz, out of a sound sleep in Massachusetts. He threw a coat over his pajamas. By the time he reached the front door, an Air Force car was pulling into his driveway. Moments later he was on his way to his office at Draper Labs.

Eyles, who had helped develop the lunar module computer programs and knew them better than anyone, listened to the problem, nodded, sat before his computer keyboard, and began a new program to eliminate the glitch in the lunar lander a quarter million miles away. He also found a way to dump the unwanted signal that would still leave the two men with their auto abort option.

Ninety minutes remained as his fingers flew across the keys. He pushed back his chair and announced, "Done." Immediately a specialist fired the program into the lunar module simulator computer in Houston. The test run flashed back in numbers.

"It's perfect!" called out the monitor for computer systems.

Flight Director Jerry Griffin barked at his crew. "Let's get it up to them!"

Antares hauled around the far side of the moon.

Alan and Ed had thirty minutes left before their lander would be over the point on the lunar surface where they must fire the descent engine for real. Or the mission went down the rat hole.

"This is CapCom, *Antares*. The new computer program has been checked. We're sending it up to you."

Electronic signals flashed at the speed of light to the spidery space vessel.

Finally: "*Antares*, transmission is completed."

Shepard felt as if he and Ed were treading a mine field. Every minute lost now was gone forever. "We have it all," he said clearly.

Alan turned to Mitchell. "This is your ball game, Ed."

Mitchell lowered the lights, stared at the bright numbers, and raced against the clock to reprogram the computers with the new descent-flight profile data. Shepard watched in silence as Ed fed sixty new sets of information, in perfect order, in the system's logic circuitry. Once again he lived up to his reputation as the best.

"It's all yours, Al,"

"Houston, we've got it," Shepard notified CapCom.

"Good show, *Antares*."

Mitchell turned from his window. "Al, we're comining up on point."

"Got it." Shepard, whose frustration had been rising, exhaled a great sigh of relief. Mitchell had completed the new computer program with barely fifteen minutes left in the bank before they would have been forced to head for a rendezvous with *Kitty Hawk*. He'd had to stand by with his hands itching to fix something. But you can't climb into a computer, buried deep inside a spaceship, with a screwdriver or a wrench. Unless you want to kill the thing. And *Antares* had needed perfect logic programming and exquisite accuracy to prevent that maddening abort signal.

Shepard swore that, no matter what happened from here on out, they were going down to the rugged surface. This wasn't just his ship. If it didn't land, Apollo 14 likely would be the last of its kind to try for the moon.

"Houston, we're commencing with the descent program." He didn't ask controllers. He told them.

"*Antares*, you have a GO."

PDI. That's what they called it, and Alan Shepard worked his controls with the precision and experience of thirty years of flying as *Antares*'s descent engine came to life with transparent flame, danced on the fire, and arced moonward.

They were still racing at 3,700 miles an hour as they moved through 46,100 feet above the surface.

Twelve minutes and thirteen seconds from landing.

They went down, thrusters working to keep them perfectly aligned along their flight path. Dull thudding sounds, the ship rocking fore and aft, side to side. Airless turbulence. Every time a thruster fired, they heard the distant hollow sound, felt the punch through their feet and hands.

No longer could they see the moon. Blind faith in *Antares*'s systems was now their lot. They fell down and backward, their eyes looking out into space.

Five miles high.

"Coming on down, just like the book says," Mitchell announced as casually as if flying any airplane on final approach.

He grinned at his partner. "No more snake bites, Al?" he chuckled.

But the snake came back hissing with a vengeance. Shepard was looking from his panel to the outside and back again when Ed gave him the bad news. Mitchell's voice had a distinct edge in it.

"Al, I'm not getting a landing radar update," he said.

Shepard didn't blink. The rule book said, "No radar, you don't land. You abort."

To hell with that.

"I'll punch it through again, Ed."

"Okay," Ed said. He paused just long enough to confirm radar function. "Nothing, Al. No update." Another pause.

"Damn snake—"

They were still "visually blind" to the moon. Their only downward-looking eye was the radar. The dead radar. Without it there was nothing to tell the computer their precise altitude.

Twenty thousand feet. It was getting sticky.

"This is Houston." That worried tone again. "We're not seeing a lock on the landing radar system."

Shepard was cool. "Roger, Houston, we're on it, trying to activate it."

"*Antares*, you're at nineteen thousand feet."

They stared, mystified. Shepard looked at Mitchell. None of this made sense.

"Houston," Alan called, "the onboard navigational system is not receiving any data. Our landing radar is out."

They were flying to the moon without their electronic seeing eye. Belly up, it was worse than useless.

Mission Control was considering the options. "We could have them pitch over before they hit ten thousand feet. They'll have a longer look at the surface that way."

"Sure, that way they can land without the radar."

"No! They'll burn too much fuel. They'd go empty before touchdown. It's too risky. We could lose them."

Alan Shepard knew the rule book. As did Ed Mitchell. Alan was doing a great job forgetting what the book said.

CapCom: "Seventeen thousand five hundred feet."

Shepard was starting to boil inside. "Roger, Houston. You guys find anything?"

"Negative, *Antares*."

Mitchell was incredibly calm as he supported Shepard. "Seventeen thousand," he said quietly.

"*Antares*," came the dreaded words, "we should go over the procedures to abort."

And there goes our landing and the Apollo program, Shepard and Mitchell thought together.

From the surface of Fra Mauro on the edge of the moon's Ocean of Storms, it wasn't yet possible to make out *Antares* against the velvet black sky, a star blazing in dazzling transparent flame with a purple glow, falling steadily, a startling traveler from outer space.

Aliens come to visit.

Except tense humans in a control center on the aliens' home world, that dazzling blue jewel suspended against the velvet blackness, were about to extinguish the flame.

Shepard called out the abort procedure. "Okay, at thirteen thousand we pitch over, activate the ascent program—" He almost choked on the words.

A morose CapCom answered, "That's affirmative, Al."

Shepard, testily, "We're aware of the ground rules, Houston."

CapCom: "Countdown to mission abort will commence at fourteen thousand feet."

The hell it will!

Shepard, startled, looked about for a moment. That voice. It wasn't CapCom. Nor Mitchell. It was Shepard himself, rebelling against the glitches and all this crap about wiring and circuitry screwing up.

He could hardly believe his "other self." The Shepard flying *Antares* by the book knew he had four minutes left for acceptance of failure.

That was it. He turned to look at his partner. "Ed," he announced, "if the radar doesn't kick in, we're going to turn her over and fly her down."

He never knew if Mitchell was surprised or not. Ed didn't say a word at that moment.

"Dammit," Shepard snapped. "We both know we can do it!"

"That's what we came here for," Ed answered after a pause. But he also had a job to do. Lay out the facts as they were. Cut through the heartache and go right to the quick. "Okay, Al. As long as we both know the risks. The light up here, the shadows, our depth perception . . . we're going to have visual disorientation. We could be on top of a crater rim before we knew what hit us."

Shepard nodded. "Yep." Then he grinned. The man talking to him was all scientist, but he also was one terrific fighter pilot. And all fighter pilots know they are the best. Alan was counting on that streak in Ed Mitchell.

"Couldn't be any worse than bringing a sick jet down at night on the deck of the *Ticonderoga*, could it?" Shepard tried to remain all business, but one corner of his mouth was twitching into that Tom Sawyer grin.

"Al, this is a mite different."

"I know. It's not pitch-black down there, and it's not plowing through the North Atlantic in forty-foot swells pitching the deck." His grin widened, and then he added: "You're right, Ed. This is different. It's easier."

"Uh-huh. Piece of cake. Maybe a thin piece, but—"

"*Antares*, you're at fifteen five."

"We copy, Houston," Shepard acknowledged. He wasn't saying a thing about abort procedures.

Deke Slayton sure as hell wasn't missing anything. He knew Alan Shepard too well; Alan wouldn't quit easily. The tone in Alan's voice made it clear to Deke the man would do whatever was necessary to plant that ship on the moon. *By God, he's gonna take her all the way down with or without his radar.*

Deke smiled and knew others in Mission Control were sharing the same thought. They knew that when the chips were down, Shepard became pure band-saw steel. They also knew there were two Alan Shepards. The warm and friendly Alan with a smile that could charm you out of your last buck. But at this moment the other Shepard was at the controls of that lunar lander. The icy, no-nonsense commander.

CapCom kept contact. "*Antares*, you're at fourteen thousand seven hundred."

"Copy, Houston. We're still trying to reactivate."

"We still see no apparent malfunction," CapCom called, a flat, puzzled tone.

"Ed," Shepard said with quiet intensity to Mitchell. "I know we can bring it down."

Mitchell never hesitated to push back the unknown. That was his whole life. He studied Shepard. "Got to admit, Al, it would be a first. Promises to be interesting."

The icy commander managed a laugh. They were a team. Two premier test pilots seeing eye to eye, fully aware of the dangers. They knew danger lurked in tumbled landscape and a bitchy electronic system.

"Fourteen thousand two hundred," CapCom announced.

No reply from *Antares*.

CapCom, now on edge, urgent, but a sense of excitement. "*Antares*, we're going to try something. We want you to reset your circuit breaker."

"Houston, we copy," Alan answered. "Pull the plug, huh?"

"Hell, it works for my toaster," Mitchell offered. "Let's do what the man says."

Shepard yanked the circuit breaker, killing the main power to the radar. He shoved the breaker back in to bring the radar back on line.

CapCom sounded like doom. "Thirteen thousand six hundred. Have you got anything yet, Antares?"

"Negative," Alan said crisply.

Jesus, he doesn't care. He's gonna land, Deke told himself.

Flame began to reflect off the higher peaks below. Mountains, boulder fields, gaping craters waited.

Gloom filled Mission Control. No way out now. Time to abort.

"Hold on, Houston!" Mitchell's voice burst from the squawk box, through headsets.

Ed glanced at Shepard. "Al, look at that."

"Houston," Al said smoothly, "we've got a radar lock."

Mission Control hung by its fingernails from one moment to the next. CapCom's voice went up sharply. "*Antares*! We're confirming incoming data on the onboard navigation system." Relief mixed with wonder. "All systems are functioning."

Shepard was Mr. Cool again. "We copy that, Houston."

"*Antares*, you have a GO to land."

"You better believe, Houston."

"Your altitude is thirteen thousand."

Shepard whooped in an exuberant war cry, slammed a gloved hand against Mitchell's back. "Hell, man, we've hung it out further than this before, Ed."

"Ten thousand two hundred," CapCom called.

"Houston, we're pitching over."

"You're looking good."

Mitchell chanted out the changing altitude and other readings from his consoles.

Seven thousand feet.

Now they had a visual on their landing site. The surface came through clearly through ghostly flame beneath them. "That's a rough runway down there," Mitchell remarked.

"What a sight!" Shepard yelled.

"Cone Crater," Mitchell sang out, "and there it is right in front of us."

Shepard stood braced at the helm, feet apart at the commander's post, and he took *Antares* down on an invisible rail toward the ancient craggy highlands of Fra Mauro.

Alan Shepard, using thirty years of pilot skills, threaded a needle between the hills and ridges along their approach path and dropped his ship down into a narrow valley, craters and boulders everywhere.

"One thousand," Mitchell called, "and we're right on the money."

"*Antares*, you're GO for a landing," CapCom announced.

Alan could be generous now. "Thank you, sir. Fantastic!"

His lunar module was a dancer balanced on flame.

Five hundred feet. Down steadily.

Shepard's eyes darted back and forth as he sought a touchdown site. *Antares* was now more a helicopter riding on fire than a ship for deep space. The bug-eyed machine glowed in shimmering white and gold as the surface reflected back light from its flame.

"Shifting course," he murmured, avoiding scattered rocks and craters.

Antares skipped and floated like a huge insect skittering along invisible water.

Shepard dodged. Thrusters banged again and again to keep them upright. Alan locked his gaze on a rocky plateau dead ahead. "We'd better move it up a little," he told Ed.

"Good move." Mitchell pointed to his right. "Right there. We can land over there . . . coming down steady, down, down, down . . . there's some dust, Al."

Fifty feet up. Fire tore into ancient lunar soil.

Mitchell spoke steadily, calmly. "Twenty feet. Descending at three feet per second . . . ten feet . . ."

A small probe beneath *Antares* jabbed into the moon.

"Contact!" Ed reported.

"Throttle's off," Shepard said.

Flame vanished.

Eerily quiet. Until Ed whooped it up. "Great! We're on the surface."

"We made a good landing, Houston," Alan called CapCom. "About the flattest place around here."

They were just sixty feet from the X they'd marked on their landing map many weeks before—within walking distance along a rugged incline to the rim of Cone Crater, their main geologic target. There they would retrieve debris blasted out of that crater by meteoric impact scientists had told them had taken place more than four billion years ago.

Mission Control, emotionally drained, was bedlam once again. They'd ridden *Antares*'s problems all the way down with Alan Shepard and Ed Mitchell, who had turned what appeared to be certain failure into a perfect lunar touchdown.

At her home, Louise Shepard shrieked with her own release of tension and joy at what her husband had accomplished. Laughing and crying, she told her family, "We can't call him Old Man Moses anymore. He's reached his Promised Land!"

The only two living beings on the moon shook hands.

Ed Mitchell gave his friend a long, hard look. The icy commander was gone, and the warm, charming Alan Shepard was on the moon beside him.

"Come on, Al. The truth now. Just between you and me." He poked a finger at Shepard, hesitated a moment. "Would you have really flown us down without the radar?"

The Tom Sawyer grin was never so wide on Alan's face. "You'll never know, Ed." He laughed. "You'll never know."

CHAPTER TWENTY TWO

MOON WALK

HE STEPPED OFF THE SMALL PORCH outside the hatch. Backward. One foot at a time. *Antares* was on the moon, but to Alan Shepard his long journey wouldn't be complete until his boots sank into lunar soil. Nine rungs down the ladder. A pause. He gripped the handrails carefully, pushed off backward. The last three and a half feet down were a lazy drop.

Moon dust flowed up and outward in a fine spray, settling quickly. He stood still for several moments. Then he turned suddenly, enthralled by where he was, by the stark, barren slopes of a landscape that had remained essentially unchanged for millions of years. Meteorite impact had blasted huge craters or simply pockmarked the surface with the fragments of disintegrating comets. Moon quakes would disturb the inner slopes of craters, sending lunar soil sliding like desert sand toward crater bottoms. Sometimes the jarring impacts of meteorites or moonquake rumblings would collapse the walls of craters, cascading dust and boulders downward.

Antares rested in absolute silence in a world that as far as Shepard could see was frozen motionless.

Immense satisfaction went through him. His first words spoken from the moon came in an inner rush of deep emotion. He had no speech prepared for posterity.

"It's been a long way, but . . . we're here!"

His words sounded a clear message to the handful of folks who believed in him, who had never lost faith with his goals, who had never wavered once in their absolute conviction that he would one day stand right where he was. In his mind's eye he saw, and shared, this moment with three people above all others. Deke Slayton, always there to back him up, to provide a granite wall of support. His wife, Louise, who knew this moment must come to pass. Bob Gilruth—from day one of Project Mercury he had held Alan above all others as the man to lead the way.

On the planet where it had all begun, smiles lit up in Mission Control. They seemed to sense the feelings of that man on the faraway moon. They were with him.

"Not bad for an old man," CapCom answered. Controllers clasped their hands over their heads in the silent signal of triumph.

Antares, Shepard reported, had touched down firmly on a slight slope that was the "flattest place around here." He had maneuvered the moon ship to a landing in a great, shallow bowl-like formation. He pushed his boots into the soft, grayish-brown dust. No living creature had ever done this before in this desolate, utterly silent field of craters. "Gazing around at the bleak landscape, it certainly is a stark place here at Fra Mauro. It's made all the more stark by the fact that the sky is completely dark. This is a very tough place."

He studied the landing pads of *Antares*. "The soil is so soft," he added, "it comes all the way to the top of the LM's footpad."

He looked across the vast expanse of Fra Mauro, turning his back to the dazzling light in the heavens. The cratered surface faded into a featureless plain, a warning that the detailed experiments they would pursue must be performed with extreme care. Their landing site offered no reference points by which they could navigate.

Shepard watched Ed Mitchell work his way down the ladder to join him in the dusty moon soil. Ed moved about quickly, testing his body reactions to the weak gravity of a world so much smaller than his own. To Mitchell this was a moment of triumph as well as immense joy. "Mobility is very great under this 'crushing' one-sixth g-load," he quipped. The moon was a bouncy playground.

The two men started their detailed work schedule. They placed samples of rocks and soil into containers for scientists back home to study. The flag went up on a staff pounded into the surface.

They carried their remote-viewing television camera sixty feet away, set it securely into the ground on its tripod so that it would capture the *Antares* and two men hard at work. They unloaded a new device for lunar transport, a modularized equipment transporter, or MET. The astronauts dubbed NASA's "super-advanced scientific development" with the inelegant title of lunar rickshaw.

The MET carried an extensive supply of tools, cameras, instruments, safety line, core tubes for digging into the lunar surface, maps and charts for Shepard and Mitchell to use to navigate their way through and around craters, gullies, and boulder fields. Included was ALSEP—an elaborate package of surface experiments.

Ed Mitchell set up a spread of geophones, placing mechanical ears against the ground. Then he walked away from *Antares* with his "thumper." He was about to give the surface a series of hard blows with small explosive charges. When they went off, with their precise power and location known, scientists could read the geophone data to determine the density of material just below the surface.

Every fifteen feet along a line extending outward for three hundred sixty feet, he pushed the "thumper" against the ground and squeezed the trigger. His equipment was less than perfect. Of twenty-one explosive charges, eight failed, and Ed had fits with the thirteen that actually fired.

"Houston, this thing's got a pretty good kick to it," he reported. He was told the blasts would be on the order of a firecracker. Firecracker, hell. "It's like firing both barrels of a twelve-gauge shotgun at the same time," he said.

He managed a good series of moon poundings, and then for three straight attempts he had successive failures. "A hair-trigger this isn't," he commented.

He gave the thumper a few good whacks of his own. Bang! The charge rattled through the surface, and back in Houston scientists nodded happily. It was quite a trick to take the moon's pulse when you're on another world.

High overhead, aboard *Kitty Hawk*, Stuart Roosa continued on his whirlwind tour of the moon, every two hours swinging completely around the cratered mass beneath him. His voice rose with excitement as he told

CapCom, "I can see *Antares* on the surface!" No question, with sunlight gleaming from the spidery moon ship and the wide area of dust splayed outward from the landing.

The oft-repeated claims that the moon's surface had remained unchanged for billions of years and would remain so for uncounted more billions were dispelled again and again by Roosa's grand photographic tour. As Shepard and Mitchell ran through their experiments at Fra Mauro, Roosa directed his attention to the crater Landsberg B. This was the aiming point where ground controllers had directed the Apollo 13 third stage to crash-land to excite Apollo 12's seismometer. Smashing into the surface at several thousand miles an hour, the huge hollow shell and its heavy engine compartment had blasted a crater two hundred feet wide in the surface.

On the far side, Roosa came up with an unexpected bonus. Swinging across the craters never seen from earth, his cameras clicking away steadily, checking his position with photos taken on earlier missions, he was astonished to find an extremely bright crater directly beneath his orbital path. Unseen, unknown on the home planet, it was the result of a meteoric impact that was only weeks or months old—another pock-mark on a world that seemed laid waste by incessant bombardment.

Deke Slayton sat before a television monitor, leaning forward, oblivious to the passing hours, watching Alan and Ed going through their paces. No question which one was Alan. He wore bright red bands on the arms and legs of his pressure suit. Deke was doing more than watching. He could sense, even feel deep inside him, the sensations Alan was experiencing, the thoughts that must be in his mind. Deke watched him walk, half float in the gossamer lunar gravity, and soon Deke found his own leg muscles beginning to move in concert with his friend. When Alan lifted his arm, Deke felt the urge to do the same.

Finally Deke closed his eyes, listening to the voices across space. Before the television monitor with its live scenes from Fra Mauro, he was one with Alan Shepard.

In that silent union Deke began to smile. Then he laughed aloud as he heard Alan bitching about the lunar dust.

The dust, Shepard described with commendable restraint to Houston, was "very fine, like talcum powder. It clings to everything."

The effect was as if the moon dust were metallic and their suits magnetic, and every time they moved and kicked up dust it rushed to collect on their boots and work upward along their suits.

Deke opened his eyes. Alan had stopped. Something in the way he stood, the manner in which he had turned, told Deke that a powerful emotional strain was running through this man.

Deke was right. At this moment Shepard, the man, had taken over Shepard, the explorer.

Alan Shepard held this moment for himself.

He stood rock still, feet braced for balance, enclosed in the elaborate pressurized garment that was his private world of life, filled with energy, with supplies of heat and cooling, water, oxygen pressure—a capsule of life created on his home world hanging in the velvety, utterly black drapery of the universe.

Alan looked long and hard at earth.

He was overwhelmed, his senses and his thoughts set afire with the miracle of what floated in ultimate darkness above him.

At this moment uncounted millions of people on earth would look into their night sky, and they would see the moon—the moon of Apollo 14—in two-thirds of what would be its fully rounded shape.

Alan looked at his home world, two-thirds encased in diamond-hard blackness.

One-third of his planet hung magically in the void, incredibly dimensional, suspended, floating, levitated. A third of a world, "yet it was breathtaking. Looking up in the black sky and seeing the brilliant planet . . . the ice caps over the poles, the white clouds, the blue water . . . gorgeous!"

It was home. "Where all my friends were . . ."

It was an incredible vista. He understood now that while on earth, it seemed almost limitless to its people with its vast oceans and upheaving mountains, where there was always a distant horizon and changing dawns and sunsets.

"But from here, from the moon," he spoke quietly to himself, "it is, in fact, very finite, very fragile . . . so incredibly fragile. That thin, thin atmosphere, the thinnest shell of air hugging the world—it can be blown away so easily! A meteor, a cataclysmic volcano, man's own uncaring outpourings of poison . . ."

And suddenly this fighter pilot, this leading test pilot, astronaut, explorer, adventurer, master of wings and rocket fire, hero to millions, wept.

Tears streaked his cheeks. He understood much of what he felt was tension draining away from the emotional travail of their flight, and then the relief of a successful landing, but all of that, magnified a thousand times over, faded before that utter, fragile beauty of "our planet."

He had the feeling that he and Mitchell, and those who had come before them, and those who would follow, were here for far more than walking through lunar dust and sending explosive booms through a scarred world's surface, and measuring magnetic fields, solar winds, and radiation levels—all that was but prologue to the real reason they had done so much and come so far.

It all condensed into this one single, long look at fragile, beautiful earth, as though he were sent here, he and the others, so they might look back at that lovely, sensitive sphere and then carry home the message that everyone there must learn to live on this planet together. It didn't matter who liked whom, or on what side a man might find himself, because they were all on the side of a world of very definite finite resources, and they would all suffer terrifying consequences if they drained to bare bones the world that had conceived the life of the human race, fostered and nurtured that life, which now threatened to contaminate and destroy—

He stopped his thoughts, forced himself out of his introspection.

Back to work! There's much to do and miles to go before I sleep my first night here under the light of the . . .

Earth.

CapCom wanted to know how Mitchell found the trampoline of diminished gravity. Ed laughed, his delight infectious. "It's easy," he announced. "Just a little push and you spring right up."

He looked like a gazelle bouncing around a grassy meadow as he worked his way through his assignments. It might be work and they might be on schedules, but to Ed Mitchell there were no boundaries, no restrictions to what he did, what he saw, what he felt.

Mitchell was swept along by a "peak experience" in which "the presence of divinity became almost palpable." In that moment he knew, beyond question or counterpoint, "that life in this universe was not just an accident based on random processes." It was knowledge gained through private, subjective awareness, every bit as real as the objective data upon which, say, the navigational program or the communications system were based.

"Clearly," he mused, "the universe has meaning and direction." He had encountered through mind that "unseen dimension" that had brought to life the intelligent design of the universe and purpose they had never remotely dreamed would be their most precious find on a world where they were the only two living souls moving, working, thinking.

A mission demands attention and labor, and Alan and Ed bent to their timelines and tasks. They walked off hundreds of feet from the security and protection of their moon ship. Observers on earth, watching the television images from Fra Mauro, learned quickly that the moon is a great harbinger of visual illusion.

What seems flat and featureless is much like an ocean surface on earth. The "flatness" is in reality a long-waved undulation of the moonscape, rising and falling in misleading, gentle swells. If this could not be discerned with a long look at that surface, the sight of Shepard and Mitchell bounding about in the distance dispelled such error in viewing.

As they approached the site for scientific experimentation, some two hundred fifty paces from their landing ship, the two men seemed to be sinking into some great bowl of moon dust. They would appear to walk along flat ground when their legs disappeared and reappeared, like small ships on a heaving sea. In reality, they strode through great shallows in the plains of Fra Mauro.

At one point CapCom was driven to expressing the surprise of Mission Control. The two men had slid from sight, risen slowly, and marched along with most of their bodies concealed by a long undulating swell.

"You're visible from the armpits up," CapCom told them, bemused.

Shepard laughed as he waded through thick layers of the moon's topsoil. "Nothing like being up to your armpits in lunar dust," he said.

Several times during their moon walk and the exertion of carrying heavy loads, especially of bending and stooping and carrying rocks, Houston found reason for alarm at their physical exertions. The physicians monitoring their progress could hear them grunting, sometimes loudly sucking in oxygen to sustain themselves. Then they realized they were pushing too hard, overloading themselves. Immediately they slowed their pace, calmed their intensity of effort. The results were everything the doctors could have hoped for.

"They're doing terrific," was the consensus in the control center.

"Their schedule calls for four hours and ten minutes on the surface," noted one doctor. "But the way they're handling things now, let them know they can have an extra half-hour beyond that time."

They stretched the extra time allotment to four hours and fifty minutes for the first moon walk so they could carry additional rocks—two the size of footballs—back to *Antares*. They used a long pulley to haul the rocks in vacuum-sealed containers into the upper stage of their moon vehicle—home and hearth while on this alien world.

They brushed and pounded and swept as much moon dust as they could from their suits, ascended the ladder back to the porch, each stopping for a final look about them, still filled with wonder at all that had happened and at the deep stirring of their souls amid their scientific work.

Mitchell took a hard look at Cone Crater and its four-hundred-foot slope, which they were to ascend on their second excursion. Indeed, the climb to the summit of Cone was the major objective of their mission. Mitchell judged it was less than a mile from *Antares*. "We shouldn't have any trouble getting up there tomorrow. There's certainly a lot of boulders on the side. I'd say some are at least twenty feet in diameter . . . I think we can make it to the rim."

Shepard and Mitchell had practiced on earth for months in rocky country, digging, picking up rock samples, learning to look for specific rock types scientists were eager to study. Geologists who had trained these men told them what they found at Cone Crater could alter much of what we believed about the formation and structure of the moon.

"You two could kick off a whole new renaissance of the moon," a scientist explained. "It's our belief that Cone was carved out of the moon's surface more than four billion years ago. One hell of a meteoric impact. But the way that crater is gouged, there's every chance it ripped away rocks from maybe three hundred feet down and that these rocks were tossed about along the crater rim. If that's what you bring back with you, then we'll be able to study material that came into existence about the same time the planets and moons of the solar system were still forming out of dust and gas, contracting into the worlds we see today throughout the system. You two will be, in every sense of the word, traveling back in time. You'll see what we're after. No mistake about that."

Shepard and Mitchell understood the importance of what the scientists were telling them, how crucial their journey was. The oldest rocks of the moon's history could be lying at their feet when they reached Cone Crater. They would reach back billions of years with no more effort than bending down to pick up rocks that looked ordinary, coated with dust, but contained within their structure the secrets of the ages.

But that would be for the next day. Their first excursion was complete.

Like young boys on all fours crawling into a tree house, the moon walkers eased their way back into the cabin, sealed the hatch behind them, pressurized their ship, and "let go." Houston admonished them to eat their fill, to drink all the water they could absorb. They replenished the pressure suit containers with oxygen and water, checked the battery packs and systems, and reveled in the pleasure of being free of their cumbersome suits.

Weary in body and mind, they slept.

They were alert, raring to go, well before the end of their scheduled ten-hour rest-and-sleep period. They had breakfast and knocked on the door of Mission Control. "Hey, we're up and running this morning," Shepard told CapCom with impatience. "The shape of the crew is excellent."

They wanted to get out ahead of their timeline, to start the most critical day of the moon walks two hours ahead of schedule. Cone Crater awaited their presence.

The medical teams nodded, the mission directors were delighted, and they passed the word to CapCom:

"Turn them loose."

Shepard and Mitchell emerged from *Antares* eager to exceed every item of their moon work list. This was the first full "geology field day" on the docket for Project Apollo.

They reloaded the MET for the trip to Cone, anticipating an easier trek than lugging about large rocks like bowling balls. Carrying a dozen of those even in lunar gravity was a back-wrenching experience.

Their MET, or rickshaw, turned out to be less than advertised. Fra Mauro was thicker and deeper in dust than the sites encountered by the earlier Apollo landings. Moving the rickshaw was like plowing through the equivalent of deep sand.

"This is ridiculous," Mitchell groused to Shepard. "Let's pick up the damn thing and carry it." Shepard agreed, and the moon walk became more and more difficult. Not only because of the dust and their increasingly heavy load of rocks, but they were once again "snared" by the undulating nature of the terrain. Optical illusion plagued them from the outset. It was much like looking at mountains in the desert of home. In clear air a mountain peak or range might seem to be just a few miles distant but in actuality could be forty or fifty miles away.

The navigation charts seemed to have been prepared for some other planet rather than the Fra Mauro area. The cartographers had done their work well enough, but they'd never encountered the "mine fields" into which Shepard and Mitchell laboriously fought their way.

Distance measurements proved to be grossly misleading. The sun angle seemed to change and twist shapes into unrecognizable features. The crystal-clear sharpness of a world without atmosphere threw off the depth perception they had honed to such sharpness as pilots. Worst of all were the gullies that had never been predicted or shown up in photographs.

The moon walk had become a fierce, slogging journey, sapping their strength and frustrating them repeatedly as they lost sight of where they were, and then found themselves unable to confirm their bearings without loss of precious time.

Every time they stopped they went through their check list, collected interesting-looking samples. At every turn they kept their major goal in mind, confirmed to each other, "There's the rim of Cone. We're getting close now."

They were gulping oxygen at a fearsome rate. Both men were drenched in perspiration; suit internal temperatures were shooting up. They stopped often to rest.

There are times when the moon can be cruel. "Damn crater," Shepard grunted through deep breaths. "It's like it's challenging us to make it to the top."

Houston was concerned about their well-being, told them to take it easy, but urged them to keep moving. Everything was reduced at this time to reaching Cone Crater, to search for those rocks more than four billion years old.

Then, finally, before them loomed a steep climb, a slope that extended four hundred feet to the rim of Cone.

To reach that goal, they would have to climb through a massive boulder field. The scene was utter devastation, rubble and smashed rocks everywhere, some the size of small houses.

They knew they were almost out of time, that soon Houston would order them to start back to *Antares*. They pushed themselves as hard as they could. The slope fought back, deep dust grabbing at their ankles. It was like climbing a steep hill of dry sand.

"You take one step up," Shepard called Houston, "and you slip back half a step."

Suddenly Alan slipped to one knee, mired in gravelly dust, trapped in bone-dry muck. He fought to get up. Quickly Mitchell came to his side, helped his moon-walking partner to his feet.

They were gasping for breath. "Let's keep going." Alan labored to get the words out.

The slope resisted every move they made. Judging their tortuous progress, Shepard was beginning to accept that Cone Crater would win, they would lose.

"I'd say the rim is at least thirty minutes away," Shepard notified Houston. He added it was now doubtful they could make it all the way unless they slowed their desperate effort, cooling down their suits and eliminating several of the important stops planned in their traverse mission after they visited Cone.

Mitchell fought not to yield to the dangerous slope. "Let's give it a whirl," he prompted. Then in controlled frustration: "Gee whiz, we can't stop without looking into Cone Crater!"

Shepard still had the energy to continue but was judging from experience now. He wanted to make the best of a bad situation and bring back the rocks coveted by the scientists, which he felt were all about them.

"I think we're looking at what we want right here," he told Mitchell. "This boulder field is the stuff that's ejected from Cone."

"But not the lowermost part," Mitchell said stubbornly, "which is what we're interested in."

They kept pushing and pulling and digging their boots and gloved fingers into the moon's soil until it was painfully clear that Cone Crater had won.

Time, oxygen, and physical strength were down to nubs. They had battled the steep climb to the rim of Cone Crater for an hour and a half of all-out physical exertion, their breathing coming in shuddering swallows, their few words heard occasionally as choking gasps.

"It's hard, hard," said a weary and despondent Shepard.

Mission Control judged the two men at the very edge of their endurance, and they were still about seventy-five feet from the top. The team leaders told CapCom, "Tell them to call it off. Get back to the ship."

"Al, Ed, you guys have already eaten into your thirty-minute extension, and you've passed that now," came the unhappy news from Houston. "We think you'd better proceed with the rock sampling."

The high rim of Cone Crater would remain unchallenged.

"I think you're finks," Mitchell told everyone.

The moon men moved to several large boulders nearby, chipping samples, recording every detail, photographing the boulders before and after whacking samples from their structure. Only scientists would later be able to tell if these samples represented the debris of impact cataclysm so many eons ago. (Later studies revealed Shepard and Mitchell had gathered rocks more than four billion years old, among the oldest found on the moon.)

Coming down the slope was immeasurably easier than clawing their way upward. Quickly their suit temperatures returned to normal, and they went with renewed will and energy to complete the moon-walk experiments. Heading back to *Antares*, they stopped at Weird and other craters, moving documented rocks into the rickshaw. They pounded core sample tubes into the surface, dug trenches to determine how surface materials had stratified over eons, and amassed data that would keep scientists busy for years.

Then, their lander was before them. They loaded their booty aboard and were ready once again to climb the nine steps to their cabin.

Well, almost.

"Houston," Alan alerted Mission Control. They were already watching every move the two men were making, but Alan Shepard was out to cop an unquestioned first for a moon visitor.

From a suit pocket he withdrew a small metal flange and carefully attached it to the long aluminum handle of the collector with which he'd picked up small rock samples. In Mission Control eyes were riveted on the screen.

What was Shepard going to do now?

"Houston," he paused for effect, "you might recognize what I have in my hand . . . the handle for the contingency sample. It just so happens to have a genuine six-iron on the bottom."

Controllers gaped.

Shepard reached into a pouch of his suit and held up a round object for the world to see.

"In my left hand I have a little white pellet that's familiar to millions of Americans."

"It's a golf ball!" came a yell from a controller.

Grins flashed throughout Mission Control.

Shepard, an avid golfer on earth, dropped the ball into the moon dust. He made a valiant effort to assume a normal two-handed stance to address the ball. No way in his bulky suit. He signed. He would attempt the first out-of-this-world golf shot with a one-handed swipe.

"I'm trying a sand-trap shot," he cracked as he swung awkwardly, the face of the six-iron spraying lunar dust and plopping the ball into a crater less than a hundred feet away.

"I got more dirt than ball," Shepard apologized.

"Looked more like a slice to me," Mitchell quipped.

Alan dropped a second ball, determined to do better. He slammed the face of the six-iron squarely into the small, white sphere, sent it sailing away against the perfect black of space in the weak lunar gravity.

"Beautiful," murmured Alan. Then, louder for his audience back on earth to hear, he called out: "There it goes! Miles and miles and miles!"

The argument would continue for many years whether it had been a shot of two hundred or four hundred yards or some spectacular distance beyond.

That was it. Both men checked to assure everything on the list to be returned to earth was aboard their ship. Shepard turned to look into the remote television camera.

"Okay, Houston, the crew of *Antares* is leaving Fra Mauro Base."

"Roger, *Antares*."

Later, the countdown timers flashed away minutes, seconds.

Flames ripped into moon dust. Gold foil tore away from the landing stage, showering outward in all directions.

The American flag whipped back and forth in the wind of rocket exhaust.

Antares climbed swiftly until, had anyone on the moon been watching, it dwindled to a sun-reflecting small star and then winked out of sight.

Shepard and Mitchell had been on the moon thirty-three hours and thirty minutes. One hour after liftoff they closed in on the waiting *Kitty Hawk*.

No docking hang-ups this time. Perfect linkup.

Everything destined for the homeward flight was floated into the command module. They closed and sealed off the tunnel between the two ships. Stu Roosa fired the LM ascent stage away from *Kitty Hawk*.

"And we say sayonara, good-bye, to *Antares*," he said in a farewell salute to the now deserted upper stage of the lunar lander, destined to crash into the moon for seismic tests. The final bell to be rung on the moon by Apollo 14.

Transparent flame ghosted behind them as they powered their way up for the return flight. Three days without any emergencies, ending in the Pacific Ocean near the waiting aircraft carrier *New Orleans*.

The legacy of Apollo 14 went far beyond the scientific probing of the moon or the dedicated and exhausting struggle of two men fighting their way upward through a hostile surface, which resisted them at every step.

All that would pass into history.

But what would not be diminished were the three remaining Apollo explorations of the moon, which had been were on the tremulous edge of cancellation. Those three flights would leave earth on the shoulders of Apollo 14.

And beyond those final Apollo voyages?

LAST APOLLO OUT

DEKE SLAYTON LOOKED UP from his desk, watching Alan Shepard easing into a chair before him. Apollo 14 was behind him, and this was Deke's first real chance to be alone with his friend since Shepard had returned from the moon.

"Remember, Al," he began eagerly, "when I got that rotten cold just before Thirteen flew?"

Alan smiled. "You were a sight."

"Yep," he agreed. "Chuck Berry turned me into a heavy vitamin kick. It did a lot more than get rid of a damn cold." Deke grinned.

Alan's interest quickened. His friend's eyes were bright. Alan nodded, waiting.

"Well," his grin grew, "it's been more than a year since then. I'm still taking those vitamins." Deke leaned back in his chair and took a deep breath. "Al, I haven't had a single skip of my heart in that whole time."

Alan studied the man who'd been grounded for nearly a decade. There was but one question to ask. "No use telling you how great that is, Deke. So the question is, what are you doing about it?"

Deke went from a broad smile to a frown, recalling his personal struggle. "I'm telling every doctor and flight surgeon I see. I pound their desks to tell them, 'hey, look, the heart's working perfectly!' And from time to time I hook myself up to one of those Holter monitors you can strap to your back and shove the recorder in your pocket and carry around with you. That way I've been logging my heart actions for a whole day at a time. The whole thing is one big pain in the ass, but now I've got records showing that in all this time my heart hasn't skipped a single beat."

Shepard didn't want to say aloud, not yet, that this could bring Deke back on line as a candidate for a launch. "What do the docs say?" he asked quietly.

Deke showed a flash of anger. Frustration from being stonewalled. "They haven't cooperated one damn bit," he growled. "Damn those people. They tell me to come down and take a battery of tests. Same old stuff I've been doing for years. All they see in me is a handy guinea pig. And whatever they do doesn't prove anything. So I stay calm, and I'm patient with them, but when I tell them I've cured my heart by these heavy doses of vitamins, most of them give me that look as if to say they know I've gone off the deep end. Crazy as a loon, as far as they're concerned. Nine out of every ten doctors I've seen say it's not possible, that it's some kind of a phantom thing. And they refuse to believe the old ticker is working perfectly."

Deke shifted in his chair and locked his gaze with Shepard's. "Look, I've got to find somebody with the guts to stick their necks out. Al, I'm okay to fly."

Shepard stayed low-key. Deke had enough fire going for both of them. He thought of the astronaut team's flight surgeon. "What about Chuck Berry?"

"Hell, Chuck's got the guts, all right," Deke said quickly. "He's on my side all the way. But not even Berry can get me restored to flight status. Not by himself. NASA insists I've got to have what they call an eminent heart authority to certify I'm healthy to fly. And that," Deke sighed, "we haven't been able to do."

Shepard weighed Deke's words. No question this man had gone the proper route from start to finish. And if his heart was performing as he'd confirmed, then it wasn't a medical problem keeping him strapped to earth. Once again "safe politics and cover your ass" were the culprits.

Alan rose from his chair; Deke stood. Shepard walked around the desk and put his arm around the man who had walked side by side with him from the first days of the manned space program. "Deke," he said carefully, "I'm going to help you every way I can. After what you did for me, that's the only way to go."

"Whatever I did for you was right," Deke said immediately.

"What counts is that you believed in me, and you did it. Look, I'm forty-seven. We both know I don't have much chance to get back into rotation for another flight. But I'll stay on at the Astronaut Office, because that's the only way I can help you knock down the walls."

Several months later, in an ancient land on the other side of the world, Dr. Chuck Berry found the solution.

Berry was then attending an international medical conference in Istanbul, Turkey. There he spent time with Dr. Harold Mankin, a world-renowned cardiologist from the Mayo Clinic in Minnesota. Berry grasped the moment to explain Deke's barriers to making his long-wanted space flight. "What can you do to help?" Berry asked.

Mankin answered, "Send Slayton up to Mayo. If nothing else, I'll be able to tell the man once and for all if he's still got his chance to"—he smiled—"get out of this world."

Deke was on his way north as soon as Chuck Berry passed him the word. Dr. Mankin lowered the medical boom on Deke, putting him through "a whole enchilada of tests. It was a pretty dynamic probing. They nearly turned me inside out. They hung me upside down on a treadmill, poked holes in me, pumped dye into my system, and examined parts of my body I didn't even know I had."

They scheduled Deke for an angiogram. He balked. Other doctors had tried to run him through an angiogram before. They had pumped him full of good cheer about how much it could tell about his heart, and then they added there was a "minimum" risk factor.

"There's a chance the angiogram will kill you," was their message. "And, if you clear the test satisfactorily, that's just the test. There's no guarantee the results will clear you for flight status."

Deke had answered in his customary manner for decisions up against the wall. "Screw that," he told the doctors. "At least give me something worthwhile for risking my life." They had nothing better to offer, and Deke had gone back trying to prove in every other way he could that there was nothing, medically, to keep him out of a spaceship.

Dr. Mankin changed Deke's mind. "All your other tests are good," he said. "If the results of the angiogram I run on you are good, Deke, then as far as I'm concerned, you get a positive reaction out of it. I'll support you all the way to fly."

"Hell, Doc, that's worth the risk," Deke responded. They gave him another "enchilada," he later recalled, grinning. "Poked a hole in my arm and ran this probe through the opening all the way to my heart."

He fretted impatiently for the final word from Dr. Mankin. The surgeon came right to the point. He called Chuck Berry. "Your man, Deke Slayton? This guy is as good as gold."

The door to upstairs had opened just a crack.

While Deke was battling the medical fraternity, the launch center on the Florida coastline was still running on all cylinders. On July 26, 1971, Apollo 15 astronauts, Dave Scott, Jim Irwin, and Al Worden, flamed into the long journey moonward. Scott and Irwin rode their lander, *Falcon*, right to the foothills of the Appenine Mountains while Worden, overhead in *Endeavour*, began an intense photographic survey for future landing sites.

Scott and Irwin stepped onto the lunar surface to stare in wonder at the Appenines, towering fifteen thousand feet above the plain where they stood. The mountains rose impressively on three sides about them, and on the fourth side plunged a mile-wide gorge, Hadley Rille. By now astronauts knew what to expect in the way of problems and had the knowledge and equipment to make great strides forward in geological surveys and scientific exploring. This was the first mission carrying a lightweight electric car, a cross between a golf cart and a dune buggy. Battery-powered electric motors drove the less than elegant go-cart at speeds on level ground up to seven miles an hour, but zipping downhill in the 77-pound buggy (462 pounds on earth) they managed the break-neck speed of eleven miles an hour. Most importantly, this lunar rover could climb and descend slopes of twenty-five degrees and carry heavy loads of tools, rocks, cameras, and the two astronauts.

With the ability to travel six miles from their lander, they increased tremendously the area to be traversed, studied, and sampled. The six miles was a safety feature. If the moon buggy broke down, the men would still have enough power and oxygen in their suits for a steady walk back to *Falcon*.

In three separate excursions during their three-day stay on the moon, the two men drove about in an exploration of nearly nineteen hours outside their lander, on the surface, climbing slopes, driving into wide, shallow craters, probing the edge of Hadley Rille.

Before returning to *Endeavour* and earth, Scott and Irwin placed on the surface a plaque honoring three more casualties of the drive to conquer space. Just four weeks before the Apollo 15 launch, three Soviet cosmonauts had died while returning to earth after a record earth orbit flight of twenty-four days. The Soyuz 11 crew, Georgi Dobrovolsky, Vladislav Volkav, and Viktor Patsayev, had been killed almost instantly during reentry when a leak developed in their spacecraft and the cabin lost pressure.

It was another reminder that space flight still had not become routine, that danger was always a companion.

Armed with the endorsement of Dr. Mankin, Flight Surgeon Chuck Berry was deep into his personal crusade to return Deke Slayton to flight status. It seemed a race between getting that reinstatement and whatever missions might be left for Deke to fly. Berry hammered on the doors of the best people in the medical field. By the time he completed his rounds there still remained a few naysayers, but the majority of the specialists, deeply impressed with the findings of Dr. Mankin, gave their blessings for Slayton to be returned to the active flying list.

Alan Shepard, who knocked on a few official doors of his own for Deke, had known this moment of triumph when he had overcome the debilitating problems of his balance system and gone on to the moon.

Now it was Deke's turn. On March 13, 1972, the doctors called him in for an intensely personal conference. It wound up with extended hands and one word: "Congratulations."

Deke Slayton's decade of agony was over.

Shepard banged him on the back. "The next step, my friend," Alan said, "is to get you a flight. And you've got a hell of a lot of competition."

Deke nodded. He wasted no time in charging into the office of Chris Kraft, the same man with whom he'd worked hand in hand for ten years.

"I'm no longer one of your managers, Chris."

The center director stared at him. What the hell was Deke talking about?

A grin spread across Deke's rugged features. "From this moment on, Chris, I'm a prime candidate for a flight crew."

Kraft pounded him on the back and shook his hand enthusiastically. Both men were willing to bypass the usual social rituals and cut to the quick. "Your news is wonderful, Deke. You know how I feel about your going all the way." He sobered quickly. "But there's also reality, my friend. Getting you a flight is not going to be easy. You can't just walk up to a spacecraft and boot someone off a crew."

"I know," Deke grunted.

"Look, the final Apollo crews are in training. We're well aware of that. Sixteen is about ready to fly, and Seventeen's been tighter than ticks the way they're working together."

"Skylab?" Deke asked with raised brows.

"We've already selected the crews for the three missions."

"Damn . . . I know."

"Well, you're in a very special position, Deke. You could exercise your seniority and bump someone off Skylab, but if I know you—"

"No, no, you're right," Deke interrupted glumly. "I wouldn't. No way would I ever cut the feet out from under another pilot."

"Well, there's the space shuttle," Kraft went on. "But it's at least six years down the line. You know that. But that ship, and its flights, are wide open."

"It's a hell of a long way down the road," Deke added.

"Tell you what. Continue with your present job. You stay as tight as you have been with Alan to wind down Apollo, and we'll see what comes along."

A month after Deke came back on active status, Apollo 16 astronauts, John Young, Charles Duke, and Ken Mattingly, swung into lunar orbit. Young and Duke left Mattingly hung in the lunar sky aboard command ship *Casper*, while they rode *Orion* to a wide plateau edging the Descartes Mountains. The second lunar rover took the two astronauts through massive boulder fields, around and through craters, their moon buggy riding over unexpected chemical rock groups that surprised geologists with their high content of aluminum basalts. During three excursions, edging to the rim of such prominent features as North Ray Crater, hauling 213 pounds of rocks and other samples back to *Orion*, they headed home with more questions for perplexed scientists than answers to old questions.

With Apollo 16 in the books, and the final moon landing gearing up for a last spectacular voyage, Deke Slayton's last-chance flight into space was beginning to take shape.

It was an old idea, long cherished but also held in great doubt. The United States had come from behind in the propaganda battle paraded as the "space race" and left the Russians choking on the exhaust of extraordinary Saturn V blastoffs to the moon. But could the fierce competition be transformed into a purpose of a higher order? Was space the new high ground where the United States and the Soviet Union might leave their angry confrontations on the earth's surface and meet in a joint mission? If this could come to pass, it might just open the doors to future cooperative programs.

The first seeds of what would become ASTP—Apollo-Soyuz Test Project—began to sprout in 1969. Thomas Paine, the NASA administrator, was aboard Air Force One carrying President Richard Nixon across the Pacific to greet the first astronauts returning from the moon.

It was the perfect opportunity for Paine to review future space programs. One choice was obvious: The Russians were the only other space power sending men above the planet. Paine postulated that both countries could benefit from a program in which either nation could rescue in an emergency the astronauts of the other. This wasn't feasible now with each nation employing not only different docking techniques in space but also using equipment that prevented linkup of American and Soviet craft. If, however, both countries could be brought to the bargaining table to launch such an effort, and would exchange scientists, engineers, and their best astronauts and cosmonauts, then in the future the planned space stations of both countries could gain an enormous safety factor when and if an emergency rescue came into demand.

Nixon saw this as a possible avenue for thawing the cold war and gave Paine the green light to proceed. Immediately on his return to Washington, Paine established contact with the Soviets through the U.S. embassy in Moscow and presented the concepts of compatible docking systems and a proving flight with both an American and a Soviet spacecraft.

Despite the high-flying aspirations of NASA and the clear superiority of the United States with the Apollo program, Russian distrust of the U.S. and its intentions held fast. They wanted more than embassy contacts and generalizations, and they balked at an agreement.

What Tom Paine had started but could not finish because of deep Russian suspicions became the task of Dr. Phillip Handler, president of the prestigious National Academy of Sciences.

On May 11–12, 1970, after weeks of preliminary discussions, Dr. Handler met with his counterpart in the USSR. The Soviet group included, as its head, academician M. V. Keldysh, president of the Soviet Academy of Sciences. Handler seized the moment to emphasize the values of a "common docking mechanism" and the enormous potential of achieving such a mutual system.

The idea was planted. The Russians moved with agonizing slowness, and not until May 24, 1972, at a Moscow summit meeting between President Nixon and Soviet leader Leonid Brezhnev, was a pact finally signed. It called for, among other space efforts, its stated purpose of "the docking of a Soviet Soyuz-type spacecraft and a United States Apollo-type spacecraft" with visits of astronauts and cosmonauts in each other's ships.

The Russian signatures on those documents did more than signify unparalleled cooperation in a vast science program.

The Russians had just punched the ticket for Deke Slayton's upcoming ride into space.

Nixon and Brezhnev inked their joint declaration just two months after medical teams removed Deke's grounding shackles.

He didn't waste a moment. Once again he sailed into the office of Chris Kraft, leaned on his desk, and stared into Kraft's eyes.

"Chris, Apollo-Soyuz is going. I expect to be a candidate for that mission. As of right now I've thrown my name into the hat."

Chris held Deke's gaze. "Give me a recommended crew, like you've always done," he replied.

Deke answered immediately. "Okay, you've got it. I'll put myself number one on the list."

Kraft nodded.

"And I'm very strong on Vance Brand," Deke said. "I've been impressed with his work as a back-up to the Skylab crews. Now, you tell me who else should be on this team."

Kraft leaned back in his seat. He ran names through his mind, considered experience, leadership, and even juggled some political considerations into the stew. "All right, Deke. I opt for Tom Stafford. He's got two Gemini flights and the Apollo 10 mission under his belt. Now, just as important for this caper is that we sent him as a representative from NASA for a cosmonaut funeral. He was there long enough to make some friendships with a lot of cosmonauts and their program officials. That could open the way for a joint project better than anything else."

Slayton accepted Kraft's recommendations. But they weren't through yet. "One more thing. I've got the seniority to command this mission. Seniority does count. I've never used it before, but I've got it and I'm using it now."

Kraft turned him down. Tom Stafford had both the experience and the all-important relationship with and trust of the Russians.

Alan Shepard was waiting for the outcome of the meeting between Deke and Kraft. Deke came into Alan's office with a mixture of exuberance and grumbling. "I'll tell you what, Al, this was what the word bittersweet is all about. As you know, I've wanted to command this damn Russian thing from the beginning. Damn if Chris didn't go and give it to Tom Stafford."

Shepard laughed. "C'mon, buddy, what'd you expect?"

Deke had a rueful grin. "You're right. I put myself in Chris's shoes, I'd make the same decision."

"That's right," Alan said quietly. "The whole thing is to fly. To get out there. Doesn't really matter what seat you're in. They're all in a row. Nobody up front, nobody in the back."

"Yeah," Deke said comfortably. "And we both know this is my only ticket. The shuttle is too far down the line. The clock runs out on me after this flight."

Apollo-Soyuz was three years away.

December 7, 1972.

Shortly after midnight on a Cape Canaveral coast mantled in darkness, thousands were convinced the sun had come up.

Light flared along the beach, spreading outward like a glowing shock wave. Nine seconds later the Saturn V booster of Apollo 17 went to full power and rose atop a blazing fireball, which split the night darkness like a great knife ripping apart the heavens. Where there had been blackness was now eye-searing light. Five hundred miles away, atop Stone Mountain in north Georgia, astonished observers saw what looked like an atomic bomb fireball accelerating ever higher above the world, its flame turning from blinding white to yellow and then crimson, and finally it became an enormous plume, ghostly, eight hundred feet of violet magic rushing away from earth.

Gene Cernan, Jack Schmitt, and Ron Evans, in their space-going vessel named *America*, were on their way for the final round of the great voyages of lunar exploration, following the first and only after-dark launch of a Saturn V.

Cernan and Schmitt, the latter the only geologist to ride the fire trail to the moon, landed their *Challenger* in the Littrow Valley of the Taurus mountain region and capped the most incredible series of expeditions in the history of the human race. They spent three days on the lunar surface, including more than twenty-two hours in a trio of stunning geological journeys, riding their lunar rover to fields of enormous boulders, to the slopes of steeply rising mountains, and along the edges of precipitous gorges from where they stood in awe of the chasms torn in the moon's surface. Before they ended the exploration, they had loaded 243 pounds of rocks and soil aboard *Challenger*, conducted dozens of scientific experiments, and stripped away many of the moon's secrets that had confounded humans for centuries. The data from Apollo 17 and the other landing missions would keep scientists of many countries intensely busy for decades studying the rocky debris distributed by the American government for world science to share.

What emerged is a picture of a moon that was born in searing heat, lived a brief life of boiling lava and shattering collisions, then died geologically in an early, primitive stage. It came into being some 4.6 billion years ago when great masses of gaseous matter called the solar nebula began condensing to form the sun, earth, and other planets and moons of the solar system. The nebula first condensed into chunks of space debris—from small pebbles to miles-wide boulders—that crashed together and fused to form celestial bodies.

Study of the Apollo data showed that, in the case of the moon, this compacting of debris generated intense heat, which turned the lunar surface into a sea of molten lava, to a depth of several miles. When the lava cooled, this became the moon's primitive crust. Debris left over from the creation of the solar system continued to bombard the moon, carving out giant craters and valleys and forming mountains by piling up large piles of rocks.

It is believed the young earth underwent the same period of meteorite bombardment and volcanism that the moon did for about a half billion years. Then the histories of the two bodies took different paths. The weak lunar gravity could not prevent volcanic gases from escaping into space. But the larger body, earth, with strong magnetic and gravity fields, held onto its volcanic gases, and they formed an atmosphere and oceans, creating conditions for the development of life. The moon became a dead body where life could not exist.

The spacecraft named *America* returned to earth on December 17.

The books were closed on Project Apollo.

The last man on the moon, Gene Cernan, had paused for a final look at the black beauty of the world about him. He had a message to send home before departing. "As I take these last steps from the surface for some time in the future to come, I'd just like to record that America's challenge of today has forged man's destiny of tomorrow. And as we leave the moon and Taurus-Littrow, we leave as we came, and, God willing, we shall return, with peace and hope for all mankind."

It's been more than two decades since he spoke those words. No American, no earth being, has yet returned to the moon. No one will again tread lunar dust until sometime in the next century.

Within a period of four years, twenty-four American astronauts, some twice, sailed through the vacuum from earth to the moon. Twelve from those twenty-four rode their landers down to the lunar surface, walked and drove through the dust and rocks of the small world.

Had the Soviet Union sustained its early lead in power and technology over the United States, the number of humans going there might have increased greatly. It was a fierce competition, and the Soviets went all-out in their desperate attempts to lead the human race to another planet, small though it might be and devoid of life. The Russians went through a series of devastating rocket explosions and suffered equally costly failures after reaching earth orbit.

Just two weeks before the last Apollo departed for the moon, the Russians were down to a last-gasp hope that their mammoth N-1 rocket, even more powerful than Wernher von Braun's spectacularly successful Saturn V, would enable them, at least, to reach the moon during the same period American astronauts were free-wheeling across its surface.

It was not to be. The fourth launch of the N-1, intended to fire a large and heavy unmanned lunar lander directly to the moon in a rehearsal for a manned flight, was ripped apart by a series of violent explosions as it climbed through the atmosphere. When the wreckage tumbled back to earth, it sounded the death knell of the Russian manned lunar effort.

Bitter and frustrated, the Soviet government insisted it had never been in the moon race. History records otherwise. Several Russian manned landers became dust collectors in remote hangars. The rocket stages and enormous fuel tanks of the leftover N-1s were hammered into storage sheds and playgrounds for children.

Whereas Russia had endured disastrous failures in its manned lunar effort, though, it had hurled its considerable power, experience, and science into a massive program of space stations in earth orbit. The Salyut stations led the way and were a spectacular success. They were followed by the expensive, complex, and extremely workable Mir station, to which the Soviets steadily added scientific and work modules. Cosmonauts, men and women, flew regularly in Soyuz spacecraft to the stations to conduct medical, science, materials processing, and military observation experiments.

A dichotomy arose in understanding the ability of the human being to endure long periods in weightlessness. Essentially, the American program was one of extreme caution, which maintained that zero-g was not only biologically deleterious but also, could lead to fatalities if astronauts remained too long in space. The Russians crafted a program to find out the effect of long exposure to weightlessness. They kept their men aboard the orbiting stations first for six months, then eight, and finally for more than a year. They suffered no long-term effects.

The American road to space in the aftermath of the stunning success of Apollo developed potholes and detours. Grandiose schemes for massive stations, a permanent base on the moon, and a grandly heralded manned expedition to Mars quickly succumbed to a disenchanted Congress and a rapidly apathetic public. Domestic problems and challenges, existing with or without a space program, stripped the national space effort to a shadow of its former glory.

No longer was there the driving force that had carried Apollo to the moon. America had been technically challenged by a cold war rival, and a sense of fear and national pride had propelled Americans to respond, to win this high-stakes battle for the high frontier. The Soviets no longer were a threat in space, and in the terms that became commonplace among the veteran ground crews, as well as the astronauts, the dreamers and builders were replaced by a new wave of NASA teams, bureaucrats who swayed with the political winds, sadly short of dreams, drive, and determination to keep forging outward beyond earth.

Slowdown was the new buzz word, a strange mixture of anemic expenditures trying to build a single answer to all needs. NASA in commendable wisdom chose the route of a system that could be used repeatedly for space missions—the STS, or Space Transportation System, known popularly as the Space Shuttle. The huge, winged, reflyable spacecraft, scientifically and engineering-wise, is a spectacular achievement. It is a technological marvel. By comparison, the predecessor Mercury, Gemini, and Apollo spacecraft were rowboats in size and performance.

NASA faltered not with its equipment, reduced in performance and reliability by fiscal pitfalls, but in the grandiose eloquence of promising that the Shuttle would be all things for all missions, that it would serve civilian and military needs, and that it would save truckloads of money in the process.

Those promises were not to be. The program escalated swiftly in cost and decelerated just as rapidly in its time schedule. Weeks became months, and projects to take months stretched into years without definite future dates that could be sustained.

Something had to fill the gap. NASA needed desperately to keep intact its management, engineering, and astronaut teams.

Enter Skylab. The space agency had rockets and spacecraft left over from the three Apollo moon missions that had fallen victim to the congressional ax. Engineers proposed modifying some of this hardware into a modest space station where astronauts could study the sun and other stars, conduct experiments seeking pure materials and medicines, and learn to live in space weightlessness for long periods in the event America one day decided to embark on a months-long manned exploration to Mars. The Soviets had a space station, some argued. Could America do less?

The cost would not be great, and Congress agreed that NASA's teams were a great national resource that should be preserved.

The third stage of a Saturn V was stripped of its engines and converted into a complete station to be hurled into orbit by the first two stages of the big booster. Skylab was a "home away from home" with racks of scientific equipment, a marvel of an astronomical laboratory, and more than thirteen thousand cubic feet for unparalleled comfort and freedom for three astronauts at a time. Gone were the cramped telephone-booth-sized craft of the past. It had cooking facilities, private quarters, showers, exercise equipment, and other "luxuries" only dreamed of before the space station roared into orbit in May 1973. Three successive missions of three astronauts each rode smaller Saturn 1B rockets and Apollo command ships to the station, and in 1974 the final "stay in space" extended to eighty-four days and proved that man suffered no ill effects from weightlessness for that length of time.

The plans to keep Skylab in orbit well into the 1990s went for naught. A constant barrage of solar wind began slowly to drag the station closer and closer to earth. Long-term NASA plans called for a Space Shuttle crew to fly to Skylab in 1978, attach a booster rocket, and fire it into a higher orbit, where it could be visited by astronaut crews for several more years. But the Shuttle languished until 1981 in a swamp of fiscal shortfalls, engineering problems, developmental snags.

In the summer of 1979, the great space station began its fall and ended its brief career with a spectacular blazing reentry through the atmosphere. Fiery chunks of its shredded body fell in the Indian Ocean and uninhabited areas of western Australia.

As the man riding herd on astronauts scheduled for space missions, Skylab in the normal course of events would have been the responsibility of Deke Slayton. But with his assignment to the upcoming Apollo-Soyuz mission, Deke was buried in training and planning for the first joint American-Russian space venture.

Once again his close friend Alan Shepard stepped into the breach. Shepard had planned to return to private life with his family when he determined he was "out of the rotation" for another space trip. That would have hung Deke up with more responsibilities than he could handle, between running the Astronaut Office and concentrating on his first space flight.

Shepard took over the reins of the Skylab astronaut needs, and Deke went with Tom Stafford and Vance Brand to Moscow.

CHAPTER
TWENTY FOUR

SALT AND VODKA

DEKE SLAYTON LEANED against the window as the aircraft continued its long descent to Moscow's main airport. Along with fellow astronauts Thomas Stafford and Vance Brand, he would soon be immersing himself both in Soviet space technology and in Russian culture and language in preparation for an unprecedented space mission.

As he watched the verdant forests and sloping hills below, he couldn't help but remember that only thirty years ago these same forests and hills had swarmed with hundreds of thousands of German troops who hammered at the very doorstep of the Soviet capital. Moscow had staged a savage defense and survived, and Russia had gone on, after suffering twenty-two million casualties, to destroy the Wehrmacht and break the back of the Nazi war machine. The Soviet Union and the United States had fought as allies then, and it was now the cold war between the world's two superpowers that beggared the imagination in its potential for wrecking civilization.

Deke had flown B-25 twin-engine bombers against the Germans and had always felt a strong, if distant, comradeship with Russian pilots who also flew B-25 bombers against the Wehrmacht. He regarded them as partners.

But from 1945 on, however, those feelings of comradeship had been eroded by growing suspicion of Soviet ambitions for political dominance of the European continent and the world at large. In those years our bombers and huge nuclear missiles had been locked on Soviet targets. And the Russians had responded in kind, heightening the tension and animosity between the two countries to the point that the possibility of nuclear war was real, at least for the politicians and military strategists in Washington and Moscow.

Deke had always found it difficult to envision nuclear war and the deployment of massive armies against an unseen target. Deke's vision of combat was more personal and remained as it had been when he first had flown against the Germans and then the Japanese in the closing days of World War II. His kind of war would always be one on one—the sword play of flashing wings and howling engines in the thin, cold air of a high-altitude struggle where man met man, where country was defended, where differences were settled, where honor was reaffirmed. The machines he had tortured through violent maneuvers over the California desert were designed to be sent against red-starred aircraft, and if the cold war ever ignited, Deke's personal view of the nature of combat would help him to strip away propaganda about the imperialistic capitalists versus the evil empire and to regard the conflict as a personal one. No matter what the bureaucrats of each nation might hurl at each other, Deke could identify his own enemy.

It would be a man, just like him. A man in a cockpit crammed with instruments and controls, wings sharply swept, a chained volcano for an engine, an array of destructive implements mounted within and beneath the machine—but it would all come down to the ultimate moment, one man against another.

That was Deke's deep and underlying strength, to be able to evaluate these strange, mysterious Russians on his own level. A fighter pilot is a fighter pilot.

Whereas only a few short years ago he might have hurtled toward Moscow in a single-seat, extreme-range fighter with two hydrogen bombs snugged up beneath his wings, he now sat within an enclosed, pressurized, comfortable jetliner cabin, and was about to alight in this strange and forbidden land to join with the Russians, to share with them, to work with them, to fly through space as a single entity.

Strange bedfellows these cosmonauts might be, but Deke and his two fellow astronauts knew full well that if they pulled off this joint orbital mission, in one fell swoop they could accomplish a tremendous easing of the tensions of the cold war.

That was terrific. But if they'd asked Deke if he minded meeting the Russians in orbit, after intensive and laborious preparation, he would have grinned and told whoever asked the question, "Do I mind? Mister, I'd get on my hands and knees and use my nose to push horseshit down a packed L.A. freeway to make this flight."

Because the pragmatic old combat pilot and test pilot and almost-astronaut knew all too well that this was his one chance to break the gravity chains of his world.

Two Russians and three Americans would make up the team that would join a Soviet spacecraft, Soyuz 19, with an Apollo spaceship and docking adapter, bringing the two together in orbit as a single space vehicle.

The problems of making the men function as a tightly knit team appeared just as formidable as the many technical issues to be resolved. For starters, the Russians didn't speak English and the Americans didn't speak Russian and, clearly, neither side could read the markings and lettering of the equipment of the other side's spacecraft.

Leading the Russian team was Alexei Arkhipovich Leonov. Deke was quick to judge the stocky, muscular cosmonaut as one of the best he'd ever run into. The longer he knew him, the more convinced he became that Leonov was as crazy as most fighter pilots. The cosmonaut veteran was, in fact, a fun-loving extrovert who'd crack jokes every chance he had, which, at first with the English language baffling him, was less often than he liked.

"Alexei was something out of a storybook," Deke said. "The guy was a top artist with his work in demand all over Russia. He went to a couple military flight schools, came out on the honor roll, and then—which proves he'd stop at nothing to qualify himself in any way he could—he made a hundred jumps as a paratrooper and became an instructor for combat jumps for the Soviet Air Force."

The Russians had had their eye on Leonov from the beginning. He was an outstanding athlete, excelled in swimming, fencing, volleyball, competition bicycling, and yachting. In 1960 he swept into the cosmonaut corps with ease.

Leonov had flown the Voshkod II flight during which he made history's first space walk. That kind of experience would be invaluable in the joint mission to come.

"I had a sort of strange association with Alexei," Deke explained. "He'd been teamed up with Oleg Makarov, and these two guys were assigned to fly the first mission to circumnavigate the moon. That was late 1968, or early '69. Well, he didn't go, of course. The Russian lunar program came unglued. And I never went to the moon, either, so we had sort of a buddy feeling between us."

The second cosmonaut for the joint mission, Valeri Nikolaievich Kubasov, was a brilliant flight engineer—but had never qualified as a fighter or test pilot. To the Russians, two pilots were superfluous, and what marked Kubasov as the perfect second man for Soyuz 19 was his experience as a cosmonaut engineer with orbital experience. He had worked on the Soviet spacecraft program from the beginning as a designer, developer, and tester of the new space vehicles. He made his first orbital mission aboard Soyuz 6 in October 1969, during which he conducted space welding, metals smelting, equipment teardown, and repair experiments. It was quite a mission. Kubasov's pilot, Georgi Shonin, rendezvoused in orbit with Soyuz 7 and 8, flew complex maneuvers and, after nearly five days in orbit, returned to earth.

"He is perfect for the joint flight of Soyuz and Apollo," Soviet officials told the American astronauts. "After all," they laughed, "if something goes wrong and equipment breaks down, you have Kubasov, who can weld together whatever has come apart."

To the Russians, the American three-man team was strangely short of space flight experience. Commanding the U.S. crew was Thomas Stafford. Forty-four years old, he was the most experienced space veteran of the five men who would meet in orbit. He'd flown the Gemini 6 and 9 missions, and then with Gene Cernan taken Apollo 10's lunar module *Snoopy* down to within nine miles of the moon's surface on a flight that came within a thin edge of crashing into the moon.

Deke Slayton, who'd waited sixteen years for his first space flight, would serve as the docking module pilot. He was the "old man" of the crew, just topping fifty years of age.

Command module pilot Vance Brand, forty-four, was, like Deke, a "space rookie," but his credentials served him well. He was a serious, hard-nosed aeronautical engineer and an experienced test pilot.

Without intending to, Stafford had the habit of cracking up Leonov as the men spent months together working to learn each other's language. There's something wooly and wild about a lanky American speaking Russian with a distinct Oklahoma twang.

A lot would ride on the Apollo-Soyuz mission scheduled for the summer of 1975. The goal of the flight was judged critical to every American and Russian: to engineer, build, and test in space a "common docking device" that on future flights could be used to "hard dock and lock" American and/or Russian spacecraft for rescue missions to a crippled spacecraft or space stations. Until now, any crew that had found itself stranded or marooned in orbit because of a propulsion failure, especially with limited oxygen and power supplies, would have stayed stranded. With the docking equipment available for spacecraft of either country, chances of rescue would take a quantum leap forward.

That was Deke's assignment: make the hardware dream come true. He must not only monitor the development of the docking system, but then prove its merit by personally bringing Apollo and Soyuz together at that critical moment when the two ships were to meet and join.

No matter how close these men became, they were never really free of the political baggage that shadowed their years of preparation. The Russians had stiff-armed all American attempts to fly a joint mission, because they were possessed of a great distrust of the United States and suffered a deep-rooted anger at having crumbled before the American advance to the moon. But the effort to test the common docking mechanisms, which would open the way for future major space projects and link the power of both nations at enormous cost and equipment savings to each, had to start somewhere, and Apollo-Soyuz was ground zero for such an effort.

Many of America's top space leaders and their powerful political allies had vivid memories of the days when NASA had stumbled badly in trying to get grapefruit-sized payloads into orbit, while the Russians were hurling satellites the size and weight of small trucks high above earth. In those early days of space flight, Moscow had taken every opportunity to trumpet Soviet superiority—leading the way with manned orbital flights, space walks, and unmanned probes of the moon and planets.

Though American technology quickly caught up to and surpassed the

Soviets, there were many who were reluctant to climb aboard the joint space flight, because they felt the Russians, having lagged miserably in space successes, would use the joint mission to gain enormous amounts of data about American space technology. By pairing off the Russian Soyuz with the American Apollo, they additionally warned, the Soviets could claim that they had achieved space parity with the United States.

Fortunately, those with the longer view of the value of space exploration prevailed. Both nations established their engineering teams and went to work, and Apollo-Soyuz became an official joint project with the signing of a space agreement on May 24, 1972, at the Moscow summit meeting between President Richard Nixon and Soviet leader Leonid Brezhnev.

Leaders of both teams took their places. Glynn Lunney, the erstwhile flight director in Houston, held the reins of the American effort. In the Soviet Union, his counterpart was Konstantin Bushuyev of the Soviet Academy of Sciences.

The Russians entered the project with equipment decidedly backward compared with the American hardware and systems. Propaganda notwithstanding, the Russians were quick to acknowledge that Apollo represented a huge technical superiority over Soyuz. The comparisons were honestly made to define the roles each team would play.

Few people in the space business were more savvy and experienced than veteran flight director Chris Kraft. "I don't think they themselves had an appreciation of their own capabilities," he said of the Russians. "I don't think people like Bushuyev had a grasp of all the things required to bring about the development of space hardware and an operational flight involving two countries. We had to work through interpreters, and that was difficult and time-consuming. . . . Initially, if we got five minutes' actual work done in a given day, we were doing extremely well."

Deke recognized the frustrations of Lunney and Kraft and the other members of the U.S. team. "Give these folks a little time," Deke told his team members. "We come from different worlds, and we don't get into the machinery of what we're doing the way they do. They've got to feel what's involved. It's like looking at the soil on a farm. You can do all sorts of technical studies, and the farmer can run the soil through his fingers and he'll know what he's facing. It's the same with these guys. As soon as they get the hang of it, they'll be galloping along with the rest of us."

Deke was right. "Before long," Chris Kraft conceded, "the teams were doing practically a full day's work in a day's time."

Deke laughed. "As I said, just give 'em their head. Right now I'd like to see the people back at our own centers watching these guys. No hollering about overtime, no problem with unions, no one crying if the bathroom doesn't work. Let's just get on with it."

Deke could be patient and insistent at the same time. "Let's get rid of this crap once and for all that the Russians were stuffing our technological secrets under their sweaters. We did not transfer carloads of technical data to them. You had to understand that the engineering of their Soyuz was twenty years old. But you also had to keep in mind that Apollo was hardly new; hell, it was a ten-year-old ship by then. We had already developed entirely new technologies for our Space Shuttle. What the Russians learned from us—if they learned anything at all—it was our system of management. That's where we really shone, and they recognized our way of doing things."

Deke stabbed a finger in the air for emphasis. "Of course, it seems nobody was aware, or they forgot, that the Russians were also developing their own new technology and their own Space Shuttle, and one hell of a booster to go with it. They just weren't telling us about it, that's all."

Deke leaned back in his chair and rubbed the top of his head. "Both sides just had to learn. And as for me, I had one overriding interest. I didn't give a damn who was ahead. Let's just do it! Apollo-Soyuz was the last train outa Dodge for ol' Deke."

You know what was the toughest part of the whole job? I'll tell you right now, it wasn't a thing to do with the nuts and bolts or technology or science. Hell, that was routine by then. The hardest part of the entire training," Deke said seriously, "was learning to speak Russian. Classroom instruction at our age makes you kinda itchy in all the worse places. You want to get the hell out from behind the desk and just fly. But we couldn't. That Russian talk was godawful tough.

"Seven hundred hours. That's how much of that stuff we had to take," he continued. "I think we would have raised all kinds of hell except for the fact the cosmonauts had to do the same thing in English! And not just the five guys who made up the prime crews. There were the back-ups, the engineers, the controllers—lots of other people who had to learn a new language in a hurry. It gets sort of desperate," he laughed, "when you don't know how to say, 'Where the hell's the toilet?'

"Now, Tom and Vance, they did a hell of a lot better than me. I learned enough to get by and that was good enough. The rule was when we got in orbit, we'd speak Russian and the cosmonauts would speak English."

Deke recalled a night when jet aircraft thundered low over Star City, thirty miles outside Moscow, the plane's roar rattling windows and shaking buildings. He had a feeling of "My God, back in the States, when we heard those big boomers go over, we'd know they were ours, and we'd say the sound you hear tonight is what lets you get a good night's sleep."

Deke sat straight up and blinked. "But now we were hearing the jet thunder that made the Russians think the same thing about themselves, because we were in Russia, and there were always Russian fighters and bombers beating up the sky. Sort of made you shiver a bit now and then.

"That wasn't the only thing that jarred you, caught you off guard. As soon as we started to relax, we'd run smack into the Russian system. The proper word is totalitarian government. Compared to the way we lived, most of those people might just as well have been working in a prison yard."

He spoke softly now, emotional about the lives of people who'd become his friends. "It wasn't just a different country. It was like being on a strange planet. Or waking up in the middle of a bad dream and you were trapped in whatever horrible place you'd been dreaming about.

"On our first trip over there," Deke recalled, "we stayed in the Intourist Hotel in downtown Moscow, near Red Square and the Kremlin. We just checked in like tourists. The place didn't have much going for it, but we weren't here to check the plumbing. Every morning the Russians showed up in buses to run us that thirty miles out to Star City, and every night they'd load us back in the buses to return us to the hotel.

"That didn't last too long," Deke laughed. "The hotel had a basement bar. An oasis in a desert of cobbled streets and concrete. It was a regular hotbed for visiting Germans, Finns, Cubans, Czechs. People from all over. One big social center for the out-of-country folk. It took us about one big, deep breath to get right into that scene, and we got wrapped up real tight with that crowd, drinking vodka, playing the tourist routine. Work all day, play much of the night. Anything to change the sterile atmosphere we were in.

"The Russians didn't like it. Their space folk were less than enthusiastic about our having a good time, to say nothing of a few hangovers the next day. But they didn't know what to do about it. At least during our first trip into Russia."

Deke grinned with the thoughts of their second visit. "Hell, we never even saw the hotel. At the airport they loaded us into buses just like before. We were ready to hit that basement bar and cut loose. Wrong! They drove us directly to Star City. They acted real mysterious, and we didn't like it one bit. It's all pine trees out there. Pines and barbed-wire fences and heavy security, and we all looked at each other and wondered what the hell was going on.

"Always count on the Russians for a surprise. While we were back in the States, they had built one hell of a private hotel for the American team. No sloppy construction, either. The damn thing was three stories tall, and it had some large apartments in it, and they equipped the place with a dining room and a special kitchen, and assigned our own cooks to us. And a bar. Can't forget that." He smiled. "And in typical Russian fashion they overdid it. They stuffed four times as much furniture as needed into every room. You could just drop anywhere into an easy chair.

"But it was still Russia. Furniture, dining room, apartments, and a bar, but they didn't have the basics like enough towels or soap or even plugs for sink and bathtub drains. If the place had been a commercial hotel, it would have gone broke in a week. Maybe not in Russia, but certainly in America.

"Now, they were real careful the way they introduced us to the place. They tried to make it a tourist special. Justified it by saying it would eliminate the daily bus rides to and from Moscow, facilitating the training. That was bullshit," Deke offered. "Their intent was to keep us under lock and key. You don't feel like a tourist behind barbed wire and armed guards and dogs and who knows what else. Easy to get in and real tough to get out. If we wanted to party downtown after training, get to Moscow to blow off steam, we had to requisition a car and driver for every time we said, 'let's get the hell outa here.' Maybe it wasn't lock and key, but it was, by God, completely under their control. When we looked past the barbed wire, we always had company. Black cars with KGB agents sitting inside. Day and night, guarding us in shifts.

"Well, what the hell, we learned to live with that. But not with everything." Deke made a face. "The food wasn't bad. Compared to what the average Russian had to eat, our meals were feasts. But the Russian beer was a goddamn disaster. We bitched about that. We bitched a lot about it. One night Tom and I were griping about the piss-poor beer, and we allowed how we wished we had something better to drink.

"That was a lesson all in itself. The next night we finished training, returned to our apartments, and every damned fridge in the building was stocked with some really great Czech beer."

Deke and Tom allowed that it didn't take a rocket scientist to figure out that the Russians had bugged the entire hotel. You couldn't sneeze without everything being listened to and recorded.

"So we put it all to good use," Deke laughed. "We learned to talk to the walls, the lamps, the mirrors—anything and everything. Because everything had ears. One of our guys said how great it would be if we had more towels and soap. Presto, we had them the next night. Another of the guys said, real casual-like, how great it would be if we had a pool table in the place. Sure enough, we came back after work, and that night we almost fell over a pool table. It was sort of archaic, the kind of table you'd find in a museum back in the States, but it was still a pool table.

"We played it carefully, not getting greedy about things. But there were a couple Russian technicians working with us we didn't like. So we talked to a lamp one night about what assholes we thought these guys were, and, oh man, we never saw those guys again. They just vanished."

What the Russians lacked in basic amenities they made up for in entertainment for their guests. Besides the astronauts, prime and back-up crews, there were another hundred technical people along. The Soviets started by throwing parties in the "Amerikanski Hotel."

But they opened up little by little, and soon the Americans were being taken into Russian homes for dinners and bottomless vodka and cognac. "These folks didn't have much to give," Deke said in admiration of their hosts, "but they shared everything they had. Sometimes the most effective things took place without even trying.

"Later, when the Russians came to Houston, we returned their hospitality. Most of them had never been outside the Soviet Union, and when we brought them to our homes, our lifestyle absolutely amazed them. They'd see a nice home, which to them was a palace, and there'd be a couple cars outside, and a bunch of bathrooms inside, and stereos and televisions all over the place. The Russians wouldn't believe the U.S. government hadn't built all the homes as showrooms to impress them. When they came to understand these were our homes—and the homes all around us were on the same level, owned by working stiffs just like us—they figured it was some sort of plot. They told us they knew we were very privileged people and the average American couldn't possibly live in such incredible luxury and freedom. Finding out that I was no different from anyone else who lived on my block or in my subdivision was a real shock to them. They also couldn't believe we came and went as we pleased. That we could just drive to an airport, we didn't need identification papers, and we went wherever we goddamned pleased. It blew their minds. Spacious ranch-styles for the most part. Many with swimming pools."

Once the Russians began to understand there were no fences, no barbed wire, no armed guards around, they were like kids turned loose in a candy store. "Hell, we welcomed them Texas-style, all right," Deke explained. "We got drunk together, we ate in restaurants that had them bug-eyed, we did saunas together, went hunting and fishing and sightseeing, went to rodeos and horse races. I'll tell you right now that just accepting them and sharing with them as friends did more good, when they got back to Russia and told their countrymen what America was really like, than all the propaganda our government had been dishing out for years. You just have to remember what Kubasov told us.

"He said that if we all continued to live together like this, the future had real promise to it. His exact words were: 'Just like you, I want my children to sleep peacefully and calmly.' Gives you a warm feeling inside, that does."

For three years the American and Russian crews, each supported by back-up teams of engineers and veteran space men, modified their spacecraft, built new equipment, learned their languages, and practiced orbital maneuvers in new simulators.

The barriers came down steadily. Russian suspicions waned visibly, and when the Soviets found themselves at Mission Control in Texas, and at the Cape Canaveral launching site in Florida, they could hardly wait to fly the American teams to their Baikonur launch complex to watch powerful Russian boosters thundering away from earth. In between the mind-numbing sessions of technical preparations and simulator practice, there were three years of moments of hot dogs and beer, borsch and vodka, swimming in the Florida surf, and a wild snowball fight at the Gagarin Cosmonaut Training Center at Star City.

Slowly the thousands of pieces of the international jigsaw puzzle came together. Astronauts and cosmonauts were independently and absolutely firm in trusting their own equipment. Now they had crossed the ultimate bridge of gaining that trust and faith in a foreign vessel they had never yet seen in space.

Deke Slayton's speciality, the new docking module, had become a continent-crossing span for the willing exchange of cultures, technical achievement, and national aspirations. Once Apollo and Soyuz were joined high above the earth, the tunnel formed by the docking module would be pressurized, hatches opened, and men who had been near enemies, and now close friends, would glide in zero gravity from one vessel to the other.

If all went well, if Deke and his revolutionary module met its promise, handshakes and bear hugs would transform all the technology and science of these men whose lives depended on mutual trust and cooperation.

What would happen in the cold, hard vacuum of space would turn on the heaters to start melting the cold war far below.

The five men who would fly Apollo-Soyuz sat about a table. Three Americans on one side, two Russians on the other. The mood was somber yet promising. Deke Slayton, Tom Stafford, and Vance Brand watched Alexei Leonov, his every move followed by Valeri Kubasov, uncork a bottle of Russian vodka.

Leonov extended his arm to offer the bottle to Tom. He nodded, accepted it, tipped it in salute to the others, hesitated a moment to swallow a pinch of salt, and then took a hefty swig of the fiery liquid. He then passed the bottle to the next man for the custom to be repeated.

When all of the five space pilots had swallowed their salt and tossed down a drink, to bring them all good fortune for their joint endeavor high above the mother planet, Leonov recorked the bottle with a twist and a slap of his powerful hand.

Again the bottle made its round, this time for each man to sign his name to the label. The somber mood faded before sudden grins and back-slapping. Leonov held the bottle high. The five vowed to uncork the same bottle again when Apollo-Soyuz was in the history books. Then they would finish the damn thing and get on to the next bottle!

They chose their words carefully. No one drank a toast to the success of the joint adventure yet to be flown and shared. To the Russians, this was bad luck.

"We toast our preparations," Kubasov explained. "That is enough." The three American pilots nodded in agreement.

The launch of two powerful rockets, one from Baikonur and one from Cape Canaveral, was now only three months away. The moment of shared salt and vodka was April 1975, with the first promise of spring in the air, a sense and smell of hope everywhere. Deke walked from the table to the window to look down on the Gagarin training center at Star City. They'd been here on and off for three years. The security fences no longer seemed threatening; everything was familiar to him. He smiled to himself. Who would have figured that an apartment in Star City would feel warm and cozy to him?

But it was. This was the apartment home of Alexei, and the stocky extrovert's wife had heaped food and drink in such quantities that the dinner table was sagging.

Deke felt good about everything that had come before this moment. The future was now primed to take care of itself. The preparations essentially were done. This was the final pre-launch training session for astronauts and cosmonauts. What waited to be finished was almost a matter of checking off items on the check lists of the boosters and spacecraft.

Seven hundred hours of language instruction crowded the heads of the Apollo-Soyuz astronauts, who had slowly improved their communications skills. Nuances of language finally became meaningful.

"That was a tough one," Deke explained. "You had to know the nitty-gritty of what the other man was saying. Our superbrains in Washington said we ought to use a computer to translate for us. That was dumb. Computers don't know nuances and idioms. It's like a computer translating from English into Russian an old favorite phrase most of us grew up with. You know, 'The spirit is willing, but the flesh is weak.' Know what you get when it goes through the computer? 'The vodka is strong, but the meat is rotten.' It don't work too good."

So they all struggled toward their avowed goal that when they floated between Apollo and Soyuz in orbit, the Russians would speak English and the Americans would speak Russian, even if Leonov and Kubasov cringed at Stafford's Oklahoma twang.

There was one final milestone to mark. It couldn't be found in the official records of the flight.

Long before launch day, they had become close friends.
That was the real beginning of the countdown for Apollo-Soyuz.

CHAPTER TWENTY FIVE

SOYUZ AND APOLLO ARE SHAKING HANDS

ALAN SHEPARD STOOD just inland from the beaches at Cape Canaveral, staring across the ocean where deep blood-red edges of first dawn began to seep across the horizon. It was a moment of false dawn, this part of the world still under the mantle of night and only beginning to yield to the sunrise. Along the horizon cumulous clouds reared up from the ocean like a row of rounded mountain ranges with misty ravines. Solar rays fanned the sky above with red-and-orange bands of the new day. The last time Alan had seen such a moment was another Florida July—long ago—when Gus Grissom had been readied to follow his suborbital path into space. On that morning the crowds at Cape Canaveral were startled to see swiftly brightening dawn paint that same huge fan. Hundreds of people taken with the sight began to cheer and applaud.

Alan turned to the huge bowl of light surrounding a giant—a Saturn 1B rocket standing high above a massive steel structure that enabled the 1B to be serviced.

All around the great rocket's gleaming light bowl, in the darker expanses of the space center, nature was in full swing. Ghostlike fog materialized with the rise in morning temperatures, spreading swirls and layers through the palmettos and scrub brush, heaving gently above the canals and waterways lacing the center. In all directions the space-port cast light, from the buildings, traffic lights, roadways, and moving cars. These sparkled through the fog, a kaleidoscope of colors and shapes. The glowing crescent of dawn diffused into rainbow hues. Orange danced lightly upon the upper level of mist as the sun began its climb above the eastern horizon, and for a long moment the space center was transformed into a glittering fantasy.

The magic faded before the brightening day. Even the searing light about the launch tower was restricted to the immediacy of the rocket complex.

Shepard nodded to himself with a long look at the great booster. Deke's bird. No other name for it. The day this man, as close to Shepard as any brother, would ride the last Apollo into space.

Alan felt deep satisfaction that Deke's long-awaited moment was almost at hand. He had an intense urge to protect Deke, even though he knew such protection was not within his ability to provide.

He and Deke had had a special relationship, but until now only Alan had lunged upward, leaving behind his friends and family. They had done the worrying, subdued their fears.

His feelings caught him by surprise. Now he had the launch jitters! He'd had them for weeks. He knew it—and so did the entire launch and control team—as he moved through their ranks like a sword ready to slice downward with the first sign of a mistake, an error, anything that wasn't done perfectly. He couldn't help that protective urge toward Deke, so Alan checked, studied, inspected, cornered, and quizzed anyone who had the least connection with the space flight to come. He would do everything he could to give Deke and his crew the safest possible flight.

Everything? That he could do?

He shook his head. He really didn't have a say in what was going on. He was now only astronaut Alan Shepard, the first American in space. Recently, with Deke well on his way to a flight, he had retired from NASA's active ranks and eased back into civilian life as a Houston businessman.

The rules said he was out of the picture. Well, to hell with that. And to hell with the rules dictating who could check and recheck preparations. If he couldn't come down personally on anyone fudging his work, he knew everyone who did have the authority to straighten out the bends on the road to space. They'd listen to him; that he knew. So, on reflection, he knew he could help, and that's exactly what he was going to do, even if everyone came to regard him as a pest and a bother and a worrywart.

That's why he was here on the grounds of the launch complex. And if strong thoughts and determination, and willing only the best to happen, might do some good, he would throw everything he had into boosting that Saturn 1B all the way up to orbit.

As his thoughts settled in his mind, he smiled suddenly. He felt a sudden and wonderful rush of confidence. Hey, the crews might call Deke the "old man" of the astronaut corps, but during these past few months Deke had been almost giddy in his attitude, happy, grinning as if he were a child about to take a ride at Disney World. Deke had reassured Alan he would be just fine. It would be a good day. A great day.

But that would be this afternoon. Right now Soyuz 19 with Alexei Leonov and Valeri Kubasov was well into its countdown, and Alan felt the urge to hurry inside the launch control center to feel and observe the sounds and the television pictures of the Russian rocket sitting on its launch pad nine thousand miles away.

On the fifteenth day of July in 1975, by the best estimates of military intelligence, some two hundred thousand rockets in the arsenals of the United States and the Soviet Union were either poised for launch or kept in pre-launch readiness. Many were small, shoulder-fired weapons. Others were clustered aboard armored vehicles and trucks. Still more were slung beneath the wings or within the bomb bays of fighters and bombers. Hundreds of heavy missiles rode beneath the ocean surface in the Sherwood Forest bays of nuclear submarines, always ready for immediate launch. And to round out the fearsome arsenal were thousands of silo-buried war rockets, some as tall as a ten-story building and crammed with multiple thermonuclear warheads.

But on this particular day only two of those rockets poised for flight carried men instead of weaponry. If all went well with Soyuz 19, then later that day the American rocket, the Saturn 1B, would begin its journey with the goal of joining it in space.

This is the Soviet Mission Control Center. Moscow time is fifteen hours fifteen minutes."

The world listened and watched the first live television ever broadcast from the launch site deep within Russia.

"Everything is ready at the Cosmodrome for the launch of the Soviet spacecraft Soyuz. Five minutes remaining for launch. Onboard systems are now under onboard control. The right control board—opposite the commander's couch—is now turned on. The cosmonauts have strapped themselves in and reported they are ready. They have lowered their face plates. The key for launch has been inserted . . . the crew is ready for launch."

Then, precisely five minutes after the ignition key was inserted, bringing the rocket to "live firing status"—

"Ignition! The engines are powered up. The launch! The booster is off! Moscow time: fifteen hours twenty minutes ten seconds. The flight is proceeding normally . . ."

Two minutes into the fiery liftoff, four liquid-fuel strap-on boosters separated from the rocket. Forty seconds later the escape tower and protective shroud blasted free of the spacecraft. Five minutes into the ascent the powerful core stage shut down and was discarded. The third and final booster blazed for another thirty seconds, then shut down. Soyuz 19 separated from the booster and eased into a perfect orbit.

In Florida the sun was bright in the mid-morning sky. Elated NASA officials awakened Stafford, Slayton, and Brand. "Your friends are upstairs. Right on schedule." It was 9:10 A.M. Cape Canaveral time.

"We're over hurdle number one," said Slayton. "Now all we have to do is get our asses up there."

Soyuz 19 was swinging around the planet on its fourth orbit when the American crew climbed into their Apollo spacecraft. It had been an interesting ride from their quarters to the launch pad. Tom Stafford filled the time practicing the Russian phrases he would use when they linked up with Leonov and Kubasov. Deke laughed aloud. Tom obviously was in his element. This would be his fourth launch. He seemed far more concerned with his linguistic pronunciation, not showing the slightest concern about the thundering blastoff they all soon would be making.

Tom's preoccupation remained unbroken even after they left their transfer vehicle and began the walk to the gantry elevator. Even Vance Brand was deep into his own thoughts, and Deke became aware that he was the only one of the three men to take the time to stop for a long, emotional study of their big booster.

"It's beautiful," Deke thought. After all his years of being grounded, of being chained to a desk, of selecting the crews that would lift off from this launch pad as he stood with his feet rooted to the ground, he was going to ride his own fire monster away from earth.

He spoke aloud to himself, oblivious to anyone who might be listening. "This mother," he said, "is going to take me upstairs." He added a "damn right" before riding the elevator to the enclosed White Room, where technicians made their final checks of the astronauts' equipment and triple-checked their fastenings, straps, and hookups within the capsule.

Every item on the check lists went with a casual smoothness that belied the complexity of the preparations. Deke thought about how leisurely everything was progressing. He recounted later that his thoughts at that moment were: "It damned well should go like all this is greased. Three missions have flown with the Apollo to link up with the Skylab space station. The only thing different about this flight is that we're going to link up with the Russians."

Another thought crossed his mind, one that made him all the more grateful for the perfect countdown. This was the last Saturn 1B booster in the stable, and this was the last Apollo—the last "throwaway spaceship" the United States ever planned to fire into orbit.

"If this sucker doesn't go," Deke thought soberly, "then I've probably shot my last bolt without ever getting into space."

He pushed away those thoughts. "Let's do it," he told himself as he'd told so many others.

Deke and the other men felt the Saturn 1B coming alive through subtle groans and creaks as fluids shifted, pressures increased, pumps were tested. It was like awakening an ancient, giant Rip Van Winkle for a brief but sensational period of life.

The final minutes ticked away; the three astronauts listened to the launch controllers passing the word "GO!" from one console after the other to the launch director. Just before Saturn 1B went to full internal power and all systems came on line, Stafford asked the launch team to send a message from the American astronaut crew to Soyuz 19:

"Alexei and Valeri, from Apollo. We'll be up there shortly."

Wonderful words to Deke. It was all coming up roses. Tom's quiet confidence was terrific.

In the moments before ignition, Deke realized suddenly that he had been working with hundreds of people, but it was strange not to have Alan Shepard right in his face, needling him, sharing the camaraderie of old test pilots. But he knew Alan was now worrying just as he had himself when Alan and his crew boomed upward on the Saturn V, right there with the control team to extend that invisible but invaluable support arm.

The countdown reached zero. Saturn 1B exploded into life.

Fiery thrust spewed in sheets over the launch platform, and it took only seconds for the astronauts to learn that their booster was no smooth and mighty Saturn V. 1B was an angry, spitting, snarling wolverine, which shook and rattled its human cargo.

The mighty booster clawed its way off the launch pad, and Deke felt as if he were riding an old pickup truck slam-banging down a rutted, country dirt road back on his Wisconsin farm. Eight engines blazed away at full thrust with a cacophony of noises—propellants pounding through lines from turbo-pumps spinning at tremendous speed, pressures surging with booming thuds throughout the stage, all to the accompaniment of teeth-rattling, eye-blurring shaking.

Deke had wanted to sit back and enjoy the trip into orbit. Instead, he felt as if he were balancing atop a long rubber balloon fighting its way through wild winds, and at its very top, where the three astronauts rode, the motions went from up and down to simultaneous spiraling. A dog shaking water from its body with a twisting, swinging motion while its legs collapsed beneath the hapless animal was Deke's description of the powered flight. Remembering Slim Pickins's "buckin' bronco" ride on the nuclear bomb dropped from his B-52 in the movie *Dr. Strangelove*, Deke could only hold on and lock his spurs into the bottom of his couch.

Punching through maximum aerodynamic pressure was another adventure as enormous forces squeezed and shook the Saturn 1B. Vibration pummeled the entire rocket and the Apollo. Then, suddenly, they were through Max Q, and they shot upward like a frightened jack rabbit. Almost at once, now into the supersonic region, engine roar and the high-pitched howl of air ripping past the rocket vanished.

While the noise of the liftoff abated, the booster complained with deep, hollow groans and the creaking of an old wooden ship wallowing in rough seas. Finally came the burnout of the first stage. For a precious moment the creaks, groans, wiggles, shaking, vibration, and other unpleasantness were gone.

Only for the moment. Explosive charges blew apart the two stages with all the velvety touch of a locomotive thundering off a high trestle to roll down a rocky slope. The second stage fired, aiming toward their target, that small doughnut in space through which they must pass to enter their planned orbit. They left the potholed country road far behind, and moments later they settled into their orbital track.

"I love it," shouted Slayton. "Goddam, I love it. Man, I tell you this was worth waiting sixteen years for."

"You liked that, huh, Deke?" Stafford grinned.

"I would like to make that ride about once a day," the fifty-one-year-old rookie astronaut laughed, throwing his arm out in triumph.

Then, instantly, it was there.

Weightlessness!

And Deke felt it. That wonderful feeling of shedding all pounds—his body as light as a falling snowflake.

"Yoweee!" Deke yelled. "This is a helluva lot of fun. I've never felt so free."

Stafford and Brand laughed and shared the enjoyment of this man who had waited so long for his chance. They quickly got out of their spacesuits, donned their flight coveralls, and settled into the comfort of weightless flight.

The Apollo crew disconnected from the spent rocket stage, maneuvered through transposition to withdraw the docking module (a procedure identical to connecting with the lunar module on the moon missions) from its parking spot atop the empty stage. They fired maneuvering thrusters to move away from the booster and began to match their orbit to that of the waiting Russian ship.

Vance Brand switched on the radio frequency for Soyuz. "*Miy nakhodit-sya na orbite!*" We are in orbit!

Deke sent a second call to the Russians, speaking in their language. "Soyuz, Apollo. How do you read me?"

They heard Kubasov answer in English. "Very well. Hello, everybody."

"Hello, Valeri," Slayton said. "How are you? Good day, Valeri."

"How are you? Good day!" Kubasov replied.

"Excellent!" Deke boomed. "I'm very happy. Good morning."

Leonov's voice came on. "Apollo, Soyuz. How do you read me?"

"Alexei, I read you excellently," Deke responded.

"I read you loud and clear," Leonov confirmed.

"Good," Slayton said. He wasn't the most loquacious pilot in the business, and that last was in English. In Russian he felt he was stumbling through their exchange. But who the hell cared?

He'd made it! Orbit!

It was time for the hunt to begin. Apollo's crew executed the first of several maneuvers to track down Soyuz over a two-day period in a game of celestial tag. Circling in a lower orbit and making precise course-changing engine burns, Apollo gradually caught up with the Russian ship, and on July 17 the two spacecraft, high over the French city of Metz, performed a stately orbital ballet as they maneuvered close to and around each other. Earthlings had a ringside seat as a television camera pointed out the Apollo window relayed pictures of Soyuz, eerie and bug-like, glowing a startling green in brilliant sun against the blackness of space and the azure blue of the earth below. "I can see your beacon in the porthole," radioed Leonov, speaking English.

"Less than five meters distance . . . three meters . . . one meter," said Stafford in Russian as he guided Apollo nearer.

The mission went smoothly with a faster than expected timeline, dispelling the naysayers who said the joint mission could never be carried off. Just fifty-two hours after Leonov and Kubasov had departed Baikonur, Stafford, Slayton, and Brand brought the two spacecraft together.

There was a slight shudder as the two craft, both traveling more than 17,400 miles per hour, docked. "We have capture," Stafford reported.

"Well done, Tom," the Soyuz commander replied. "It was a good show. Soyuz and Apollo are shaking hands."

At that moment history had a new page to be read for future centuries.

Throughout the world, millions of fascinated, even astonished, viewers watched the American and Russian spacemen clear the hatches between their respective vessels and push themselves in a weightless ballet from one spacecraft to the other.

Television observers shook their heads in disbelief at the reality of three Americans speaking Russian and two Russians speaking English, congratulating themselves high above the earth while the cold war was put temporarily on hold far below.

"*Tovarich*," ("friend") Stafford called out as he shook hands with Leonov. "Very, very happy to see you. How are things?" Leonov asked, giving Stafford and Slayton the traditional Russian bear hug. They then floated into the Soyuz cabin to be greeted by Kubasov. The spacemen traded gifts, including flags of their nations and commemorative plaques. Leonov, a gifted artist, gave the astronauts sketches he had made of them during training.

The visiting Americans gathered around a green metal table in Soyuz for a meal and to toast the success of the mission with apple juice. They feasted on reconstituted strawberries, Roquefort cheese, sticks of apples and plums, and tubes of borsch, which the cosmonauts had mischievously labeled vodka. Later, aboard Apollo, the cosmonauts were treated to potato soup, bread, more strawberries, and grilled steak. Space flyers have never considered their food the best, prompting Leonov to remark, "As the philosophers say, the best part of a good lunch is not what you eat, but with whom you eat."

During the forty-seven hours the two ships were linked, astronauts and cosmonauts executed their own brand of shuttle diplomacy. Leonov said that the flight had been made possible by the "climate of detente" that had been begun by President Richard Nixon and termed it "a first step on the endless road of space exploration." Slayton said that through space flight men of various nations could gain a greater sense of understanding.

President Gerald Ford could hardly wait for his chance to talk to the crew aboard Apollo while the two were still docked. He asked for and spoke directly to Deke Slayton.

"As the world's oldest space rookie, do you have any advice for young people who hope to fly on future space missions?" the president asked.

Deke answered in his own pragmatic style. The best advice he could offer the youngsters of the world, he said, is to "decide what you want to do, and then never give up until you've done it."

Before the hatches between the craft were closed for the last time, the two commanders held a brief farewell ceremony in Soyuz. Each crew had brought along halves of two commemorative aluminum and steel medallions, which Stafford and Leonov joined together. They also signed four forms for the Federation Aeronautique Internationale, a body that verifies achievements in aviation and space. In addition, they traded small boxes of seeds from their countries—the Americans contributed a hybrid white spruce; the Russians, Scotch pine, Siberian larch, and Nordmann's fir—for planting on return to earth. In the Soviet Union this ancient tradition of a visitor planting a tree is considered a gesture of true friendship.

As both ground teams did, at Mission Control near Houston and at Star City.

It was incredible. Alan Shepard had more than once found himself holding his breath during the most critical elements of the flight, the fiery ascent, separation of rocket stages, insertion into orbit, flying the exquisitely demanding shifts in power and height above the earth to ride the rails of orbital mechanics in order to meet up with Soyuz, the docking.

Yet Alan quickly distanced himself from the technical aspects. His mind soared to capture a vision of Deke testing the delicious sensation of zero gravity, of knowing finally that incredible feeling of weightlessness.

And he thought of Deke floating on the edge of the heavens, staring out at the moon and planets, individual stars and whole galaxies, and he knew Deke was now one of those privileged few who with a single glance could drink in the beauty of the gleaming earth below, lift his head, and become part of the universe.

Alan Shepard smiled warmly.

He knew it couldn't get any better for Deke Slayton.

The docking of the two spacecraft had until now been strictly an American show. It was time to prove the competence and quality of Russian space science and the skill of the cosmonauts. The two spacecraft separated, and Apollo adopted the passive role. Leonov flew Soyuz 19 in with consummate ease. As if he'd been flying this maneuver for years, he slid the two ships together for a perfect hard dock.

There remained no question that either nation in an emergency now had the capability of rescuing men and women in crisis high above the planet.

Precedent continued to be broken as the mission progressed. Only a few years before, the idea of a joint American-Soviet space mission had been judged unthinkable. Now five spacemen from both countries, gathered 140 miles above the planet with multi-national spacecraft, held press conferences with reporters from several nations on earth.

That was the lasting foundation and the heart of the Apollo-Soyuz mission. World news media hailed the flight as a stunning high-water mark in space exploration. That was the least of it. Of all the aims and achievements of Apollo-Soyuz, nothing could have mattered less than spacecraft or technical achievement or scientific goals.

That representatives of the two most powerful military nations in history, bristling with antagonism and weaponry, met peacefully, in full cooperation, hiding none of their equipment, was the "salt and bread" of this space flight.

Finally the two spacecraft separated for the last time. Deke Slayton worked the thruster controls of Apollo with a deft and sure hand, backing off from Soyuz and then flying dazzling maneuvers about the Soviet craft. They drifted together awhile, then moved away from each other.

As the ships began to part, Alexei Leonov declared, "Mission accomplished." From Tom Stafford came the reply, "Good show." Apollo and Soyuz went their separate ways, conducting independent experiments.

On July 21, after six days in space, Leonov and Kubasov fired retrorockets to slip out of orbit and start back to earth. Soviet television cameras provided the world with the first-ever live coverage of a Russian spacecraft returning home, picking up Soyuz as it descended through cloudy skies, dangling beneath a single red-and-white parachute. Eight feet above a flat, featureless wheat field in central Russia, Soyuz triggered braking rockets to cushion the landing and stirred up a massive cloud of dust, which briefly engulfed the craft. A helicopter landed beside Soyuz within thirty seconds, and recovery workers quickly opened the hatch. The cosmonauts stepped out smiling, hugged recovery team members, and waved at cameramen. A crowd of several hundred persons, including farmers and peasants, clapped and cheered loudly from a short distance away.

Applause also came from the Apollo crew still in orbit. Stafford, Slayton, and Brand told Mission Control to give Leonov and Kubasov their best and to say they were glad they had made a safe landing.

Soyuz was home.

The last Apollo's work was not yet done. Three more days of orbital experiments awaited the astronauts as they moved into the final phase of their flight plan.

CHAPTER TWENTY SIX

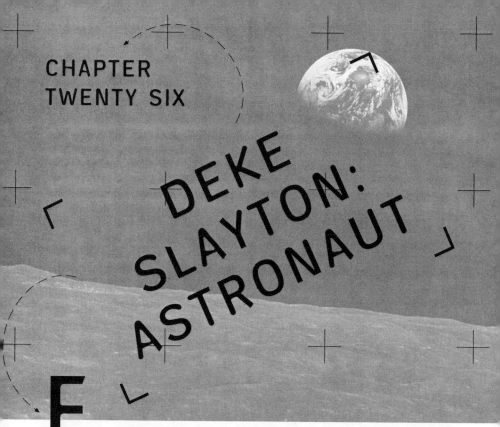

DEKE SLAYTON: ASTRONAUT

FROM THE SURFACE of the Gulf of Alaska, looking through scattered clouds, one star in the night moved. It first appeared just above the horizon, a sharp pinpoint of slivered light traveling with a sense of unreality across the top of the world.

In absolute silence it sped closer until it was at its brightest, directly overhead, sweeping in a long, sinking fall toward the southeastern curve of the planet over Canada, a bright messenger in the heavens. Directly ahead lay the western shores of Lake Superior, no longer in the dark of night but washed by the deep orange glow of dawn.

Deke Slayton floated weightlessly to a viewing port of his Apollo spacecraft, braced with one arm and his leg against his seat. He looked downward, still awed by the swelling light racing across the horizon, turning the huge expanse of Lake Superior orange, quickly advancing to a pale crimson and then resplendent with the sun climbing quickly upward above the flanks of the world.

His practiced eye, skilled at searching for landmarks from high above the earth through a lifetime of flight, checked the southern shore of the lake against the sparkling lights of Duluth. Now he followed the Mississippi River winding southward, dawn reflecting off the ribbons. He looked for the confluence of the Mississippi with the Wisconsin River. There. Reflected dawn showed him clearly where they joined. Now he looked barely north to where the Mississippi widened. There was La Crosse, unmistakable with its night lights still glowing. From 140 miles high, the town of Sparta, east from La Crosse, was vis-ible.

Below Sparta, five miles on a map, wrapped together to the human eye from this height, the countryside flowed along hills and valleys. Deke knew this Wisconsin farmland better than any place on the planet below. A hundred and sixty acres of that land was where he had spent his boy-hood years, farmland with a heart-touching similarity to his family's ori-gins in Norway. But for more than a century his family had lived on that land over which he now sped at five miles every second. What was visi-ble as early morning countryside, without country roads too distant to view, without any boundaries save the two rivers, without a specific marking of the Slayton family farm, he saw clearly in the memories of his mind's eye.

Then, with a silent sigh of time it was gone, flowing far behind the hurtling spacecraft.

Deke Slayton felt at home in two places at that same time. Down below, a forever place in his memory, and his immediate physical sur-roundings—within the interior of the spaceship he had waited sixteen years to fly.

Deke floated away from the view port, glancing at Tom Stafford and Vance Brand engrossed in flight plan chores. The green-and-white Russian Soyuz with Alexei Leonov and Valeri Kubasov had already returned to earth following their historic orbital linkup. Soon Deke and his crewmates would end their three additional days of flight. Apollo would then begin its fiery descent through atmosphere.

But not yet. There was still time to revel in the wonder of weightless-ness and the mind-numbing display of celestial fireworks draped across the velvety black curtain of space.

Deke Slayton, Tom Stafford, and Vance Brand, and the two Soviet cosmonauts had demonstrated that all terrestrial boundaries were artifi-cial, invisible, and meaningless from 140 miles high, that one flight could point the way to a safe and sane future.

Most flyers had long known a great truth. Boundaries between states and nations did not exist. When they looked down from their winged craft, borders appeared only as rivers and coastlines and mountains and valleys, and it was impossible to tell what belonged to whom. From fifty thousand feet high, national borders could be found only on maps. To the eye, a single homogeneous world flowed together.

Now, at last, in the summer of 1975, Deke saw Mother Earth from space, the high throne of orbit, and he could raise his eyes to look with awe at his stunning yet unfathomable universe. He sailed in an enclosed cylinder of magic, silently, his own planet's turbulent, distorting atmosphere well below him.

Deke could not escape his problem of sleeplessness. No physical ailment disturbed him. His enemy was time. After a while his body demanded the rest and his mind the escape from responsibilities within Apollo, but there was so much to see! So much at which he marveled, drunk with the wonder of what was displayed before him on this incredible celestial stage. He begrudged every moment he could not look at the earth and beyond. He fought not to miss a single moment, another angle, a variation in view.

Even with a sunrise and a sunset fitted neatly within a time span of every ninety minutes, a double explosion of waxing and waning brilliance and colors as the sun raced above the horizon or fell in silent escape as the rim of the earth ascended, he knew he could see only a part of the glory before him. Each sunrise, each sunset, differed from all those that had gone before. The clouds changed the colors and the intensity. Smoke and dust and moisture altered the colors and their shapes and the quiet rush of kaleidoscopic hues across the curving surface of the world. He found himself holding his breath, then gasping for air, as if this might help to grasp—and freeze the moment until he could study it carefully— each race between the gods of sun and earth.

It was a struggle he knew he couldn't win and was grateful that he would always find the unexpected when he looked out the Apollo window.

When heavy cloud decks enveloped the planet, they created a whole new surface that had never before existed, of high mountain ranges, tumbling ravines. Sometimes the clouds would create huge cliffs, sheer walls miles high into which shadows fell to give them a startling sense of solidity, as though the whiteness below was some Antarctic winter mountain scene now spread across all the visible world. No oceans, no land surface, only that startling, shifting panorama, and then, suddenly, it became something else.

Ethereal clouds, misty, as fine as a woman's hair, wispy and ghostlike. They appeared everywhere or strangely vanished, then showed up again, brushing the edges of islands and the shores of continents. They were members of the cloud family, a living race dancing and floating above the planetary surface.

Astonished, awed, he had the strangest thought that perhaps this is what the angels could see. . . .

Deke gloried in the freedom of weightlessness, able to float and turn lazy pirouettes within the cabin, and what mesmerized him at night offered him a variety of detail when they sailed through daylight washed across the earth. He sought out everything, and he remembered what Alan Shepard had told him to look for, how to pick out the telltale features of life activity on his globe.

He marveled at those moments when they drifted over oceans and seas with their surfaces glassy and undisturbed by wind or waves, for at such times he could see clearly the huge V shapes of wakes from ships plowing through those waters.

Once he saw a shining streak, a single wirelike gleam that ran for uncounted miles. A different sign of life on the world below—sunlight reflecting off a long, straight railroad track.

Dust storms raised great brown mists off the deserts; snow turned high mountains into zebra-like stripes, differentiating between white snow and dark rock.

And there were moments when the angles were just right, and the sun reflecting off the blue ocean formed a huge, eye-stabbing bowl too bright for him to watch with unprotected eyes.

Green fields, vast tracts of farmland, cities that were darker patches on the surface, huge rivers that from his vantage point were ribbons winding through dark green.

But it was the night that held him in its wonder.

When the sun fell behind them and night swept across the planet, Deke knew he was sailing into a world utterly different from what was bathed by that same sun. Nightfall brought with it what was invisible during the day.

Cities sprang to life with coruscating multicolored lights, a swarming of neon illuminations, brilliantly lit streets, buildings ablaze from neat rows of glowing windows, manmade oases of color and brightness connected by long tendrils of highways marked with headlights. Deke could make out where cities started and ended, and if there were haze or thin clouds the light shimmered and expanded its reach, soft glowing bowls unmistakable even from that high platform in orbit. Oil well and gas well fires shone in darkened deserts. Within the vast reaches of invisible forests and jungles was other light, this time huge blazes devouring trees and grasslands.

The signs of an inhabited planet stirring, surging, moving, living. Good or bad was not judged from Apollo; it was there for Deke to see.

There were other lights not of storm or kindled by men. Silently, magically, glowing colors would rush down from arctic regions, the aurora borealis, electrical charges ignited by the sun in the upper atmosphere of earth. In seconds they would cover the distance of half a planet, greens and reds and yellows, oranges and blues, painted with the softest of brushes, swirling, flashing, and spinning within their own tremulous structure.

But of all that he saw, what gripped him the most were the light flashes in the darkened atmosphere that he had seen before—but always high, high above him. Meteors blazing through the atmosphere, shooting stars beneath him!

The fireflies of space dashing blindly through cremation.

Then came the moment.

Deke would never forget it. He became part of a wonder that opened all space to him.

Meteors flashed in greater number than he had yet seen, the spattered debris of ancient planetary formation and collisions of rock consumed by the atmosphere of the home world.

Something he could not measure in size, but unquestionably large, perhaps even huge, rushed at earth with tremendous velocity. The meteor hurtled in toward earth, but at an angle that would send it skimming along the upper reaches of the atmosphere, almost parallel with the planet surface below. Deke first saw the intruder when it punched deep enough into earth's air ocean, grazing the edges of the atmosphere with a speed he could not judge, except that it was a rogue body, gravity-whipped to tremendous velocity.

It tore into thin air; instantly its outer surface began to burn, its front edges blazing like a giant welding torch gone mad. It skipped along the atmosphere and gained an upward thrusting lift, like a flat rock hurled across smooth water. Deke gazed in wonder at the sight and watched the burning invader continue its journey along the atmosphere and then flash beyond. Away now from the clutches of air, still burning, it left behind an ionized trail of particles and superheated gases. Now away from earth, it lofted high and far until it raced beyond earth's shadow. Sunlight flashed through the ionized trail, and the departing mass created its own record of passage, enduring long enough for Deke to watch until the last flicker, the final gleam, was gone.

He felt he should not lower his gaze. His vision moved along the arrowing path of the now invisible wanderer of the solar system, and Deke stared, unblinking, as the mass of stars in his own galaxy shone down on him, an uncountable array of suns, stars he knew were smaller than his own sun, many vastly greater in size and energy, but all members of the great pinwheeled Milky Way of which Deke and his world were one tiny member.

He was a man humbled, awed, grateful for what had happened, what he had been given to see, and he was now seeing clearly into tomorrow, focusing on life and time itself. Understanding that life was indifferent. Understanding that time is a dimension measured only within the mind.

He let the vision soar into a future. Not one he could measure specifically, but a future restricted to a definite road. That world below a cornucopia of life. It gave up wonders so that the human race could evolve and grow and turn its gaze skyward, wondering about stars and other worlds.

Here was the beginning of life, its present, and its end. The bounty of earth was finite. Its supply of energy, foodstuffs, clean atmosphere, pristine waters—all were finite. And whatever age to come was being poisoned, tortured, shortened by the myopic, uncaring abuses of mankind.

One phrase remained in Deke's mind from the moment he had read the words written at the turn of the century by a Russian scientist-schoolteacher, Konstantin E. Tsiolkovsky, who was the first human to envision and draw up concepts for the use of rockets in space travel. In a simple but wonderfully elegant turn of words, Tsiolkovsky surveyed the future and saw what the human race must do and where it must go.

"Earth is the cradle of the mind," wrote the self-taught man reaching for tomorrow, "but one cannot live in the cradle forever."

There it was. Finite. If Deke, and the others before him and those who would follow, were successful, then man had taken his first faltering steps not merely to other worlds close by, but to far distant stars and worlds revolving about those alien suns. Deke looked down on his world in a mixture of sadness and hope, for he knew this planet earth one day would pass into history. No one knew when, but all knew it was inevitable.

Down there, man consumed in a mindless orgy the limited resources of Spaceship Earth. It might be that long before the wells were dry and the fields dusty, an agitated sun could yet turn earth into a cinder. Or instability would ratchet its way through the interior of our sun and diminish its energy output, and earth would slip into a deep freeze. Either would toll the bell for the end of man.

Now, at least, they were doing something about that inevitable tomorrow. All that had been done, from Gagarin to Shepard, from Leonov to Armstrong, all the space travelers right to this moment with Deke himself carrying the torch, would against the long march of history be no more than the blink of an eye.

Well, he reasoned, brief as it might be, the flutter of an eyelash against the many pages of history, he had done his part.

He knew that if one day men were successful in journeying to distant stars and populating the planets of those far away stellar engines, then the race of man was safe. A star might go nova, obliterate an entire solar system, but if man populated many solar systems . . . he smiled to himself. Then life would go on.

That was the gift to the future of all those among us who made it possible. Shepard, Armstrong, Conrad, Grissom, Gagarin, Leonov, Stafford, Scott, Irwin, Young, Cernan, Glenn, Cooper, Schirra, Carpenter, and— and so many more.

And on that roll call there was now this name to remember.

Always.

Deke Slayton.

Astronaut.

INDEX

INDEX

Gemini program (Project Gemini), 176-91: and Apollo program, 157, 167, 175; goals, 167; preparation for first manned flight, 168, 176-78; Slayton grounded during, 9, 16; success of, 190, 194, 209. *See also* Gemini spacecraft *and entries for specific Gemini missions, above*

Gemini spacecraft, 177; ban on naming of, 179, 180; boosters (Titan II), 157, 168, 179, 282; compared to other craft, 177, 209, 221, 332; hatch design, 181; oxygen pressurization, 199-200

General Dynamics, Convair division, 71, 143-44

German rocket program, 31-32

Gibbs, Col. Asa, 34

Gilruth, Robert (Bob), 76-77, 78-80, 81, 89, 129, 144, 154, 162-63, 209, 210, 214, 215, 228, 279, 287, 307

Gleaves, James, 203

Glenn, John, 11, 12, 64, 98, 132, 134-35, 156, 161, 278, 365; as backup for first manned Mercury flights, 79, 86, 89, 140, 144; *Friendship Seven* mission, 80, 144-52, 154, 157, 198; peers' evaluation of, 77-78, 81; and Shepard's *Freedom Seven* mission, 81, 102, 105, 106

Glennan, T. Keith, 64

g-loads, 53, 116, 122-23; centrifuge training, 72-73, 74, 123

global weather forecasting, 12

Gordon, Dick: Apollo 12 mission, 255, 256, 257; Gemini 11 mission, 189, 190, 229

Grand Bahama Island (GBI), 120, 122, 123-24

gravity. See g-loads; weightlessness

Griffin, Jerry, 298

Grissom, Betty, 66, 206, 211

Grissom, Mark, 66

Grissom, Scott, 66

Grissom, Virgil ("Gus"), 12, 64, 77, 84-85, 86, 89, 134, 138-40, 166, 182, 211-12, 276, 278, 365; Apollo 1 mission, 192-94, 195-96, 197-98, 200, 201, 202-3, 205, 207, 211, 215, 216, 348; Gemini 3 (*Molly Brown*) mission, 168, 176-77, 178-80, 181, 240; *Liberty Bell Seven* mission, 78-79, 80, 136, 140-42, 144, 154, 179, 196; and Shepard's *Freedom Seven* mission, 102, 104, 124

Grumman Aircraft Company, 216, 239, 273-74

Grumman F11F Tiger, 54-55

Grumman Gulfstream, 214

Gumdrop (Apollo 9 command module), 240

—H—

Hadley Rille, Moon, 323, 324

Haise, Fred: Apollo 13 mission, 260, 265, 266, 267, 268, 271-72, 273

Ham (space chimpanzee), 90

Handler, Dr. Phillip, 327

Harris, Gordon, 41

Harter, Alan, 205

hatch design: Apollo, 195-96, 202, 216; Gemini, 181; Mercury, 142, 196

Hawk (Vostok V), 166

Hitler, Adolf, 31, 87

Holiday Inn, Cocoa Beach, 87-88, 159-60, 206

House, Dr. William, 251, 252-53

INDEX

–O–

Oak Ridge Institute of Nuclear Studies, 210
Ocean of Storms, Moon, 255, 256, 293, 301
Odyssey (Apollo 13 command module), 260, 264, 265, 267, 269, 270, 271, 272
Office of Naval Research, 35
O'Hara, Dee, 141
O'Malley, Tom (T.J.), 143, 145
Operation Paperclip, 31
orbital flights: first, see Vostok I; first manned maneuver to change orbit, 179; first U.S., see Friendship Seven
Orbiter project, 35
Orion (Apollo 16 lunar module), 325
oxygen pressurization, 194-95, 196-200

–P–

Paine, Thomas O., 221, 229, 326-27
Patrick Air Force Base, 140
Patsayev, Viktor, 324
Patuxent River Naval Air Base, 42, 57
Peenemünde, Germany, 31
Pentagon: charge to von Braun, 34; and first U.S. satellite launch, 35, 36, 46-47
Petrone, Rocco, 197, 198, 200-201, 204-5, 220
Phillips, Maj. Gen. Samuel C. (Sam), 210-11, 215, 221, 241
Pickering, Dr. William, 46, 47
Pierce, Jim, 204
Pilotless Aircraft Program, Langley Air Force Base, 76
pilots, 49-60. See also test pilots; fighter pilots
"pilot seances," 78
"plugs out" testing, 196
Polaris missiles, 164
Popovich, Pavel R., 230
Powers, Col. John ("Shorty"), 97
Pravda, 174
Probst, Gary, 202
"procedures trainer," 86
Project Apollo. See Apollo program
Project Gemini. See Gemini program
Project Mercury. See Mercury program
Project Orbiter, 35
Project Vanguard, 35, 40, 44-45
Proton rocket, 226

–R–

R-7 rocket, 37-38, 226
Rathmann, Jim, 84-85, 138-39, 140
Reasoner, Harry, 45
Redstone Arsenal, Huntsville, Alabama, 30-31, 33, 36, 40
Redstone rocket: adapted as Jupiter-C, 35, 40; compared to R-7, 37; compared to Saturn V, 277; development of, 33-34; as Mercury booster, 79, 80, 102, 103, 104-5,

INDEX

Service Medal award, 128-29; naval training, 42, 57, 66; request for Apollo 13 mission, 253-54, 263; response to news of Gagarin's flight, 96-97; response to Sputnik, 42-43, 44; retirement from NASA, 349; return to flight status, 253; selected for first Gemini flight, 168; and Slayton, 11, 22, 28, 62, 63, 64, 65, 108, 156, 168, 169-70, 178, 307, 320-22, 324, 328, 349, 353, 357, 362; and Skylab, 333; as a test pilot, 42, 54-56; White House reception and parade for, 127-29, 131-32

Shepard, Alice, 66, 67, 68

Shepard, Julie, 66, 67, 68

Shepard, Laura, 66, 67, 68

Shepard, Louise, 66-68, 81, 103-4, 106-7, 109-11, 115, 127-28, 131, 251-52, 281, 305, 307

Shinkle, John, 199

Shonin, Georgi, 337

Sigma Seven mission (Schirra), 157

simulators: Apollo 1, 193; Apollo 13, 262-63, 264-65; Apollo 14, 298

Sinatra, Frank, 211-12

Sjoberg, Sig, 291

Skylab, 325, 327, 332-33, 352

Slayton, Deke, 9, 12, 43-44, 70-71, 79, 134, 135, 366; and Apollo program, 212-13, 217, 221; Apollo 1, 192, 195, 197, 200-201, 205, 206-7, 211, 212; Apollo 7, 222; Apollo 8, 227-28, 229; Apollo 10, 241; Apollo 11, 16-17, 19, 22, 25, 28, 237-38, 239; Apollo 13, 253-54, 262, 264, 265-67, 269, 270; Apollo 14, 275-78, 280, 287, 288, 298, 302, 303, 307, 309; Apollo-Soyuz Test Project mission, 9, 327, 333, 334-36, 338, 339-46, 348, 349, 351-56, 358, 359-65; assessment of Mercury Seven peers, 77-78; astronaut pin, 211, 247; as Chief of Flight Crew Operations, 9, 11, 169; as Coordinator of Astronaut Activities, 156, 157; crew selection method, 237, 238, 254; and Gemini program, 168, 169-70, 176, 177, 180, 181, 188; and Grissom, 84-85, 205, 276; heartbeat irregularity and medical grounding, 9, 16, 72-74, 153-54, 156, 178, 211, 253, 257, 258-59, 320-22; introduced as Mercury Seven astronaut, 62-66; memorial service for, 10, 11, 13; and Mercury missions: *Aurora Seven*, 154; *Faith Seven*, 157-58, 159, 161; *Freedom Seven*, 76, 77, 78, 79, 108, 111-12, 114-15, 116, 118, 119, 120, 121, 122-24; and Mercury Seven training, 70-71, 74, 87; reinstatement offlight status, 324; and Shepard, 22, 28, 62, 63, 64, 65, 108, 168, 169-70, 178, 252, 253-54, 276, 307, 309-10, 320-22, 328, 349, 353; slated for second orbital flight, 80, 152-53; as a test pilot, 43, 50-51, 52-54, 63, 257; at White House reception and parade, 129, 132; as a World War II fighter pilot, 51, 65, 334, 335

Slayton, Kent, 65, 66

Slayton, Marge, 65-66, 73

Snoopy (Apollo 10 lunar module), 241, 242-43, 337

solar powered flight, 217

sonic booms, 121-22

South Korea, 33

Soviet Academy of Sciences, 327, 339

Soviet Air Force, cosmonauts from, 93, 172, 336

Soviet "firsts" in space, 9; man in space, 92-98; satellite in orbit, 37-42; space walk, 171-74; woman in space, 166

Soviet Union/Russia, 8-9; arguments for yielding space race to, 98-99, 130; joint mission with U.S., *see* Apollo-Soyuz Test Project; long-endurance flights, 166, 331; Luna 15 project, 248-49; manned lunar project, 12, 226, 235-36, 330-31; rocketry program, 32, 36, 98; *see also* N-1; R-1; spacecraft, *see* Voshkod; Vostok; Soyuz; space port, 36-

–T–

INDEX

Thor missiles, 34, 164

Time magazine, 47

Titan rocket, 135; as Gemini booster (Titan II), 157, 168, 179, 282; as ICBM, 34, 143, 164

Titov, Gherman Stepanovich, 93, 142, 154

translunar injection, 285

"trench," at Mission Control, 19

Truman, Harry, 33, 141, 210

Tsiolkovsky, Konstantin E., 364-65

–U–

United States, 8-9; commitment to a manned lunar landing, *see under* Kennedy; competition with Soviet Union, *see* U.S.-Soviet competition; first manned space flight, *see* *Freedom Seven*; first manned orbital flight, *see Friendship Seven*; first satellite launch, 46-48; joint mission with Soviets, *see* Apollo-Soyuz Test Project; postwar rocketry program, 31-36. *See also* entries at U.S.

Unsinkable Molly Brown, The (film), 179

urination systems, 107, 141

U.S. Air Force: Astronaut Maneuvering Unit (AMU), 186-88; astronauts, 64, 79, 134; facilities, 164, 166; lighthouse "launch" film, 83; missile development programs, 34, 143, 210; rescue planes, 186; review of Slayton's health status, 153

U.S. Army: campaign to launch a U.S. satellite, 35-36, 40-41, 45-46; communications satellite system, 135; Operation Paperclip, 31; rocketry program, 30-31, 32-33, 34-35, 41, 135

U.S. Congress: Kennedy's address to, 133; Manned Spacecraft Center site selection, 163, 164-65; and Mercury program, 71, 72; NASA funding, 32, 135-36, 142, 165, 279, 332; opposition to space program, 32, 331; and Gemini 3 space sandwich, 180

U.S. "firsts" in space: docking in orbit, 183-84; manual control of spacecraft, 118-19; manual maneuvers in orbit, 179; manned lunarlanding and return, *see* Apollo 11; manned lunar orbit, *see* Apollo 8

U.S. Marines, 127, 128; astronaut, 64, 79

U.S. Naval Academy, 57

U.S. Navy: astronauts, 56, 57-60, 64, 79, 129, 134, 255; recovery teams, 112, 120, 179, 186

USS *Coastal Sentry Queen*, 185

USS *Iwo Jima*, 271, 273

USS *Lake Champlain*, 125

USS *Missouri*, 115

USS *New Orleans*, 319

USS *Noa*, 152

USS *Oriskany*, 57, 59

U.S.-Soviet competition, 8-9, 12; and distrust of joint mission, 326-27, 338-39; end of, and reduction in U.S. space program, 331-32 ; after *Friendship Seven* flight, 152; long-endurance flight, 157, 166, 183; manned flight, 86, 91; lunar landing, 12, 17, 174, 208, 209, 226-27, 235-36, 237, 247-48, 330-31; manual control of spacecraft, 118-19; orbital flight, 130, 142, 143, 152; satellite in orbit, 42; rendezvous and docking in orbit, 190; rocketry and boost capability, 36, 129, 131, 133, 166; space walking, 174, 190-91

U.S.-Soviet space treaty, 208

ABOUT THE AUTHORS

ALAN SHEPARD was selected by NASA as one of the United States' original Mercury Seven astronauts in 1959. He became the first American to rocket into space on May 5, 1961, making a fifteen-minute suborbital trip, which took him 116 miles high and 302 miles down the Atlantic tracking range in his *Freedom Seven* capsule. After serving for several years as chief of the Astronaut Office, he commanded the Apollo 14 mission as it embarked for the moon, where he made two surface excursions totaling nine hours. In 1974 he retired from both NASA and the Navy, where he held the rank of Rear Admiral. Shepard currently serves as president of the Mercury Seven Foundation, which raises scholarship money for college science and engineering students. He and his wife, Louise, live in Houston.

DEKE SLAYTON was also one of America's original Mercury Seven astronauts. He was assigned to fly the second Project Mercury orbital mission but was grounded by an irregular heartbeat. He stayed with NASA to supervise the astronaut corps and was restored to flight status in 1972. Three years later on July 17, 1975, Slayton made it into space after sixteen years as an astronaut. Flying for the Apollo-Soyuz Test Program, he rode an Apollo capsule that linked up with a Soviet Soyuz spacecraft carrying two cosmonauts. They conducted joint experiments for forty-seven hours. Slayton and his two Apollo crew members remained in orbit for nine days. For the next two years Slayton was manager of the space shuttle approach and landing tests at Edwards Air Force Base. From 1977 until he retired from NASA in 1982, he was manager for orbital flight tests. Deke Slayton, with his wife Bobbie at his side, died of cancer in his home in League City, Texas, in June 1993.

JAY BARBREE, a finalist to be the first Journalist in Space, is the author of four books and has been an NBC News space correspondent for the entire history of NASA. He shared an Emmy for NBC's Apollo lunar landing coverage and is the recipient of a lifetime award from the NBC Space Veterans Association. He lives in Merritt Island, Florida.

HOWARD BENEDICT was senior aerospace writer for the Associated Press for more than thirty years. He covered more than two thousand missile and rocket launches and wrote the main stories on the first sixty-five U.S. astronaut flights. He has received numerous awards for his writing and has written two books on NASA. Since 1990 he has served as executive director of the Mercury Seven Foundation, working with the surviving members of the United States' original seven astronauts. He lives in Florida.

ACKNOWLEDGMENTS

Turner Publishing would like to thank the staff at NASA Media Services, with special thanks to Mike Gentry, for their assistance in providing photographs for MOON SHOT. Additional photographs were provided courtesy of the Shepard and Slayton families and Turner Publishing acknowledges this contribution with gratitude.